T0270991

CAMBRIDGE TRACTS IN MATHEMATICS

General Editors

B. BOLLOBAS, F. KIRWAN, P. SARNAK, C.T.C. WALL

134 Birational Geometry of Algebraic Varieties

János Kollár
University of Utah

Shigefumi Mori
RIMS, Kyoto University

With the collaboration of

C. H. Clemens
University of Utah

A. Corti
University of Cambridge

Birational Geometry of Algebraic Varieties

CAMBRIDGE
UNIVERSITY PRESS

PUBLISHED BY THE PRESS SYNDICATE OF THE UNIVERSITY OF CAMBRIDGE
The Pitt Building, Trumpington Street, Cambridge CB2 1RP, United Kingdom

CAMBRIDGE UNIVERSITY PRESS
The Edinburgh Building, Cambridge CB2 2RU, United Kingdom
40 West 20th Street, New York, NY 10011-4211, USA
10 Stamford Road, Oakleigh, Melbourne 3166, Australia

First published 1998

Typeset in Computer Modern 10/13pt

A catalogue record for this book is available from the British Library

Library of Congress Cataloguing in Publication data

Kollár, János.
  Birational geometry of algebraic varieties / János Kollár and
  Shigefumi Mori, with collaboration of C.H. Clemens and A. Corti.
    p.   cm. – (Cambridge tracts in mathematics; 134)
  Includes bibliographical references and index.
  ISBN 0 521 63277 3
  1. Surfaces, Algebraic.   2. Algebraic varieties.   I. Mori,
  Shigefumi.   II. Title.   III. Series.
  QA571.K65   1998
  516.3′5–dc21   98-24732 CIP

ISBN 0 521 63277 3 hardback

Transferred to digital printing 2002

Contents

Preface

One of the major discoveries of the last two decades in algebraic geometry is the realization that the theory of minimal models of surfaces can be generalized to higher dimensional varieties. This generalization is called the minimal model program or Mori's program. While originally the program was conceived with the sole aim of constructing higher dimensional analogues of minimal models of surfaces, by now it has developed into a powerful tool with applications to diverse questions in algebraic geometry and beyond.

So far the program is complete only in dimension 3, but large parts are known to work in all dimensions.

The aim of this book is to introduce the reader to the circle of ideas developed around the minimal model program, relying only on knowledge of basic algebraic geometry.

In order to achieve this goal, considerable effort was devoted to make the book as self-contained as possible. We managed to simplify many of the proofs, but in some cases a compromise seemed a better alternative. There are quite a few cases where a theorem which is local in nature is much easier to prove for projective varieties. For these, we state the general theorem and then prove the projective version, giving references for the general cases. Most of the applications of the minimal model program ultimately concern projective varieties, and for these the proofs in this book are complete

Acknowledgments

The present form of this book owes a lot to the contributions of our two collaborators.

H. Clemens was our coauthor in [CKM88]. Sections 1.1–3, 2.1, 2.2, 2.4 and 3.1–5 are revised versions of sections of [CKM88]. We owe special thanks to H. Clemens and to Astérisque for allowing us to use this material.

A. Corti showed us his unpublished note 'Semi-stable 3–fold flips' which utilized the reduction ideas of [Sho92] (as explained in [K⁺92, Chap. 18]) to semi-stable flips. This allowed him to eliminate from the proof the use of the classification of surface quotient singularities. For 3–folds this is a relatively small difference, but each time we eliminate the use of a special low-dimensional result, we are hopefully a step closer to flips in higher dimensions. Corti's note formed the basis of our treatment of semi-stable flips in section 7.4.

In the past five years, we have given lecture series about the material presented in this book at the Regional Geometry Institute at Park City Utah, at RIMS Kyoto University and at the University of Utah. We would like to thank our audience, colleagues and students who contributed many observations and improvements to the present form of this book. We received especially helpful comments from D. Abramovich, A. Bertram, J. Cheah, O. Fujino, Y. Kawamata, H. Kley, S. Kovács, T. Kuwata, Y. Lee, R. Mayer, Y. Miyaoka, R. Morelli, N. Nakayama, Th. Peternell, E. Szabó and N. Tziolas.

We also thank the financial support of the NSF (grant number DMS-9622394), the University of Utah and the Japanese Ministry of Education, Science and Culture under a Grant-in-Aid for International Scientific Research (Joint Research Program 08044078) and Scientific Research (B) Program 09440010.

Introduction

From the beginnings of algebraic geometry it has been understood that birationally equivalent varieties have many properties in common. Thus it is natural to attempt to find in each birational equivalence class a variety which is simplest in some sense, and then study these varieties in detail.

Each irreducible curve is birational to a unique smooth projective curve, thus the investigation of smooth projective curves is equivalent to the study of all curves up to birational equivalence.

For surfaces the situation is more complicated. Each irreducible surface is birational to infinitely many smooth projective surfaces. The theory of minimal models of surfaces, developed by the Italian algebraic geometers at the beginning of the twentieth century, aims to choose a unique smooth projective surface from each birational equivalence class. The recipe is quite simple. If a smooth projective surface contains a smooth rational curve with self-intersection -1, then it can be contracted to a point and we obtain another smooth projective surface. Repeating this procedure as many times as possible, we usually obtain a unique 'minimal model'. In a few cases we obtain a model that is not unique, but these cases can be described very explicitly.

A search for a higher dimensional analogue of this method started quite late. One reason is that some examples indicated that a similar approach fails in higher dimensions.

The works of Reid and Mori in the early 1980s raised the possibility that a higher dimensional theory of minimal models may be possible if we allow not just smooth varieties but also varieties with certain mild singularities. This approach is called the Minimal Model Program or Mori's Program. After many contributions by Benveniste, Kawamata,

Kollár, Reid, Shokurov, Tsunoda, Viehweg and others, the program was completed in dimension three by Mori in 1988.

Since then this program has grown into a method which can be applied successfully to many problems in algebraic geometry.

The aim of this book is to provide an introduction to the techniques and ideas of the minimal model program.

Chapter 1 gives an introduction to the whole program through a geometric approach. Most of these results are not used later, but they provide a useful conceptual foundation.

Chapter 2 is still introductory, discussing some aspects of singularities and the relevant generalizations of the Kodaira Vanishing Theorem.

The first major part of the program, the Cone Theorem, is proved in Chapter 3. These results work in all dimensions.

The rest of the book is essentially devoted to the study of 3–dimensional flips and flops. Flips and flops are new types of birational transformations which first appear in dimension 3. Most major differences between the theory of surfaces and 3–folds can be traced back to flips and flops.

Chapter 4 is devoted to the classification of certain surface singularities. These results are needed in further work on the 3–dimensional theory.

The singularities appearing in the course of the minimal model program are investigated in Chapter 5. The results are again rather complete in all dimensions.

Flops are studied in Chapter 6. Flops are easier to understand than flips, and, at least in dimension 3, their description is rather satisfactory.

Chapter 7 is devoted to 3–dimensional flips. The general theory is still too complicated and long to be included in a textbook, thus we restrict ourselves to the study of a special class, the so-called semi-stable flips. We have succeeded in simplifying the proofs in this case considerably. Semi-stable flips appear naturally in many contexts, and they are sufficient for several of the applications.

A more detailed description of the contents of each chapter is given at its beginning.

Sections 4.5 and 5.5 are each a side direction, rather than being part of the main line of arguments. In each case we felt that the available references do not adequately cover some results we need, and that our presentation may be of interest to the reader.

Prerequisites

We assume that the reader is familiar with basic algebraic geometry, at the level of [Har77].

There are a few other results that we use without proof.

In the proof of (1.10) we need an estimate for the dimension of the deformation space of a morphism. This result, whose proof is rather technical, is fundamental for much of sections 1.1 and 1.2. These theorems are, however, not used in subsequent sections.

In section 1.5 we recall the basic properties of intersection numbers of divisors and a weak form of Riemann–Roch that we need frequently.

In section 2.4 we state and use the basic comparison theorem of algebraic and analytic cohomologies and also a special case of the Hodge decomposition of the singular cohomology.

In all these cases we need only the stated results, not the techniques involved in their proofs.

A few times we need the Leray spectral sequence (see [God58, 4.17], [HS71, VIII.9] or [Bre97, IV.6] for proofs):

Theorem 0.1. *Let* $f : X \to Y$ *be a morphism of schemes and* F *a quasi-coherent sheaf on* X. *Then there is a spectral sequence*

$$E_2^{i,j} = H^i(Y, R^j f_* F) \Rightarrow H^{i+j}(X, F).$$

We also use resolution of singularities from [Hir64] on many occasions. We need two versions of this result as follows:

Theorem 0.2. *Let* X *be an irreducible reduced algebraic variety over* $\mathbb{C}$ *(or a suitably small neighbourhood of a compact set of an irreducible reduced analytic space) and* $I \subset \mathcal{O}_X$ *a coherent sheaf of ideals defining a closed subscheme (or subspace)* Z. *Then there are a smooth variety (or analytic space)* Y *and a projective morphism* $f : Y \to X$ *such that*

(1) f *is an isomorphism over* $X \setminus (\mathrm{Sing}(X) \cup \mathrm{Supp}\, Z)$,
(2) $f^* I \subset \mathcal{O}_Y$ *is an invertible sheaf* $\mathcal{O}_Y(-D)$ *and*
(3) $\mathrm{Ex}(f) \cup D$ *is an snc divisor.*

This follows from the Main Theorems I and II (or I′ and II′ in the analytic case) of [Hir64]. The result without the assertion (1) is called the Weak Hironaka Theorem, which is all we need in this book. Very short proofs of the Weak Hironaka Theorem for quasi-projective X are given in [AdJ97], [BP96], [Par98]. All these papers reduce the Weak Hironaka Theorem to the torus embedding theory of [KKMSD73]. (They state only that D is an snc divisor, but the proofs work for the full snc statement (3).)

The relative version of resolution is the following:

Theorem 0.3. *Let $f : X \to C$ be a flat morphism of a reduced algebraic variety over $\mathbb{C}$ (or a suitably small neighbourhood of a compact set of a reduced analytic space) to a non-singular curve and $B \subset X$ a divisor. Then there exists a projective birational morphism $g : Y \to X$ from a non-singular Y such that $\mathrm{Ex}(g) + g^*B + (f \circ g)^*(c)$ is an snc divisor for all $c \in C$.*

This follows from the Main Theorem II (or II′) of [Hir64]. It is used only in Chapter 7 with $\dim X = 3$. If C and X are projective, this is a special case of [AK97, Thm. 2.1]. The latter paper also ignores $\mathrm{Ex}(g)$ but the proof again can be modified to yield the full snc statement.

Notation 0.4. In order to avoid possible misunderstanding, here is a list of some of the standard notation we use.

(1) Let X be a normal scheme. A *prime divisor* is an irreducible and reduced subscheme of codimension one. A *divisor* on X is a formal linear combination $D = \sum d_i D_i$ of prime divisors where $d_i \in \mathbb{Z}$. In using this notation we assume that the D_i are distinct. A $\mathbb{Q}$-*divisor* is a formal linear combination $D = \sum d_i D_i$ of prime divisors where $d_i \in \mathbb{Q}$. D is called *effective* if $d_i \geq 0$ for every i. For $\mathbb{Q}$-divisors A, B, we write $A \geq B$ or $B \leq A$ if $A - B$ is effective. (This notation will not be used extensively since it can be easily confused with $A - B$ being nef.) A divisor (or $\mathbb{Q}$-divisor) D is called $\mathbb{Q}$-*Cartier* if mD is Cartier for some $0 \neq m \in \mathbb{Z}$. X is called $\mathbb{Q}$-*factorial* if every $\mathbb{Q}$-divisor is $\mathbb{Q}$-Cartier. The *support* of $D = \sum d_i D_i$, denoted by $\mathrm{Supp}\, D$, is the subscheme $\cup_{d_i \neq 0} D_i$.

(2) *Linear equivalence* of two divisors D_1, D_2 is denoted by $D_1 \sim D_2$; *numerical equivalence* of two $\mathbb{Q}$-divisors D_1, D_2 is denoted by $D_1 \equiv D_2$. (We do not define linear equivalence of $\mathbb{Q}$-divisors.) D is said to be *trivial* (resp. *numerically trivial*) if $D \sim 0$ (resp. $D \equiv 0$).

(3) A $\mathbb{Q}$-Cartier divisor D on a proper scheme is called *nef* if $(D \cdot C) \geq 0$ for every irreducible curve $C \subset X$.

(4) A *morphism* of schemes is everywhere defined. It is denoted by a solid arrow $f : X \to Y$. A *map* of schemes is defined on a dense

open set; it is denoted by a dotted arrow $f : X \dashrightarrow Y$. In many books this is called a rational map.

(5) Let $f : X \to Y$ be a morphism and D_1, D_2 two divisors on X. We say that they are *linearly f-equivalent* (denoted by $D_1 \sim_f D_2$) iff there is a Cartier divisor B on Y such that $D_1 \sim D_2 + f^*B$. Two $\mathbb{Q}$-divisors are called *numerically f-equivalent* (denoted by $D_1 \equiv_f D_2$) iff there is a $\mathbb{Q}$-Cartier $\mathbb{Q}$-divisor B on Y such that $D_1 \equiv D_2 + f^*B$. D is said to be *(linearly) f-trivial* (resp. *numerically f-trivial*) if $D \sim_f 0$ (resp. $D \equiv_f 0$).

(6) For a scheme X, red X denotes the unique reduced subscheme with the same support as X.

(7) For a birational morphism $f : X \to Y$, the *exceptional set* $\mathrm{Ex}(f) \subset X$ is the set of points $\{x \in X\}$ where f is not biregular (that is f^{-1} is not a morphism at $f(x)$). We usually view $\mathrm{Ex}(f)$ as a subscheme with the induced reduced structure.

(8) Let X be a smooth variety and $D = \sum d_i D_i$ a $\mathbb{Q}$-divisor on X. We say that D is a *simple normal crossing* divisor (abbreviated as *snc*) if each D_i is smooth and they intersect everywhere transversally.

(9) Let X be a scheme. A *resolution* of X is a proper birational morphism $g : Y \to X$ such that Y is smooth.

(10) Let X be a scheme and $D = \sum d_i D_i$ a $\mathbb{Q}$-divisor on X. A *log resolution* of (X, D) is a proper birational morphism $g : Y \to X$ such that Y is smooth, $\mathrm{Ex}(g)$ is a divisor and $\mathrm{Ex}(g) \cup g^{-1}(\mathrm{Supp}\, D)$ is a snc divisor. Log resolutions exist for varieties over a field of characteristic zero by (0.2).

(11) Let $f : X \dashrightarrow Y$ be a map of schemes. Let $Z \subset X$ be a subscheme such that f is defined on a dense open subset $Z^0 \subset Z$. The closure of $f(Z^0)$ is called the *birational transform* of Z. (This is sometimes also called the proper or strict transform.) It is denoted by $f_*(Z)$. If $g : Y \to X$ is birational then we obtain the somewhat unusual looking notation $g_*^{-1}(Z)$. The same notation is used for divisors.

(12) For a real number d, its *round down* is the largest integer $\leq d$. It is denoted by $\lfloor d \rfloor$. The *round up* is the smallest integer $\geq d$. It is denoted by $\lceil d \rceil$. The *fractional part* is $d - \lfloor d \rfloor$ and often denoted by $\{d\}$. If $D = \sum d_i D_i$ is a divisor with real coefficients and the D_i are distinct prime divisors, then we define the *round down* of D as $\lfloor D \rfloor := \sum \lfloor d_i \rfloor D_i$, the *round up* of D as $\lceil D \rceil := \sum \lceil d_i \rceil D_i$ and the *fractional part* of D as $\{D\} := \sum \{d_i\} D_i$.

(13) If X is an analytic space, we usually take an arbitrary compact set $Z \subset X$ and work on a suitable small open neighbourhood $U \supset Z$. We may shrink U if it is convenient, without mentioning this explicitly. U is often called the germ of X around Z. If $g : Y \to X$ is a proper morphism of analytic spaces, we usually work over U as above. With these settings, the arguments for algebraic varieties often work and the notation introduced above can be used similarly. Meromorphic maps and bimeromorphic maps are simply called maps and birational maps.

(14) $P := R$ indicates that the new symbol P is defined to be equal to the old expression R.

(15) $\mathbb{Z}_{>0}$ denotes the set of positive integers, and similarly $\mathbb{R}_{\geq 0}$ denotes the set of non-negative real numbers.

1
Rational Curves and the Canonical Class

In this chapter we explore the relationship between the canonical class K_X of a smooth projective variety X and rational curves on X.

The first section considers the case when $-K_X$ is ample; these are called Fano varieties. The main result shows that X contains a rational curve $C \subset X$ which has low degree with respect to $-K_X$. This result, due to [Mor79], is one of the starting points of the minimal model theory. It is quite interesting that even for varieties over $\mathbb{C}$, the proof proceeds through positive characteristic.

In section 2 we generalize these results to the case when $-K_X$ is no longer ample, but it has positive intersection number with some curve. The proofs are very similar to the earlier ones, we just have to keep track of some additional information carefully.

This leads to the geometric proof of the Cone Theorem for smooth projective varieties in section 3, due to [Mor82]. Unfortunately, for most applications this is not strong enough, and we prove a more general Cone Theorem in Chapter 3 with very different methods.

In section 4 we illustrate the use of the Cone Theorem by using it to construct minimal models of surfaces. The rest of the book is essentially devoted to generalizing these results to higher dimensions.

The last section contains the proof of some of the basic ampleness criteria.

Unfortunately, the methods of this chapter are not sufficient to complete the minimal model program in higher dimensions. In fact, they are not used in subsequent chapters. Nonetheless, we feel that these results provide a very clear geometric picture, which guides the later, more technical works.

The geometric ideas explained in this chapter provided the impetus for much further research. Many of these results are described in [Kol96].

1.1 Finding Rational Curves when K_X is Negative

This section will serve as a warm-up. In it we explore the general theme:

1.1. How do rational curves on a variety influence the birational geometry of that variety?

We will see that the absence of rational curves has some very pleasant consequences. Later this will be turned around, and we will see that certain complications of birational geometry of a variety X are caused precisely by certain special rational curves on X.

The simplest example is Castelnuovo's Theorem on (-1)-curves in the theory of surfaces (cf. [Har77, V.5.7]):

Theorem 1.2. *If X is a smooth proper surface, then there is a nontrivial birational morphism $f : X \to Y$ to a smooth surface Y iff X contains a smooth rational curve with self-intersection -1.*

One side of this is easy to generalize as follows:

Proposition 1.3. *[Abh56, Prop. 4] Let X be smooth of any dimension and $f : Y \to X$ a proper birational morphism. For any $x \in X$, either $f^{-1}(x)$ is a point or $f^{-1}(x)$ is covered by rational curves.*

Proof: Let us consider first the case when X is a surface.

We resolve the indeterminacies of f^{-1} by successively blowing-up points of X. At each step we introduce a $\mathbb{P}^1$. Thus every $f^{-1}(x)$ is dominated by a union of some of these $\mathbb{P}^1$-s. By Lüroth's Theorem, every $f^{-1}(x)$ is a union of rational curves.

The general case can be proved the same way provided we know how to resolve indeterminacies of maps. However a much weaker version of resolution is sufficient. Since we will use (1.3) later only when X is a surface, we only sketch the proof in the higher-dimensional case.

We may assume that Y is normal. The exceptional set of f is of pure codimension one by [Sha94, II.4.4] (see also (2.63)). Let $E \subset Y$ be an irreducible component of the exceptional set. At a general point $e \in E$, (Y, E) is isomorphic to a succession of blow ups with smooth centers by (2.45).

Thus there is a rational curve $C \subset E$ that passes through e such that $f(C)$ is a point. Since a rational curve can specialize only to unions of rational curves, there is a rational curve through every point of E.   □

Using Zariski's Main Theorem [Har77, V.5.2] this implies the following.

Corollary 1.4. *Let Z be a smooth variety and $g : Z \dashrightarrow X$ a rational map. Let $Y \subset X \times Z$ be the closure of the graph of g, and let q and p be the first and the second projections. Let $S \subset Z$ be the set of points where g is not a morphism. Then for every $z \in q(p^{-1}S)$ there is a rational curve $z \in C_z \subset q(p^{-1}S)$.* □

Corollary 1.5. *Let X and Z be algebraic varieties, Z smooth and X proper. If there is a rational map $g : Z \dashrightarrow X$ which is not everywhere defined, then X contains a rational curve.* □

The simplest situation where one could apply this corollary is when Z is a surface which we obtain as a family of curves. In some cases one can assert that a map g as in (1.5) cannot be a morphism.

Lemma 1.6 (Rigidity Lemma). *Let Y be an irreducible variety and $f : Y \to Z$ a proper and surjective morphism. Assume that every fiber of f is connected and of dimension n. Let $g : Y \to X$ be a morphism such that $g(f^{-1}(z_0))$ is a point for some $z_0 \in Z$. Then $g(f^{-1}(z))$ is a point for every $z \in Z$.*

Proof: Set $W = \mathrm{im}(f \times g) \subset Z \times X$. We obtain proper morphisms

$$f : Y \xrightarrow{h} W \xrightarrow{p} Z.$$

$p^{-1}(z) = h(f^{-1}(z))$ and $\dim p^{-1}(z_0) = 0$. By the upper semi-continuity of fiber dimensions, there is an open set $z_0 \in U \subset Z$ such that $\dim p^{-1}(z) = 0$ for every $z \in U$. Thus h has fiber dimension n over $p^{-1}(U)$, hence h has fiber dimension at least n everywhere. For any $w \in W$, $h^{-1}(w) \subset f^{-1}(p(w))$, $\dim h^{-1}(w) \geq n$ and $\dim f^{-1}(p(w)) = n$. Therefore $h^{-1}(w)$ is a union of irreducible components of $f^{-1}(p(w))$, and so $h(f^{-1}(p(w))) = p^{-1}(p(w))$ is finite. It is a single point since $f^{-1}(p(w))$ is connected. □

Corollary 1.7 (Bend and Break I). *Let X be a proper variety, C a smooth proper curve, $p \in C$ a point, and $g_0 : C \to X$ a non-constant morphism. Assume that there is a smooth, connected (possibly non-proper) pointed curve $0 \in D$ and a morphism $G : C \times D \to X$ such that:*

(1) *$G|_{C \times \{0\}} = g_0$,*
(2) *$G(\{p\} \times D) = g_0(p)$, and*
(3) *$G|_{C \times \{t\}}$ is different from g_0 for general $t \in D$.*

Then there is a (possibly constant) morphism $g_1 : C \to X$, and a linear combination of rational curves $Z = \sum a_i Z_i$ where $a_i > 0$ and $Z_i \subset X$, such that

(1) *$(g_0)_*(C)$ is algebraically equivalent to $(g_1)_*(C) + Z$, and*

(2) *$g_0(p) \in \cup_i Z_i$.*

In particular, X contains a rational curve through $g_0(p)$.

We frequently refer to this result in the following imprecise but suggestive form:

If g_0 moves with a point fixed, then it degenerates into a sum of rational curves and another map $g_1 : C \to X$.

Proof: We compactify D to a proper curve $\bar{D}$, and so we have a rational map $\bar{G} : C \times \bar{D} \dashrightarrow X$. We claim that $\bar{G}$ is undefined somewhere along $\{p\} \times \bar{D}$. To see this apply (1.6) to the projection map $f : U \times \bar{D} \to U$ for a neighbourhood U of p in C. $f^{-1}(p)$ is mapped to a single point; thus the same holds for every fiber, and $G(z, t) = g_0(z)$ for general $t \in D$, a contradiction.

Let S denote the normalization of the closure of the graph of $\bar{G}$. Let $\pi : S \to C \times \bar{D}$ and $G_S : S \to X$ be the projections and define $h : S \to C \times \bar{D} \to \bar{D}$ as the composite of π and the second projection of the product.

As we remarked, there is a point $(p, d) \in C \times \bar{D}$ such that π is not an isomorphism above (p, d). We can write $h^{-1}(d) = C' + E$ where $C \cong C' \subset S$ is the birational transform of $C \times \{d\}$ and E is π-exceptional. Let $g_1 : C \to X$ be the restriction of G_S to C' and set $Z = G_S(E)$.

g_0 can be identified with $G_S|_{h^{-1}(0)}$, thus $(g_0)_*(C)$ is algebraically equivalent to $(g_1)_*(C) + Z$. By (1.3) E is a union of rational curves, hence so is Z. Using (1.4) we see that there is a rational curve through $g_0(p)$.   □

It is interesting to note that the algebraicity assumption is essential:

Example 1.8. [Bla56] Let E be an elliptic curve and M a line bundle of degree ≥ 2 with generating sections s and t. In $V = M \oplus M$, the sections $(s, t), (\sqrt{-1}s, -\sqrt{-1}t), (t, -s), (\sqrt{-1}t, \sqrt{-1}s)$ are everywhere independent over $\mathbb{R}$, thus they generate a 'lattice bundle' L over E. Let $X = V/L$ and $C =$ the zero section in V/L. Then C must move leaving a point fixed by the positivity of the bundle V, yet V/L has no rational curves.

Conclusion: The family of deformations of the mapping of C into X (leaving a point of C fixed) has no non-trivial compactifiable subvarieties.

If the curve C in (1.7) is rational, then we can take $g_1 := g_0$ and so the conclusions do not yield anything new. The following variant of (1.7) shows how to get non-trivial degenerations of rational curves.

Lemma 1.9 (Bend and Break II). *Let X be a projective variety and $g_0 : \mathbb{P}^1 \to X$ a non-constant morphism. Assume that there is a smooth connected (possibly non-proper) pointed curve $0_D \in D$ and a morphism $G : \mathbb{P}^1 \times D \to X$ such that*

(1) $G|_{\mathbb{P}^1 \times \{0_D\}} = g_0$,
(2) $G(\{0\} \times D) = g_0(0)$, $G(\{\infty\} \times D) = g_0(\infty)$ *and*
(3) $G(\mathbb{P}^1 \times D)$ *is a surface.*

Then $(g_0)_*(\mathbb{P}^1)$ is algebraically equivalent on X to either a reducible curve or a multiple curve (i.e. of the form aC for some $a > 1$).

Proof. Let $\bar{D}$ be a smooth compactification of D and $q : S \to \bar{D}$ a $\mathbb{P}^1$-bundle containing $\mathbb{P}^1 \times D$ as an open set such that q is compatible with the second projection $\mathbb{P}^1 \times D \to D$. Let $\bar{G} : S \dashrightarrow X$ be the rational map extending G. Let $\tilde{r} : \tilde{S} \to S$ be a sequence of blow ups such that the induced map $\tilde{G} : \tilde{S} \to X$ is a morphism. We prove the lemma by induction on the number of blow ups in $\tilde{r}$.

First we treat the case when $\bar{G}$ is a morphism. Let H be ample on X and $C_0, C_\infty \subset S$ the two sections extending $\{0\} \times D, \{\infty\} \times D$. Then $((\bar{G}^*H)^2) > 0$ and $(C_0 \cdot \bar{G}^*H) = 0 = (C_\infty \cdot \bar{G}^*H)$. By the Hodge Index Theorem, $(C_0^2) < 0$, $(C_\infty^2) < 0$ and so $\bar{G}^*H, C_0, C_\infty$ are linearly independent elements of the Néron–Severi group of S. On the other hand, the Néron–Severi group of S has rank 2 since S is a $\mathbb{P}^1$-bundle over $\bar{D}$, a contradiction.

Suppose $\bar{G}$ is not a morphism and let $\tilde{S} \xrightarrow{r'} S' \xrightarrow{r} S$ be the blow up at a point $P \in q^{-1}(y)$ needed to eliminate the indeterminacy. Let F_1 denote the exceptional curve of r. Then $(q \circ r)^*(y) = F_1 + F_2$ where F_1, F_2 are (-1)-curves which intersect at a single point $Q = F_1 \cap F_2$. Let $\bar{G}' : S' \dashrightarrow X$ be the induced map.

We claim that $\bar{G}'$ is a morphism along F_2. Note that $(g_0)_*(\mathbb{P}^1)$ is algebraically equivalent to $\tilde{G}_*((q \circ \tilde{r})^*(y))$, which we may assume irreducible and reduced as a 1-cycle because otherwise we are done. If G is not defined at another point $P \neq P' \in q^{-1}(y)$ then

$$\tilde{G}_*((q \circ \tilde{r})^*(y)) = \tilde{G}_* \operatorname{red}(\tilde{r}^{-1}(P)) + \tilde{G}_* \operatorname{red}(\tilde{r}^{-1}(P')) + (\text{effective cycle}),$$

a contradiction. We are left to show that $S' \dashrightarrow X$ is defined at Q. Every irreducible component of $\operatorname{red}((r')^{-1}(Q))$ has multiplicity at least 2 in $(q \circ \tilde{r})^*(y)$ and so needs to be contracted by $\tilde{G}$. Thus the claim is proved.

If $S' \to S''$ is the contraction of the (-1)-curve F_2, then $S'' \dashrightarrow X$ needs one less blow up for the elimination of indeterminacy. We are done by induction. □

We are ready to formulate and prove the first main result about the existence of rational curves. This theorem is of independent interest, even after we consider a later variant which is, in some aspects, considerably sharper.

Theorem 1.10. *[Mor82] Let X be a smooth projective variety such that $-K_X$ is ample. Then X contains a rational curve. In fact, through every point of X there is a rational curve D such that*

$$0 < -(D \cdot K_X) \leq \dim X + 1.$$

Proof: This is done in several steps.

Step 1. We intend to apply (1.7). Thus we have to find a morphism $f : C \to X$ which we will be able to deform. Pick any curve C. If we want to find a rational curve through a given point $x \in X$, then we require C to pass through x and pick $0 \in C$ such that its image is x.

Step 2. We need the following non-trivial result [Mor82] from deformation theory, treated in many books on the subject, for instance [Kol96, II.1]:

Morphisms f of C into X have a deformation space of dimension

$$\geq h^0(C, f^*T_X) - h^1(C, f^*T_X) = -(f_*(C) \cdot K_X) + (1 - g(C)) \cdot \dim X.$$

We use this result through the following consequence.

Having a deformation space of dimension m implies that there is an m-dimensional pointed irreducible affine variety $0 \in Z$ and a morphism $F : C \times Z \to X$ such that $F|_{C \times \{0\}} = f$ and $F|_{C \times \{z\}} \neq F|_{C \times \{0\}}$ for $0 \neq z \in Z$.

Since $\dim X$ conditions are required to fix the image of the basepoint

0 under f, morphisms f of C into X sending 0 to x have a deformation space of dimension

$$\geq h^0(C, f^*T_X) - h^1(C, f^*T_X) - \dim X = -(f_*(C) \cdot K_X) - g(C) \cdot \dim X.$$

Thus whenever the quantity $-(f_*(C) \cdot K_X) - g(C) \cdot \dim X$ is positive there must be an non-trivial one-parameter family of deformations of the map $f : C \to X$ keeping the image of 0 fixed. By (1.6) therefore, we obtain a rational curve in X through x. We remark that this part of the proof works also for Kähler manifolds, but by (1.8) it fails for arbitrary compact complex manifolds.

Step 3. We show how to get $-(f_*(C) \cdot K_X) - g(C) \cdot \dim X > 0$. To do this, we need to get $-(f_*(C) \cdot K_X)$ big enough. We consider three cases:

(1) $g(C) = 0$. If $-(f_*(C) \cdot K_X) > 0$, then C moves in X, but we already knew that X has a rational curve through x.

(2) $g(C) = 1$. If $-(f_*(C) \cdot K_X) > 0$, compose f with the endomorphism of C given by multiplication by the integer n. Then

$$-((f \circ n)_*(C) \cdot K_X) - \dim X = -n^2(f_*(C) \cdot K_X) - \dim X,$$

so this time some multiple of C moves (so that one point of some sheet over the image stays fixed).

(3) $g(C) \geq 2$. A curve of genus ≥ 2 has no endomorphisms of degree greater than 1. Thus we can try to consider another curve $h : C' \to C$ and deform $f \circ h$. The problem here is that although $((f \circ h)_*(C') \cdot K_X = \deg h \cdot (f_*(C) \cdot K_X)$, the genus also changes. Thus, for example, if we try to move an m-sheeted unbranched cover of C, we are only guaranteed a deformation space of dimension

$$m[-(f_*(C) \cdot K_X) - g(C) \cdot \dim X] + (m - 1) \dim X.$$

This does not necessarily get positive by making m large, even when $-(f_*(C) \cdot K_X) > 0$.

Step 4. Thus we are in trouble in the case $g(C) > 1$ because C does not admit endomorphisms of high degree. However, there is a situation in which a curve C does in fact admit endomorphisms of high degree, namely, in finite characteristic. The Frobenius morphism is such an endomorphism. We next see how to pass from our original situation to one over a field of characteristic $p > 0$.

Step 5. Take a curve C and a smooth variety X. First suppose that both C and X are defined by equations with integral coefficients:

$$h_1(x_0, \ldots, x_n), \ldots, h_r(x_0, \ldots, x_n) \qquad \text{define} \qquad X,$$
$$c_1(y_0, \ldots, y_m), \ldots, c_s(y_0, \ldots, y_m) \qquad \text{define} \qquad C.$$

Let $\mathbb{F}_p$ be the field with p elements and $\bar{\mathbb{F}}_p$ its algebraic closure. Then the equations c_j and h_i above define varieties C_p and X_p in the projective spaces $\bar{\mathbb{F}}_p \mathbb{P}^m$ and $\bar{\mathbb{F}}_p \mathbb{P}^n$ respectively.

These varieties are non-singular, and $\dim C_p = 1$, for almost all p (that is, for all p except for a finite number of exceptions). The mapping

$$(y_0, \ldots, y_m) \to (y_0^p, \ldots, y_m^p)$$

gives an endomorphism F_p of C_p, called the *Frobenius* endomorphism. Although F_p is injective set-theoretically, it is a morphism of degree p.

By 'generic flatness' over $\operatorname{Spec} \mathbb{Z}$, the values $((f_p)_*(C_p) \cdot K_{X_p})$, $g(C_p)$, and $\chi(T_X|_{C_p})$, are constant for almost all p. The 'base-pointed' deformation space of the composite morphism

$$C_p \xrightarrow{F_p^m} C_p \xrightarrow{f_p} X_p$$

has dimension bounded below by

$$-p^m((f_p)_*(C_p) \cdot K_{X_p}) - g(C_p) \cdot \dim X.$$

$((f_p)_*(C_p) \cdot K_{X_p})$ is negative and independent of p, so the above expression $-p^m((f_p)_*(C_p) \cdot K_{X_p}) - g(C_p) \cdot \dim X$ is positive for $m \gg 1$ (for almost all p we could use $m = 1$). Then, as in Step 2, we produce a rational curve A_p on X_p for almost all p.

Step 6. Suppose now that we are in the general case in which the coefficients of the h_j (defining X in $\mathbb{P}^n$), the c_j (defining C in $\mathbb{P}^m$), the b_j (defining the graph of the map in $\mathbb{P}^n \times \mathbb{P}^m$) and the coordinates of $0 \in C \subset \mathbb{P}^m$ are not necessarily integers. In any case, these coefficients generate a finitely generated ring $R \supset \mathbb{Z}$. Let P be any maximal ideal in R. Then R/P is a finite field (since an infinite field cannot be finitely generated as a ring over $\mathbb{Z}$). So R/P is isomorphic to $\mathbb{F}_{p^k}$, the finite field with p^k elements for some p. In this case, our Frobenius morphism is given by raising the homogeneous coordinates $(x_0, \ldots, x_m)$ of $\mathbb{F}_{p^k} \mathbb{P}^m$ to the p^k-th power. The rest of the argument proceeds as above, giving us a rational curve A_P, for all closed points P in some Zariski open dense set of $\operatorname{Spec} R$.

Step 7. Now we assume that $-K_X$ is ample and that X is embedded by $|-mK_X|$ for some positive integer m. In this step, we wish to replace A_p with a rational curve B_p with $-(K_{X_p} \cdot B_p) \leq \dim X + 1$.

To do this, notice that, if $-(K_{X_p} \cdot A_p) > \dim X + 1$, then the morphism $A_p \to X_p$ deforms with two points q, q' fixed in at least a two-parameter family. Since $\mathbb{P}^1$ has only a one-dimensional family of automorphisms leaving two points fixed, the image of $A_p \subset X_p$ must move. By (1.9), one sees that A_p decomposes into a sum of rational curves of lower degrees. So we must be able to find a rational curve of lower degree as long as $-(A_p \cdot K_X) > \dim X + 1$.

Step 8. In this last step, we must conclude the existence of a rational curve on the variety X of characteristic zero from the existence of the bounded degree rational curve B_p for almost all p. (The general case using $P \in \operatorname{Spec} R$ is analogous.)

Principle 1.11. *If a homogeneous system of algebraic equations with integral coefficients has a non-trivial solution in $\overline{\mathbb{F}}_p$ for infinitely many p (in the general case, for a Zariski dense subset of $\operatorname{Spec} R$), then it has a non-trivial solution in any algebraically closed field.*

Traditional proof: By elimination theory, the existence of a common solution to a system of equations is given by the vanishing of a series of determinants of matrices whose entries are polynomials (with integral coefficients) in the coefficients of the equations (see, for instance, [vdW91, 16.5]). A determinant vanishes if it vanishes mod p for an infinite number of primes p.

Modern proof: The equations define a closed subscheme $Z \subset \mathbb{P}^N_{\operatorname{Spec} R}$. The projection $\pi : \mathbb{P}^N_{\operatorname{Spec} R} \to \operatorname{Spec} R$ is proper, so $\pi(Z) \subset \operatorname{Spec} R$ is closed. If $\pi(Z)$ contains a Zariski dense set of closed points, it also contains the generic point.   □

In our situation, for most p we have homogeneous forms

$$(g_{p,0}, \ldots, g_{p,n})$$

of degree $m(\dim X + 1)$ in (t_0, t_1) giving the map $\mathbb{P}^1 \to X \subset \mathbb{P}^n$ such that

$$h_i(g_{p,0}, \ldots, g_{p,n}) = 0$$

identically in (t_0, t_1) for all i. This condition can be expressed as a system of equations in the coefficients of the g_k. Since this system has a solution for a Zariski dense subset of the primes p, it has a solution in any algebraically closed field by the above principle.

Step 9. Finally, we should remark that Steps 2 and 7 allow the construction of a rational curve of degree $\leq \dim X + 1$ through any pre-given point of X. So, if $-K_X$ is positive, X must be covered by an algebraic family of rational curves of degree $\leq \dim X + 1$. $\square$

1.2 Finding Rational Curves when K_X is not Nef

1.12. Now let us weaken our hypotheses about X. Namely, from now on we only assume that, for our fixed f,

$$-(f(C) \cdot K_X) > 0,$$

rather than assuming the positivity of $-K_X$. We also fix an ample divisor H on X.

The main result is the following.

Theorem 1.13. *[Mor82] Let X be a smooth projective variety and H an ample divisor on X. Assume that there is an irreducible curve $C' \subset X$ such that $-(C' \cdot K_X) > 0$. Then there is a rational curve $E \subset X$ such that*

$$\dim X + 1 \geq -(E \cdot K_X) > 0 \quad and \quad \frac{-(E \cdot K_X)}{(E \cdot H)} \geq \frac{-(C' \cdot K_X)}{(C' \cdot H)}.$$

Proof: The proof proceeds along the lines of section 1, except we have to keep track of the resulting curves more carefully.

Step 1. Let $f : C \to X$ be the normalization of C'. If

$$-(f(C) \cdot K_X) - g(C) \cdot \dim X > 0, \tag{1.1}$$

then C deforms with one point fixed. As before, this family must degenerate to $f'(C) +$ (sum of rational curves).

Step 2. In order to achieve (1.1), we pass to finite characteristic, and compose f with the m-th power of the Frobenius morphism. For $m \gg 0$, we are able to degenerate $p^m \cdot f(C_p)$ to

$$f_m^1(C_p) + Z_{p,m}^1 \sim p^m \cdot f(C_p), \tag{1.2}$$

where $Z_{p,m}^1$ is a sum of rational curves. Set $M = -(C' \cdot K_X)/(C' \cdot H)$. The ratio $-(f(C_p) \cdot K_{X_p})/(f(C_p) \cdot H_p)$ equals M for almost all p and it does not change if we replace f with its composition with a power of a Frobenius morphism. If

$$-(f_m^1(C_p) \cdot K_{X_p}) - g(C_p) \cdot \dim X_p > 0,$$

we can move $f_m^1(C_p)$ as before (without composing again with the Frobenius morphism). We iterate these moves. Each time the intersection number of H_p with $f_m^j(C_p)$ goes down, so the process must stop. Thus we reach an algebraic equivalence

$$f_m^s(C_p) + Z_{p,m}^s \sim p^m \cdot f(C_p),$$

such that

$$-(f_m^s(C_p) \cdot K_{X_p}) \leq g(C_p) \cdot \dim X_p.$$

Let $a = -(f_m^s(C_p) \cdot K_{X_p})$; $b = -(Z_{p,m}^s \cdot K_{X_p})$; $c = (f_m^s(C_p) \cdot H_p)$; $d = (Z_{p,m}^s \cdot H_p)$.

For large m, $(c + d)$ is large, $(a + b)/(c + d) = M$, so $(a + b)$ must be large. But a is bounded, so b must get large.

Step 3. We claim the following.

For any $\epsilon > 0$, if $m \gg 0$ then there exists an irreducible component E_p of $Z_{p,m}^s$ such that

$$\frac{-(E_p \cdot K_{X_p})}{(E_p \cdot H_p)} > M - \epsilon. \tag{1.3}$$

First if $a/c < M$, then $b/d \geq M$ by (1.14). Hence by (1.14) again, E_p exists as claimed. For large m, if c gets large then we eventually get $a/c < M$ and we are done as above. If c stays bounded, then a, c stay bounded and b, d get large and $b/d + \epsilon > (a + b)/(c + d) = M$. Hence for $m \gg 0$, one has

$$\frac{-(Z_{p,m}^s \cdot K_{X_p})}{(Z_{p,m}^s \cdot H_p)} > M - \epsilon.$$

Now again by (1.14), the required E_p exists as above.

Lemma 1.14. *If $c, d > 0$, then $\frac{a+b}{c+d} \leq \max\{\frac{a}{c}, \frac{b}{d}\}$.*   □

Step 4. Suppose now that $-(E_p \cdot K_{X_p}) > (\dim X + 1)$. Then, as in (1.10, Step 2), we can move the rational curve E_p with two points fixed and the moving curve must degenerate into a sum of two or more (not necessarily distinct) rational curves. We use (1.14) again to conclude that the inequality (1.3) must hold for at least one of the components E_p' of the degeneration. If

$$-(E_p' \cdot K_{X_p}) > (\dim X + 1),$$

E_p' moves and as above we find E_p'' for which (1.3) holds. This process cannot continue indefinitely, since at each step $(E_p \cdot H_p)$ goes down. So eventually we arrive at a curve (which we again call E_p) such that

$$\dim X + 1 \geq -(E_p \cdot K_{X_p}) > 0, \quad \text{and} \quad \frac{-(E_p \cdot K_{X_p})}{(E_p \cdot H)} \geq M - \epsilon. (1.4)$$

This implies that $(E_p \cdot H_p) \leq (\dim X + 1)/(M - \epsilon)$. Hence all the fractions occurring above have bounded denominators, so once ϵ is sufficiently small, we can take it to be zero.

Furthermore, $(E_p \cdot H_p)$ is bounded independently of p. We can reason as in (1.10, Step 8) to conclude the existence of a rational curve E on the complex projective manifold X which satisfies the requirements of (1.13).   □

Remark: This argument does not allow us to say anything about the position of the rational curves on X. A different argument, however, shows that, through any point of C' there is a rational curve (see [MM86] or [Kol96, II.5]).

1.3 The Cone of Curves of Smooth Varieties

Our main goal in this section is to prove the Cone Theorem for smooth projective varieties. This result, proved in [Mor82], was the first major step of Mori's program. The final form of the theorem is established in Chapter 3.

Definition 1.15. Let $K = \mathbb{Q}$ or $K = \mathbb{R}$ and V a K-vector space. A subset $N \subset V$ is called a *cone* if $0 \in N$ and N is closed under multiplication by positive scalars.

A subcone $M \subset N$ is called *extremal* if $u, v \in N$, $u + v \in M$ imply that $u, v \in M$. M is also called an *extremal face* of N. A 1-dimensional extremal subcone is called an *extremal ray*.

It is not hard to see that any closed convex cone is the convex hull of its extremal rays.

Definition 1.16. Let X be a proper variety. A *1-cycle* is a formal linear combination of irreducible, reduced and proper curves $C = \sum a_i C_i$. A 1-cycle is called *effective* if $a_i \geq 0$ for every i. Two 1-cycles C, C' are called *numerically equivalent* if $(C \cdot D) = (C' \cdot D)$ for any Cartier divisor D. 1-cycles with real coefficients modulo numerical equivalence form an $\mathbb{R}$-vector space; it is denoted by $N_1(X)$. The class of a 1-cycle C is denoted by $[C]$.

Let $NS(X)$ denote the Néron–Severi group of X (cf. [Har77]). Intersection of curves and divisors gives a perfect pairing

$$N_1(X) \times (NS(X) \otimes_{\mathbb{Z}} \mathbb{R}) \to \mathbb{R}.$$

The Theorem of the Base of Néron–Severi asserts that $N_1(X)$ is finite dimensional; see [Kle66] for a proof. Its dimension is called the *Picard number* of X and denoted by $\rho(X)$.

If X is smooth over $\mathbb{C}$, then there is an injection $N_1(X) \to H_2(X, \mathbb{R})$ by the Lefschetz theorem on $(1,1)$-classes; see [GH78, p.161]. This shows that $N_1(X)$ is finite dimensional when X is smooth and defined over $\mathbb{C}$.

Definition 1.17. Let X be a proper variety. Set

$$
\begin{aligned}
NE_{\mathbb{Q}}(X) &= \{\textstyle\sum a_i[C_i] : C_i \subset X, 0 \le a_i \in \mathbb{Q}\} \subset N_1(X); \\
NE(X) &= \{\textstyle\sum a_i[C_i] : C_i \subset X, 0 \le a_i \in \mathbb{R}\} \subset N_1(X), \quad \text{and} \\
\overline{NE}(X) &= \text{the closure of } NE(X) \text{ in } N_1(X),
\end{aligned}
$$

where the C_i are irreducible curves on X. Clearly $NE_{\mathbb{Q}}(X)$ is dense in $\overline{NE}(X)$.

For any divisor D, set $D_{\ge 0} := \{x \in N_1(X) : (x \cdot D) \ge 0\}$ (similarly for $> 0, \le 0, < 0$) and $D^\perp := \{x : (x \cdot D) = 0\}$. We also use the notation

$$\overline{NE}(X)_{D \ge 0} := \overline{NE}(X) \cap D_{\ge 0},$$

and similarly for $> 0, \le 0, < 0$.

It is also natural to consider the cones of divisors, but we do not treat them in this book. The closure of the cone of nef divisors is dual to the cone of effective 1-cycles. This cone was first studied by [Hir60].

Next we give some general results and work out a few examples of cones of curves. One of the basic results is due to [Kle66]. A proof is given in section 1.5.

Theorem 1.18 (Kleiman's Ampleness Criterion). *Let X be a projective variety and D a Cartier divisor on X. Then D is ample iff*

$$D_{>0} \supset \overline{NE}(X) \setminus \{0\}.$$

The situation of ample D is illustrated by the following picture.

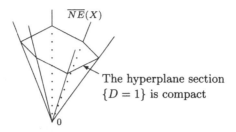

The hyperplane section $\{D = 1\}$ is compact

Corollary 1.19. *Let X be a projective variety and H an ample divisor. Then:*

(1) $\overline{NE}(X)$ *does not contain a straight line.*

(2) *For any $C > 0$ the set $\{z \in \overline{NE}(X) : (z \cdot H) \leq C\}$ is compact.*

(3) *For any $C > 0$ there are only finitely many numerical equivalence classes of effective 1-cycles $Z = \sum a_i Z_i$ with integral coefficients such that $(Z \cdot H) \leq C$.*

Proof. (1) follows from the fact that a linear function cannot be positive on a straight line minus the origin.

Fix a norm $\|\ \|$ on $N_1(X)$ and assume that $W_C := \{z \in \overline{NE}(X) : (z \cdot H) \leq C\}$ is not compact. Then there is a sequence $z_i \in W_C$ such that $\|z_i\| \to \infty$. $z_i/\|z_i\|$ is a bounded sequence, hence a suitable subsequence converges to a point $y \in \overline{NE}(X) \setminus \{0\}$ and $(y \cdot H) = \lim(z_i \cdot H)/\|z_i\| = 0$. Thus H is not ample, a contradiction.

Finally, 1-cycles with integral coefficients correspond to a discrete set in $N_1(X)$, and so it has only finitely many points in any compact set. $\square$

Lemma 1.20. *If D is a divisor on an irreducible and proper surface X with $(D^2) > 0$, then either $|nD| \neq \emptyset$ or $|-nD| \neq \emptyset$ for $n \gg 0$.*

Proof: By the Riemann–Roch Theorem,

$$h^0(nD) + h^0(K_X - nD) \geq \frac{n^2}{2}(D^2) - \frac{n}{2}(D \cdot K_X) + \chi(\mathcal{O}_X),$$

$$h^0(-nD) + h^0(K_X + nD) \geq \frac{n^2}{2}(D^2) + \frac{n}{2}(D \cdot K_X) + \chi(\mathcal{O}_X).$$

Letting n get large, we notice that the right-hand-side of each equation gets big. But it cannot be true that both $h^0(K_X - nD)$ and $h^0(K_X + nD)$ get big, since the two divisors sum to a fixed linear system $|2K_X|$. Thus $h^0(nD)$ or $h^0(-nD)$ grows quadratically with n. $\square$

Corollary 1.21. *Let X be an irreducible and projective surface with an ample divisor H. The set $Q := \{z \in N_1(X) : (z^2) > 0\}$ has two connected components*

$$Q^+ := \{z \in Q : (z \cdot H) > 0\} \quad and \quad Q^- := \{z \in Q : (z \cdot H) < 0\}.$$

Furthermore, $Q^+ \subset \overline{NE}(X)$.

Proof: By the Hodge Index Theorem the intersection form on $N_1(X)$ has exactly one positive eigenvalue. Thus in a suitable basis it can be written as $x_1^2 - \sum_{i \geq 2} x_i^2$, and we can choose the basis such that $[H] = (\sqrt{(H \cdot H)}, 0, \ldots, 0)$. This gives the two connected components

$$Q^+ = (x_1 > (\sum_{i \geq 2} x_i^2)^{1/2}) \quad and \quad Q^- = (x_1 < -(\sum_{i \geq 2} x_i^2)^{1/2}).$$

For any $[D] \in Q$, either D or $-D$ is effective. An effective curve has positive intersection with H. Thus the effective curves in Q are precisely the ones in Q^+. $\qquad\square$

Lemma 1.22. *Let X be an irreducible and projective surface and $C \subset X$ an irreducible curve. If $(C^2) \leq 0$, then $[C]$ is in the boundary of $\overline{NE}(X)$. If $(C^2) < 0$ then $[C]$ is extremal in $\overline{NE}(X)$.*

Proof: If $D \subset X$ is an irreducible curve such that $(D \cdot C) < 0$, then $D = C$. If $(C^2) = 0$ then $D \mapsto (D \cdot C)$ is a linear function which is non-negative on $\overline{NE}(X)$ and zero on C.

In general, $\overline{NE}(X)$ is spanned by $\mathbb{R}_{\geq 0}[C]$ and $\overline{NE}(X)_{C \geq 0}$, because the class of every irreducible curve $D \neq C$ is in $\overline{NE}(X)_{C \geq 0}$. If $(C^2) < 0$, then $[C] \notin \overline{NE}(X)_{C \geq 0}$. Thus $[C]$ generates an extremal ray. $\qquad\square$

Example 1.23. Let us now look at our series of examples:

(1) Suppose X is a minimal ruled surface over a curve B. Then $\overline{NE}(X)$ is a cone in $\mathbb{R}^2$. By (1.19), $\overline{NE}(X)$ is generated by its two edges. Let f be the homology class of the fiber of $X \to B$ and s the other edge. By the adjunction formula, $(f \cdot K_X) = -2 < 0$. By (1.21), $s^2 \leq 0$. If $s^2 < 0$, take a sequence D_n of effective 1-cycles converging to a point of $\mathbb{R}_{\geq 0}[s]$, and notice that, for $n \gg 0$, $(D_n^2) < 0$. There is an irreducible component E_n of $\operatorname{Supp} D_n$ such that $(E_n^2) < 0$, hence by (1.22) above, $[E_n] \in \mathbb{R}_{\geq 0}[s]$. If $s^2 = 0$, fix any irreducible D other than a fiber. Then $[D]$ and f span $N_1(X)$. Write $(xf + yD)^2 = 2xy(f \cdot D) + y^2(D \cdot D) = 0$. Then s is a solution to $2x(f \cdot D) + y(D \cdot D) = 0$, so s must have a

rational slope. If the genus of B is 0 or 1, from the classification of ruled surfaces one can see that there is an irreducible curve $S \subset X$ with $[S] \in \mathbb{R}_{\geq 0}[s]$. If $g(B) \geq 2$ then there may not be such a curve, though the construction of such an example is not straightforward.

For instance, let $\rho : \pi_1(B) \to SU(2, \mathbb{C})$ be a representation which remains Zariski dense on any finite index subgroup of $\pi_1(B)$. This corresponds to a rank two vector bundle on B. Let X be the associated ruled surface. See [Har70] for details.

(2) Let A be an abelian surface with an ample divisor H. Since the self-intersection of any curve on an abelian surface is non-negative, $\overline{NE}(A) = \bar{Q}^+$ by (1.20). If $\dim N_1(A) \geq 3$ (e.g. $A = E \times E$ for some elliptic curve E), then $\overline{NE}(A)$ is a 'round' cone.

Every point on the boundary of $\overline{NE}(A)$ is extremal. Most of those points have irrational coordinates, thus they do not correspond to any curve on A.

(3) Cubic surfaces $X \subset \mathbb{P}^3$.

Here $\dim N_1(X) = 7$ and X contains 27 lines $L_1, \ldots, L_{27}$. By (1.22), these generate 27 extremal rays.

We shall see that $NE(X) = \mathbb{R}_{\geq 0}[L_1] + \cdots + \mathbb{R}_{\geq 0}[L_{27}]$. In particular, $NE(X) = \overline{NE}(X)$ is a cone over a finite polyhedron. A similar picture emerges for all Del Pezzo surfaces.

(4) [Nag60] Let X be obtained from $\mathbb{P}^2$ by blowing up at the nine base points of a pencil of cubic curves, all of whose members are irreducible. Choosing one of the nine points as the zero section, we get an infinite group of automorphisms of X generated by the other eight sections. So X has infinitely many (-1)-curves, all of which span an extremal ray of $\overline{NE}(X)$. $|-K_X|$ is the elliptic pencil, thus $-K_X$ is nef, but not ample.

With these examples in mind, we are ready to state the first result of Mori for smooth varieties of arbitrary dimension. The singular case is treated in Chapter 3. The proof in the smooth case is more geometric, and it is worthwhile to see it first.

Theorem 1.24 (Cone Theorem). *[Mor82] Let X be a non-singular projective variety.*

(1) *There are countably many rational curves $C_i \subset X$ such that $0 < -(C_i \cdot K_X) \leq \dim X + 1$, and*

$$\overline{NE}(X) = \overline{NE}(X)_{K_X \geq 0} + \sum_i \mathbb{R}_{\geq 0}[C_i].$$

(2) *For any $\epsilon > 0$ and ample divisor H,*

$$\overline{NE}(X) = \overline{NE}(X)_{(K_X + \epsilon H) \geq 0} + \underset{\text{finite}}{\sum} \mathbb{R}_{\geq 0}[C_i].$$

The meaning of the theorem is illustrated by the following picture.

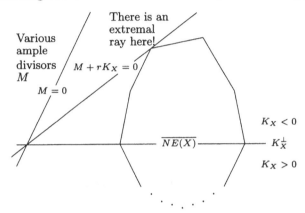

Proof: Up to numerical equivalence, there are only countably many curves on X. From each numerical equivalence class such that $0 < -(C \cdot K_X) \leq \dim X + 1$ we pick a rational curve C_i (if there is such). Set

$$W := \text{closure of } (\overline{NE}(X)_{K_X \geq 0} + \sum_i \mathbb{R}_{\geq 0}[C_i]).$$

The main part of the proof is to show that $\overline{NE}(X) = W$.

Assume the contrary. Then there is a divisor D such that the corresponding linear function is positive on $W \setminus \{0\}$ but negative somewhere on $\overline{NE}(X)$.

Let H be an ample divisor on X and $\mu > 0$ the largest number such that $H + \mu D$ is nef. (As a consequence of (1.24) we see that μ is rational, but for now all we know is that μ is some real number.) Choose $0 \neq z \in \overline{NE}(X)$ such that $(z \cdot (H + \mu D)) = 0$. Then $(z \cdot K_X) < 0$ since $\overline{NE}(X)_{K_X \geq 0} \subset W$.

By definition of $\overline{NE}(X)$, there are effective 1-cycles $Z_k = \sum_j a_{ij} Z_{kj}$ (with real or rational coefficients) on X such that $[Z_k] \to z$. The situation is illustrated by the following picture.

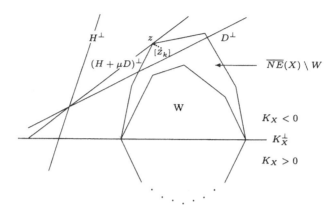

For any rational $\mu' < \mu$, the $\mathbb{Q}$-divisor $H + \mu'D$ is ample on X. By (1.14),

$$\max_j \frac{-(Z_{kj} \cdot K_X)}{(Z_{kj} \cdot (H + \mu'D))} \geq \frac{-(Z_k \cdot K_X)}{(Z_k \cdot (H + \mu'D))}.$$

We rearrange the indices so that the maximum is achieved for Z_{k0}. Next we apply (1.13) to Z_{k0} and the ample divisor $H + \mu'D$. We obtain that there is a rational curve $E_{i(k)}$ such that $\dim X + 1 \geq -(E_{i(k)} \cdot K_X) > 0$ and

$$\frac{-(E_{i(k)} \cdot K_X)}{(E_{i(k)} \cdot (H + \mu'D))} \geq \frac{-(Z_{k0} \cdot K_X)}{(Z_{k0} \cdot H + \mu'D)} \geq \frac{-(Z_k \cdot K_X)}{(Z_k \cdot (H + \mu'D))}.$$

Up to numerical equivalence, $E_{i(k)}$ coincides with one of the curves $\{C_i\}$, in particular $(E_{i(k)} \cdot D) \geq 0$. These give the inequalities

$$\frac{-(E_{i(k)} \cdot K_X)}{(E_{i(k)} \cdot H)} \geq \frac{-(E_{i(k)} \cdot K_X)}{(E_{i(k)} \cdot (H + \mu'D))} \geq \frac{-(Z_k \cdot K_X)}{(Z_k \cdot (H + \mu'D))}.$$

There is a constant M such that $MH + K_X$ is ample. This means that $M \geq -(E_{i(k)} \cdot K_X)/(E_{i(k)} \cdot H)$, which gives the inequality

$$M \geq \frac{-(E_{i(k)} \cdot K_X)}{(E_{i(k)} \cdot H)} \geq \frac{-(Z_k \cdot K_X)}{(Z_k \cdot (H + \mu'D))}.$$

Let now $i \to \infty$ and $\mu' \to \mu$. Then

$$M \geq \frac{-(Z_k \cdot K_X)}{(Z_k \cdot (H + \mu'D))} \to \frac{-(z \cdot K_X)}{(z \cdot (H + \mu D))} = \frac{-(z \cdot K_X)}{0} = +\infty.$$

This is a contradiction, thus $\overline{NE}(X) = W$.

Fix now H and ϵ. If $(C_i \cdot (K_X + \epsilon H)) < 0$, then

$$(C_i \cdot H) \leq \frac{-(C_i \cdot K_X)}{\epsilon} \leq \frac{\dim X + 1}{\epsilon}.$$

By (1.19) there are only finitely many such C_i. Set

$$V_{H,\epsilon} := \overline{NE}(X)_{(K_X + \epsilon H) \geq 0} + \sum_{(C_i \cdot (K_X + \epsilon H)) < 0} \mathbb{R}_{\geq 0}[C_i].$$

$V_{H,\epsilon}$ is a closed cone, and it contains W. We have proved that $W = \overline{NE}(X)$, hence part (2) of the Cone Theorem follows.

The only missing ingredient is to show that the right hand side of (1.24.1) is closed. This is a formal argument about cones in $\mathbb{R}^m$, cf. [Kol96, p.161]. We do not need this part later, so we skip its proof. $\square$

While the Cone Theorem reveals the interesting structure of the cone of curves, it is not immediately clear that this structure can be related to more traditional geometric properties of a variety. The key step is the observation that extremal rays correspond to morphisms of a variety:

Definition 1.25. Let X be a projective variety and $F \subset \overline{NE}(X)$ an extremal face. A morphism $\mathrm{cont}_F : X \to Z$ is called the *contraction* of F if the following hold:

(1) $\mathrm{cont}_F(C) =$ point for an irreducible curve $C \subset X$ iff $[C] \in F$,
(2) $(\mathrm{cont}_F)_* \mathcal{O}_X = \mathcal{O}_Z$.

Remark 1.26. In general not every extremal face can be contracted, cf. (1.27). One of our main aims is to find conditions which guarantee the existence of cont_F. Low dimensional smooth cases are discussed in the next section, but the general result is only proved in Chapter 3.

Condition (1) specifies the fibers of cont_F set theoretically, and (2) says that cont_F is its own Stein factorization. These imply that cont_F is uniquely determined by F.

Let $g : X \to Z$ be a morphism between projective varieties such that $g_* \mathcal{O}_X = \mathcal{O}_Z$. Let F be the closed cone spanned by all $[C]$ such that $g(C)$ is a point. Then $\mathrm{cont}_F = g$, thus g is the contraction of an extremal face.

Example 1.27. [Zar62] We begin by letting $g : X \to \mathbb{P}^2$ be the blow up of $\mathbb{P}^2$ at 12 points $p_1, \ldots, p_{12}$ on a smooth cubic plane curve D. Let $C \subset X$ be the birational transform of the plane cubic. $C^2 = -3$, so that C can be contracted via an analytic morphism $f : X \to Y$ to an analytic surface Y (cf. [Gra62]). However Y cannot be projective if the 12 points are in general position. To see this, suppose M is any line bundle on Y. Then, f^*M is linearly equivalent to $g^*\mathcal{O}_{\mathbb{P}^2}(b) + \sum a_i E_i$, where E_i is the exceptional curve above p_i. $f^*M|_C$ is linearly equivalent to 0. So we would have a linear equivalence $\mathcal{O}_D(b) + \sum a_i p_i \sim 0$ on D, which is clearly impossible for general choice of the p_i.

However, if the p_i are the points of intersection of a quartic curve Q with D, then the linear system $|M|$ spanned by Q and by the quartics of the form $C + (\text{line})$ is birationally transformed to a free linear system $g_*^{-1}|M|$ and it realizes $f : X \to Y$ as a morphism into a projective space.

1.4 Minimal Models of Surfaces

The aim of this section is to present the classical theory of minimal models of surfaces from the point of view of extremal ray theory. This illustrates extremal ray theory in a very simple setting. Then we review the first steps generalizing these results to threefolds.

In this section all varieties are over an algebraically closed field k.

Theorem 1.28. *Let X be a smooth projective surface and $R \subset \overline{NE}(X)$ an extremal ray such that $(R \cdot K_X) < 0$. Then the contraction morphism (1.25) $\mathrm{cont}_R : X \to Z$ exists and is one of the following types:*

(1) *Z is a smooth surface and X is obtained from Z by blowing up a closed point; $\rho(Z) = \rho(X) - 1$.*

(2) *Z is a smooth curve and X is a minimal ruled surface over Z; $\rho(X) = 2$.*

(3) *Z is a point, $\rho(X) = 1$ and $-K_X$ is ample. (In fact, $X \cong \mathbb{P}^2$ but this is harder to prove.)*

Proof. Let $C \subset X$ be an irreducible curve such that $[C] \in R$. We show that the above three cases correspond to the sign of (C^2).

Assume first that $(C^2) > 0$. By (1.21), $[C]$ is an interior point of $\overline{NE}(X)$. $[C]$ also generates an extremal ray, thus $N_1(X) \cong \mathbb{R}$. By assumption $(C \cdot K_X) < 0$, thus K_X is negative on $\overline{NE}(X) \setminus \{0\}$ and so $-K_X$ is ample by (1.18).

Next consider the case when $(C^2) = 0$. We prove that $|mC|$ gives the contraction morphism for $m \gg 1$. Since C is effective, $H^2(X, \mathcal{O}(mC)) = 0$ for $m \gg 1$. By assumption $K_X \cdot C < 0$, thus

$$h^0(X, \mathcal{O}(mC)) \geq \chi(X, \mathcal{O}(mC)) = \frac{-K_X \cdot C}{2} m + \chi(\mathcal{O}_X) \geq 2 \quad \text{for } m \gg 1.$$

Any fixed component is a multiple of C, thus there is an $m' > 0$ such that $|m'C|$ has no fixed components. Since $(C^2) = 0$, the general member is disjoint from C, thus $|m'C|$ is base point free. Let $\text{cont}_R : X \to Z$ be the Stein factorization of the corresponding morphism.

Let $\sum a_i C_i$ be a fiber of cont_R. Then $\sum a_i[C_i] = [C] \in R$. Since R is an extremal ray, this implies that $[C_i] \in R$ for every i. Thus $(C_i^2) = 0$ and $(C_i \cdot K_X) < 0$. By the adjunction formula this implies that $C_i \cong \mathbb{P}^1$ and $(C_i \cdot K_X) = -2$. Thus

$$-2 = (C \cdot K_X) = \left(\sum a_i C_i \cdot K_X\right) = -2 \sum a_i,$$

which shows that $\sum a_i C_i$ is an irreducible and reduced curve, isomorphic to $\mathbb{P}^1$. Thus X is a minimal ruled surface over Z.

Finally assume that $(C^2) < 0$. By the adjunction formula C is a (-1)-curve and (1.28.1) reduces to the Castelnuovo contraction theorem (1.2). □

The cases (1.28.2–3) provide a structure theorem for X, while (1.28.1) introduces a new surface Z. We can apply (1.28) to Z and continue if possible. At each step the Picard number drops by one, and so this process terminates. This gives the following:

Theorem 1.29. *Let X be a smooth projective surface. There is a sequence of contractions $X \to X_1 \to \cdots \to X_n = X'$ such that X' satisfies exactly one of the following conditions:*

(1) $K_{X'}$ *is nef;*
(2) X' *is a minimal ruled surface over a curve C;*
(3) $X' \cong \mathbb{P}^2$. □

Definition 1.30. If $K_{X'}$ is nef above, then X' is called a *minimal model* of X. It turns out that in this case the morphism $X \to X'$ is unique, thus X' is determined by X.

Summary 1.31. Our work so far gives the following step by step approach to the minimal models of surfaces:

Step 0. We start with a smooth projective surface X.

Step 1. If K_X is nef, then we go to Step 5. If K_X is not nef, then by the Cone Theorem we can choose a K_X-negative extremal ray $R \subset \overline{NE}(X)$.

Step 2. By (1.28), the contraction morphism $\text{cont}_R : X \to Z$ exists. There are two possibilities.

Step 3. If $\dim Z = \dim X$ then go back to Step 0 with Z replacing X.

Step 4. If $\dim Z < 2 = \dim X$ then (1.28.2–3) gives a structure theorem for X.

Step 5. If K_X is nef, then we stop. These surfaces should be investigated by other methods.

The aim of Mori's program is to provide a similar step by step approach to higher dimensional varieties. The part corresponding to Step 1 is the Cone Theorem, though we have not proved the contractibility part yet.

The next step is to develop an analogue of (1.28) for higher dimensional varieties. There are already numerous cases for smooth 3-folds over $\mathbb{C}$ [Mor82]. The list is essentially the same for smooth 3-folds over any algebraically closed field [Kol91a]. The case of smooth 4-folds over $\mathbb{C}$ was worked out by many authors, see [AW97] for an overview.

Theorem 1.32. *[Mor82] Let X be a non-singular projective threefold over $\mathbb{C}$ and $\mathrm{cont}_R : X \to Y$ the contraction of a K_X-negative extremal ray $R \subset \overline{NE}(X)$. The following is a list of all possibilities for cont_R:*

    E : *(Exceptional)* $\dim Y = 3$, cont_R *is birational and there are five types of local behaviour near the contracted surface:*

        $E1$: cont_R *is the (inverse of the) blow-up of a smooth curve in the smooth threefold Y.*

        $E2$: cont_R *is the (inverse of the) blow-up of a smooth point of the smooth threefold Y.*

        $E3$: cont_R *is the (inverse of the) blow-up of an ordinary double point of Y. (Locally analytically, an ordinary double point is given by the equation $x^2 + y^2 + z^2 + w^2 = 0$.)*

        $E4$: cont_R *is the (inverse of the) blow-up of a point of Y which is locally analytically given by the equation $x^2 + y^2 + z^2 + w^3 = 0$.*

        $E5$: cont_R *contracts a smooth $\mathbb{CP}^2$ with normal bundle $\mathcal{O}(-2)$ to a point of multiplicity 4 on Y which is locally analytically the quotient of $\mathbb{C}^3$ by the involution $(x, y, z) \mapsto (-x, -y, -z)$.*

    C : *(Conic bundle)* $\dim Y = 2$ *and cont_R is a fibration whose fibers are plane conics. (General fibers are, of course, smooth.)*

    D : *(Del Pezzo fibration)* $\dim Y = 1$ *and general fibers of cont_R are Del Pezzo surfaces.*

    F : *(Fano variety)* $\dim Y = 0$, *$-K_X$ is ample and hence X is a Fano variety.* $\square$

The cases C and D can be viewed as quite satisfactory structure theorems. Case F may appear less complete, but a complete list of the occurring Fano threefolds has been worked out; see, for instance [Isk80].

The cases E1 and E2 are close analogues of (1.28.1) and we can again apply (1.32) to Y. The unpleasant surprise is the existence of cases E3, E4 and E5. Here Y is singular, and we cannot apply (1.32) to Y.

In the next chapter we outline a program, called Mori's program or the Minimal Model Program, which generalizes (1.31) to higher dimensions. There are many surprises and still unsolved questions along the way.

1.5 Ampleness Criteria

This section is intended to supplement section 1.3, giving a quick introduction to the fundamental ampleness criteria (1.18) and (1.44). We apply the criteria only to projective varieties in this book, so, for simplicity of treatment, we often consider only this case. For more detailed accounts we refer to [Har70, Kol96].

We assume that the reader is somewhat familiar with intersection numbers of divisors. First we fix our notation, then we outline various approaches to the definition and finally we give a complete list of the basic properties that we use.

Notation 1.33. Let X be a proper scheme over a field k, $Z \subset X$ a closed subscheme of dimension d and $L_1, \cdots, L_d$ Cartier divisors on X. $(L_1 \cdots L_d \cdot Z)$ denotes the *intersection number* of the divisors $L_1, \ldots, L_d$ on Z. If $L_1 = \cdots = L_d = L$ then we also use the notation $(L^d \cdot Z)$. When $Z = X$ we sometimes use $(L_1 \cdots L_d)$ if no confusion is likely.

1.34 (Definitions of the intersection number).
There are at least four ways of defining intersection numbers:

(Classical approach) If $L_1, \cdots, L_d$ and Z intersect only at smooth points with independent tangent planes then let $(L_1 \cdots L_d \cdot Z)$ be the number of intersection points (over the algebraic closure of k). From here we extend to arbitrary divisors by linearity. See, for instance, [Sha94, Chap. IV].

(Cohomological approach) Here one proves that

$$\chi(Z, \mathcal{O}_Z(m_1 L_1 + \cdots + m_d L_d)) = C \cdot m_1 \cdots m_d + \text{(other terms)},$$

and then set $(L_1 \cdots L_d \cdot Z) := C$. This is the method of [Kle66, Kol96].

(General intersection theory) One can define the intersection of a

Cartier divisor with any closed subscheme and then let $(L_1 \cdots L_d \cdot Z)$ be the degree of the zero cycle obtained by intersecting Z first with L_d, then intersecting the result with L_{d-1} and so on. See [Ful84, 2.3] or [Har77, I.7].

(Topological approach) If X is defined over $\mathbb{C}$ then $\mathcal{O}_X(L)$ has a first Chern class $c_1(L) \in H^2(X, \mathbb{Z})$ and $Z \subset X$ has a fundamental class $[Z] \in H_{2k}(X, \mathbb{Z})$. If X is connected then [point] $\in H_0(X, \mathbb{Z})$ is well defined and set

$$(L_1 \cdots L_d \cdot Z) \cdot [\text{point}] := (c_1(L_1) \cup \cdots \cup c_1(L_d)) \cap [Z].$$

See, for instance, [GH78].

With all of these approaches many things need to be checked and their equivalence is also not obvious. The next proposition lists the basic properties of intersection numbers that we use. For proofs we refer to the above mentioned works.

Proposition 1.35. *The intersection numbers have the following properties:*

(1) *If $X \supset Y \supset Z$ and the L'_i are Cartier divisors on Y such that $\mathcal{O}_X(L_i)|_Y \cong \mathcal{O}_Y(L'_i)$ for every i, then*
$$(L_1 \cdots L_d \cdot Z) = (L'_1 \cdots L'_d \cdot Z).$$
(2) *$(L_1 \cdots L_d \cdot Z)$ is symmetric and multilinear in the L_i.*
(3) *If $X \subset \mathbb{P}^N$ and $\mathcal{O}_X(L) = \mathcal{O}_X(1)$ then $(L^d \cdot Z) = \deg Z$.*
(4) *If $L_1 \equiv 0$ then $(L_1 \cdots L_d \cdot Z) = 0$.*
(5) *If $D \subset Z$ is a Cartier divisor and $\mathcal{O}_X(L_d)|_Z \cong \mathcal{O}_Z(D)$ then*
$$(L_1 \cdots L_d \cdot Z) = (L_1 \cdots L_{d-1} \cdot D).$$
(6) *If Z, Z' are integral and $f : Z' \to Z$ is birational then*
$$(L_1 \cdots L_d \cdot Z) = (f^*L_1 \cdots f^*L_d \cdot Z').$$

We also use the following form of Riemann–Roch:

Theorem 1.36 (Very Weak Riemann–Roch). *Let X be a proper scheme of dimension n over a field and $L_1, \cdots, L_r$ Cartier divisors on X. Then $\chi(X, \mathcal{O}_X(k_1 L_1 + \cdots + k_r L_r))$ is a polynomial of degree at most n in $(k_1, \cdots, k_r)$, and*

$$\chi(X, \mathcal{O}_X(\textstyle\sum_{i=1}^r k_i L_i)) = \frac{((\sum_{i=1}^r k_i L_i)^n)}{n!} + (\text{terms of degree} < n),$$

where $((\sum_{i=1}^r k_i L_i)^n)$ is the intersection number explained above.

This result is much weaker than the usual Riemann–Roch (cf. [Har77, App. A]) which identifies the coefficients. The cohomological approach is the easiest way to get (1.36). Reduce it to the case where the L_i are all very ample and then apply the argument of [Har77, III.5, Ex.5.2.(a)] using induction on dim X and r. The proof is essentially the same as Step 1 in the proof of (1.37).

We are now ready to prove the first ampleness criterion:

Theorem 1.37 (Nakai–Moishezon criterion). *Let X be a proper scheme over a field and L a Cartier divisor on X. Then L is ample on X iff $(L^{\dim Z} \cdot Z) > 0$ for every closed integral subscheme $Z \subset X$.*

Proof. We prove the theorem only when X is projective.

If L is ample, we may assume that $X \subset \mathbb{P}^N$ and $\mathcal{O}_X(mL) = \mathcal{O}_X(1)$. Then $m^d(L^d \cdot Z) = \deg Z > 0$ (1.35.3).

To prove the converse, we use Noetherian induction. Thus we may assume that $L|_Z$ is ample for every closed subscheme $Z \subsetneq X$. We may also assume $n = \dim X > 0$.

Step 1. $h^0(X, \mathcal{O}_X(kL)) > 0$ for some $k > 0$.

Let B be a very ample effective Cartier divisor on X such that $L + B$ is very ample. Then a general member $A \in |L + B|$ is an effective Cartier divisor and $L \sim A - B$. In the exact sequence

$$0 \to \mathcal{O}_X(kL - B) \to \mathcal{O}_X(kL) \to \mathcal{O}_B(kL) \to 0,$$

$\mathcal{O}_B(L)$ is ample by induction, and so $h^i(B, \mathcal{O}_B(kL)) = 0$ for $k \gg 0$ and $i \geq 1$. Hence $h^i(X, \mathcal{O}_X(kL - B)) = h^i(X, \mathcal{O}_X(kL))$ for $k \gg 0$ and $i \geq 2$. From the exact sequence

$$0 \to \mathcal{O}_X(kL - B) \to \mathcal{O}_X((k+1)L) \to \mathcal{O}_A((k+1)L) \to 0$$

we get similar equalities: $h^i(X, \mathcal{O}_X(kL - B)) = h^i(X, \mathcal{O}_X((k+1)L))$ for $k \gg 0$ and $i \geq 2$. Hence $h^i(X, \mathcal{O}_X((k+1)L)) = h^i(X, \mathcal{O}_X(kL))$, and so $h^i(X, \mathcal{O}_X(kL))$ is a constant for $i \geq 2$ and $k \gg 0$. Thus

$$\begin{aligned} h^0(X, \mathcal{O}_X(kL)) &\geq h^0(X, \mathcal{O}_X(kL)) - h^1(X, \mathcal{O}_X(kL)) \\ &= \chi(X, \mathcal{O}_X(kL)) + (\text{constant}) \\ &= (L^n)/n! \cdot k^n + (\text{lower order terms}). \end{aligned}$$

Step 2. $\mathcal{O}_X(kL)$ is generated by global sections for some $k > 0$.

By Step 1 there is a non-zero $s \in H^0(X, \mathcal{O}_X(mL))$ for some $m > 0$.

Let $D := (s = 0) \subsetneq X$. Then

$$\mathcal{O}_Y \cong \mathrm{im}[\mathcal{O}_X \xrightarrow{s} \mathcal{O}_X(mL)] = \ker[\mathcal{O}_X(mL) \to \mathcal{O}_D(mL)],$$

for some closed subscheme Y of X. We note that $Y \neq X$ is possible since s may vanish on some component of X. We have an exact sequence

$$0 \to \mathcal{O}_Y((k-1)mL) \to \mathcal{O}_X(kmL) \to \mathcal{O}_D(kmL) \to 0.$$

Since $\mathcal{O}_D(kmL)$ is very ample for $k \gg 0$, it is enough to show that $H^0(X, \mathcal{O}_X(kmL)) \twoheadrightarrow H^0(D, \mathcal{O}_D(kmL))$ is surjective for $k \gg 0$.

If $Y \subsetneq X$, then $H^1(Y, \mathcal{O}_Y(kmL)) = 0$ for $k \gg 0$ by induction, and we obtain the surjection. So we may assume $Y = X$, that is, D is a Cartier divisor. $H^1(D, \mathcal{O}_D(kmL)) = 0$ for $k \gg 0$ by induction. As in Step 1 we see that $h^1(X, \mathcal{O}_X(kD))$ is non–increasing in k for $k \gg 0$, and so it is constant for $k \gg 0$. This shows that $H^0(X, \mathcal{O}_X(kmL)) \twoheadrightarrow H^0(D, \mathcal{O}_D(kmL))$ is surjective for $k \gg 0$.

Step 3. Let $\phi : X \to \mathbb{P}^N$ be the morphism induced by $|kL|$ for some $k > 0$. Then ϕ is finite because if $C \subset X$ is a curve such that $f(C)$ is a point then $\mathcal{O}_X(kL)|_C \cong \mathcal{O}_C$ and $(L \cdot C) = 0$ by (1.35.4) and (1.35.1). $\square$

We also need the following property of nef divisors, due to [Kle66].

Theorem 1.38. *Let X be a proper variety and L a nef Cartier divisor. Then $(L^{\dim Z} \cdot Z) \geq 0$ for every integral closed subscheme $Z \subset X$.*

Proof. We prove this for X projective. The general case is easily reduced to the projective case by Chow's Lemma and (1.35.6).

Let $n = \dim X$. We prove $(L^{\dim Z} \cdot Z) \geq 0$ by induction on n. If $Z \subsetneq X$ then $(L^{\dim Z} \cdot Z) \geq 0$ by the induction hypothesis, so it remains to prove $(L^n) \geq 0$.

Let A be a very ample divisor on X.

$$f(x, y) := ((xL + yA)^n) = \sum_{i=0}^{n} \binom{n}{i} (L^{n-i} \cdot A^i) x^{n-i} y^i$$

is a homogeneous polynomial of degree n in x, y. Pick general $A_j \in |A|$. Then $(L^{n-i} \cdot A^i) = (L^{n-i} \cdot (A_1 \cap \cdots \cap A_i))$ by (1.35.5), hence by induction, $(L^{n-i} \cdot A^i) \geq 0$ for $i = 1, \cdots, n-1$. $(A^n) > 0$ by (1.35.3).

All the coefficients of $f(1, t)$ are non-negative, except possibly the constant term. Thus $f(1, t)$ is a strictly increasing function for $t \geq 0$.

If $f(1,0) = (L^n) < 0$ then there exists a unique $t_0 > 0$ such that $f(1, t_0) = 0$. We derive a contradiction from this assumption.

$$
\begin{aligned}
0 = f(1, t_0) &= ((L + t_0 A)^n) \\
&= (L \cdot (L + t_0 A)^{n-1}) + t_0 (A \cdot (L + t_0 A)^{n-1}). \quad (1.5)
\end{aligned}
$$

In order to estimate the first term on the right, let $s = a/b > t_0$ be a rational number, $a, b > 0$. Then $((bL + aA)^n) = b^n f(1, s) > 0$ and

$$
\begin{aligned}
((bL + aA)^d \cdot Z) &= \sum_{i=0}^{d} \binom{d}{i} (L^{d-i} \cdot A^i \cdot Z) b^{d-i} a^i \\
&\geq (A^d \cdot Z) a^d > 0 \quad \text{if } Z \subsetneq X, \quad (1.6)
\end{aligned}
$$

by induction. Hence $bL + aA$ is ample by (1.37). For $m \gg 0$, there exist $H_1, \cdots, H_{n-1} \sim m(bL + aA)$ which intersect transversally. Then by (1.35.5),

$$
m^{n-1}(L \cdot (bL + aA)^{n-1}) = (L \cdot H_1 \cap \cdots \cap H_{n-1}) \geq 0.
$$

Hence

$$
(L \cdot (L + t_0 A)^{n-1}) = \lim_{s \to t_0} (L \cdot (L + sA)^{n-1}) \geq 0.
$$

The inequality (1.6) for $Z = A$ shows that $(A \cdot (L + t_0 A)^{n-1}) \geq t_0^{n-1}(A^n) > 0$. Substituting into (1.5) we get a contradiction. $\qquad \square$

1.39 (Proof of (1.18)). Assume D is ample. One can include it in a basis $D = D_1, \cdots, D_{\rho(X)}$ of $N_1(X)^* \otimes \mathbb{Q}$. We can further assume that D_i and $2D - D_i$ are all ample $\mathbb{Q}$-divisors. Then $n : N_1(X) \to \mathbb{R}$ defined by $n(C) := \sum_i |(C \cdot D_i)|$ is a norm on $N_1(X)$. Since $2\rho(X)(C \cdot D) - n(C) = (C \cdot \sum(2D - D_i)) \geq 0$ for every $C \in \overline{NE}(X) \setminus \{0\}$, one has $(C \cdot D) \geq n(C)/(2\rho(X)) > 0$.

To see the converse, let A be an ample divisor on X. D is strictly positive on $\overline{NE}(X) \setminus \{0\}$, thus $L := tD - A$ is nef for some integer $t \gg 0$. By (1.38), $tD = A + L$ satisfies $(D^{\dim Z} \cdot Z) \geq (A^{\dim Z} \cdot Z) > 0$ for every $Z \subset X$. Thus tD is ample by (1.37). $\qquad \square$

Remark 1.40. Theorem (1.18) still holds if X is proper and smooth (or at least $\mathbb{Q}$-factorial) but it fails for arbitrary proper varieties. See [Kle66, Har70, Kol96] for proofs.

The above results can be generalized to the relative case by the following proposition.

Proposition 1.41. *Let $f : X \to Y$ be a proper morphism and D a Cartier divisor on X. Let $y \in Y$ be a point and X_y the fiber of f over y. If $\mathcal{O}_{X_y}(D)$ is ample, then D is ample over some open set $U \ni y$ of Y.*

Proof. As a first step, we claim the vanishing:

(∗) $R^i f_* F(\nu D) = 0$ near y for $i > 0, \nu \gg 0$ for any coherent sheaf F
    defined near $f^{-1}(y)$.

We prove (∗) by descending induction on i. The vanishing is known for $i > \dim f^{-1}(y)$, thus assume (∗) for some $i > 1$. Let $u_1, \cdots, u_r$ be generators of the maximal ideal $m_{y,Y}$ and $s : F^{\oplus r} \to F$ the homomorphism $s(a_1, \cdots, a_r) := \sum_i u_i a_i$ defined near $f^{-1}(y)$. Then we have an exact sequence:

$$F(\nu D)^{\oplus r} \to F(\nu D) \to \mathcal{O}_{X_y} \otimes F(\nu D) \to 0.$$

$R^i f_*(\ker s)(\nu D) = R^i f_*(\operatorname{im} s)(\nu D) = 0$ by the inductive hypothesis, thus we get an exact sequence near y for $\nu \gg 0$:

$$R^{i-1} f_* F(\nu D)^{\oplus r} \to R^{i-1} f_* F(\nu D) \to R^{i-1}(f_y)_*(\mathcal{O}_{X_y} \otimes F)(\nu D) = 0.$$

Thus $R^{i-1} f_* F(\nu D) = m_{y,Y} R^{i-1} f_* F(\nu D)$, and Nakayama's Lemma implies that $R^{i-1} f_* F(\nu D) = 0$. Thus vanishing holds for $i - 1$, proving (∗).

Let I be the ideal sheaf of X_y. Applying (∗) to I, we obtain that $f_* \mathcal{O}_X(\nu D) \twoheadrightarrow \mathcal{O}_{X_y}(\nu D)$ for $\nu \gg 1$. $\mathcal{O}_{X_y}(D)$ is ample, thus $\mathcal{O}_{X_y}(\nu D)$ is generated by global sections for $\nu \gg 1$. Thus the composite

$$f^* f_* \mathcal{O}_X(\nu D) \to \mathcal{O}_X(\nu D) \to \mathcal{O}_{X_y}(\nu D)$$

is surjective for $\nu \gg 1$. By the Nakayama Lemma, $f^* f_* \mathcal{O}_X(\nu D) \twoheadrightarrow \mathcal{O}_X(\nu D)$ over an affine open set $U \ni y$ of Y, thus $\nu D|_{f^{-1}(U)}$ induces a finite morphism near $f^{-1}(y)$. This proves (1.41). $\qquad\square$

The following is a relative version of the Nakai–Moishezon criterion (1.37), and it follows immediately from (1.37) via (1.41).

Theorem 1.42 (Nakai–Moishezon criterion). *Let $f : X \to Y$ be a proper morphism and L a Cartier divisor on X. Then L is f-ample on X iff $(L^{\dim Z} \cdot Z) > 0$ for every closed integral subscheme $Z \subset X$ such that $f(Z)$ is a point.* $\qquad\square$

The following is an immediate corollary to (1.38). We just list it for comparison.

Corollary 1.43. *Let $f : X \to Y$ be a proper morphism and L an f-nef Cartier divisor on X. Then $(L^{\dim Z} \cdot Z) \geq 0$ for every integral closed subscheme $Z \subset X$ such that $f(Z)$ is a point.* $\square$

The following is the relative version of Kleiman's criterion (1.18).

Theorem 1.44 (Kleiman's Ampleness Criterion). *Let $f : X \to Y$ be a projective morphism and D a Cartier divisor on X. Then D is f-ample iff*

$$D_{>0} \supset \overline{NE}(X/Y) \setminus \{0\}.$$

This follows from the preceding two results as in (1.39) except for the terminology in the relative setting, which will be defined in (2.16).

We also need the following (cf. [Har77, II.7.10]):

Proposition 1.45. *Let $f : X \to Y$ be a morphism of projective varieties with M an ample divisor on Y. If L is an f-ample Cartier divisor on Y then $L + \nu f^* M$ is ample for $\nu \gg 0$.* $\square$

It is worthwhile to note that a similar statement does not hold for f-nef divisors.

Example 1.46. Let $0 \in E$ be an elliptic curve. Set $X := E \times E$, $Y = E$ and let $f : X \to Y$ be the first projection. Let $\Gamma_0 := \{(x, 0) \mid x \in E\}$ and $\Gamma_1 := \{(x, x) \mid x \in E\}$ be sections of f and $D := \Gamma_1 - \Gamma_0$. Then D is f-nef, even numerically f-tivial. $((D + f^* A)^2) = (D^2) = -2$, thus $D + f^* A$ is not nef for any A.

We can get a similar example where f is birational as follows. Let L be an ample invertible sheaf on E, $\pi : \mathbb{P}_E(L \oplus \mathcal{O}_E) \to E$ the $\mathbb{P}^1$-bundle and $g : \mathbb{P}_E(L \oplus \mathcal{O}_E) \to V$ the contraction of the negative section S of π. Note that

$$h = id_E \times g : E \times \mathbb{P}_E(L \oplus \mathcal{O}_E) \to E \times V$$

is a birational morphism and it induces f on $X \cong E \times S$. Then $G := (id_E \times \pi)^* D$ is h-nef, and $G + h^* B$ is not nef for any B, since $G + h^* B|_X = D + f^*(B|_E)$.

2

Introduction to the Minimal Model Program

This chapter provides the first glimpse of the general minimal model program and it also collects some preparatory material.

Section 1 explains the aims and methods of the minimal model program, still at an informal level. One of the fundamental observations is that, starting with dimension three, the minimal model program leads us out of the class of smooth varieties. Therefore, any precise explanation of the minimal model program has to be preceded by a study of the resulting singularities.

Section 2 is an aside; it considers various generalizations of the minimal model program. In applications these are very useful, but they do not introduce new conceptual difficulties.

For us the most useful is the study of the so-called log category. Here one considers pairs (X, D) where X is a variety and D a formal linear combination of irreducible divisors. It seems that for the minimal model program, this provides the natural setting.

Various classes of singularities of such pairs (X, D) are considered in section 3. These are somewhat technical, but indispensable for the later developments. A more detailed study of the log category can be found in [Kol97].

Sections 4 and 5 are devoted to proving the vanishing theorems which are used in subsequent chapters. We prove just as much as needed later, and so we restrict ourselves to the case of smooth projective varieties. In this case the proofs are rather simple and they reveal the relationship of vanishing theorems with the topology of varieties. There are several approaches to vanishing theorems; see [KMM87, EV92, Kol95, Kol97] for other treatments.

2.1 Introduction to Mori's Program

As we noted after (1.32), because of the singularities occurring in cases E3, E4 and E5, we cannot apply (1.32) to the resulting variety Y. Thus we need to go back and see if one can prove a version of (1.32) when we allow X to have certain singularities. It turns out that the singularities in E3 and E4 do not cause much trouble. More generally, most of the proof goes through with minor changes if we allow X to have isolated hypersurface singularities, cf. [Cut88a].

Unfortunately the singularity in case E5 is not a hypersurface singularity, and it is much harder to give a direct geometric analysis of varieties with such singularities. Even before [Mor82], such singularities were known to cause problems. For instance, [Uen75] observed the following.

Example 2.1. Let A be an Abelian threefold and $X := A/\pm$ the quotient of A by the involution $x \mapsto -x$. The 2-torsion points are fixed by the involution and they give 64 singular points on X which are locally analytically isomorphic to $\mathbb{C}^3/\pm$. [Uen75, 16.17] noted that X is not birational to any smooth projective variety with nef canonical class. This shows that there is no 3-dimensional analogue of (1.29) if one insists on staying within the framework of smooth varieties.

2.2. It seems that a more conceptual approach is required. As with many questions, there are two opposing points of view:

• Minimalist: We should try to identify the *smallest* class of singularities which we encounter starting with smooth varieties and applying (1.32) and its generalizations.
• Maximalist: We should try to identify the *largest* class of singularities where an analogue of (1.29) is possible.

For the beginning of the section we try to follow the maximalist approach. This is most suitable for general discussion. Unfortunately, essentially none of the results hold without rather strong restrictions on the singularities. Thus, when it comes to proofs, we have to get close to the minimalist approach.

2.3. In (1.31.1) we first ask if K_X is nef. For this question to make any sense, the intersection numbers of K_X with curves have to be defined. These numbers make sense if K_X is Cartier, or more generally, when K_X is $\mathbb{Q}$-Cartier. It seems that, starting with dimension three, nothing can be done without this assumption.

Next we have to establish the Cone Theorem for X. For smooth varieties this was done in (1.24). The main use of smoothness is through the formula (1.10. Step 2), which frequently fails when X is singular. There are, however, large classes of singularities where (1.10) holds, and this leads to the Cone Theorem for many singular varieties. It turns out that the Cone Theorem holds for any normal threefold [Kol92, p.295], but it fails in higher dimensions [Kol92, 5.5.2.2.1]. Unfortunately, the geometric approach so far failed to prove a Cone Theorem which is strong enough for the whole program in dimensions 4 and up. A different approach to the Cone Theorem is discussed in Chapter 3.

2.4. In (1.31.2) we need to contract the extremal ray R. It seems that this is a much more delicate result, which holds only for rather special singularities. The proof of this result, presented in Chapter 3, in fact reverses the above order: the contraction of extremal rays is proved before the Cone Theorem.

For the moment we mention the following result which classifies the resulting extremal contractions into three types. The subsequent discussion shows that these three cases have distinct features.

Proposition 2.5. *Let X be a normal projective variety. Assume that X is $\mathbb{Q}$-factorial. Let $f : X \to Y$ be the contraction of an extremal ray $R \subset \overline{NE}(X)$. Then we have one of the following cases:*

(1) *(Fiber type contraction)* $\dim X > \dim Y$.
(2) *(Divisorial contraction)* f *is birational and* $\mathrm{Ex}(f)$ *is an irreducible divisor.*
(3) *(Small contraction)* f *is birational and* $\mathrm{Ex}(f)$ *has codimension ≥ 2.*

Proof. The only content is the assertion that if f is birational and the exceptional set contains an irreducible divisor E then in fact $E = \mathrm{Ex}(f)$. By assumption E is $\mathbb{Q}$-Cartier, and $(E \cdot R) < 0$, as we see in (3.39). If $C \subset X$ is a curve such that $f(C)$ is a point then $[C] \in R$, thus $(E \cdot C) < 0$. This implies that $C \subset E$, thus $E = \mathrm{Ex}(f)$.   □

2.6. The three cases of (2.5) do behave very differently from the point of view of our program:

Case 1. Fiber type contractions correspond to (1.31.4). If $(R \cdot K_X) < 0$, then the general fiber F of f is an algebraic variety where $-K_F = -K_X|_F$ is ample. Thus, at least in principle, we reduce the problem of understanding X to the study of the lower dimensional variety Y and the fibers of f. Moreover the fibers are of a very special kind – they are analogues of $\mathbb{CP}^1$ and of Del Pezzo surfaces.

Case 2. Divisorial contractions should correspond to (1.31.2). For this to be true we need to establish that K_Y is $\mathbb{Q}$-Cartier, but preferably that Y is $\mathbb{Q}$-factorial. If this holds, then Y can be considered to be 'simpler' than X, since $\rho(Y) \leq \rho(X) - 1$ (in fact it is easy to see that $\rho(Y) = \rho(X) - 1$).

Let $D \subset Y$ be any Weil divisor and $D' \subset X$ its birational transform. We can choose $a(D)$ such that $(R \cdot (D' + a(D)E)) = 0$. Finally choose m such that $D'' := m(D' + a(D)E)$ is Cartier. D'' has a good chance of being the pull back of a Cartier divisor $\bar{D}$ from Y. If this is indeed the case then $mD \sim \bar{D}$ and so D is $\mathbb{Q}$-Cartier.

Case 3. Small contractions. This is a new situation. It could never happen for surfaces for dimensional reasons, and it did not happen for smooth threefolds X. We claim that in this case, Y has 'very bad' singularities where no multiple of K_Y is Cartier.

Indeed, assume that mK_Y and mK_X are both Cartier. Then mK_X and $f^*(mK_Y)$ are two Cartier divisors on X which are linearly equivalent outside the codimension two set $\text{Ex}(f)$. This implies that they are linearly equivalent on X. This is, however, impossible, since $(R \cdot K_X) < 0$ and $(R \cdot f^*(mK_Y)) = 0$.

So we were led out of the class of varieties that we can control. In order to continue at this point, we have to introduce a new operation called a *flip*. This is the algebraic analogue of topological surgery:

Instead of contracting the (codimension at least two) subvariety $E = \text{Ex}(f) \subset X$, we remove it, and then compactify $X \setminus E$ by adding another (codimension at least two) subvariety E^+. (For the moment, it is not at all clear that this operation exists or that it is well defined, let alone that it improves things.)

Example 2.7 (Example of flips). We first study an example of this situation. In the example, the flip removes the curve $C \cong \mathbb{CP}^1$ from the singular variety X_n and replaces it with $D \cong \mathbb{CP}^1$ to achieve the 'improved' variety X_n^+ (which in this case is non-singular).

We start with an auxiliary construction. Let us consider $Y = (xy - uv = 0) \subset \mathbb{C}^4$. This has an isolated singularity at the origin. If we blow it up, we get $\tilde{X} = B_0 Y$. The exceptional set $Q \subset \tilde{X}$ is the projective quadric $(xy - uv = 0) \subset \mathbb{P}^3$. This has two families of lines: $x = cv$, $y =$

$c^{-1}u$ and $x = cu$, $y = c^{-1}v$. These two families can be blown down to smooth threefolds X resp. X^+. X resp. X^+ can also be obtained alternatively by blowing up the ideals (x, v) resp. (x, u). Let $C \subset X$, resp. $C^+ \subset X^+$ be the exceptional curves of $X \to Y$, resp. $X^+ \to Y$. Thus we have the left hand square of the following diagram:

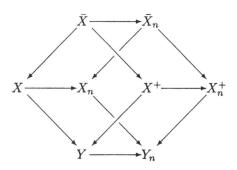

Consider the action of the cyclic group μ_n: $(x, y, u, v) \mapsto (\zeta x, y, \zeta u, v)$. This defines an action on all of the above varieties. The corresponding quotients are denoted by a subscript n.

The fixed point set of the action (i.e. the set of points fixed by some $g \in \mu_n \setminus \{1\}$) on Y is the 2-plane $(x = u = 0)$. On the projective quadric Q the action has two fixed lines: $(x = u = 0)$ corresponding to the above fixed 2-plane and $(y = v = 0)$ corresponding to the ζ-eigenspace. On X therefore the fixed point set has two components: the birational transform of the $(x = u = 0)$ plane and the image of the $(y = v = 0)$ line, this latter being an isolated fixed point. It is easy to see that $(x, v' = vx^{-1}, u)$ give local coordinates at the isolated fixed point. The group action is $(x, v', u) \mapsto (\zeta x, \zeta^{-1}v', \zeta u)$.

On X^+ the fixed point set is irreducible and it contains the exceptional curve C^+. X_n^+ is smooth.

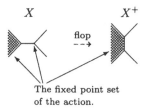

X X^+

flop

The fixed point set
of the action.

The group quotient of a flop induces a flip. The codim ≥ 2 part of the fixed point set creates the singularities, and the codim 1 part creates the sign of K.

It is not hard to compute the intersection numbers of the canonical classes with the exceptional curves. We obtain that

$$C_n \cdot K_{X_n} = -\frac{n-1}{n} \quad \text{and} \quad C_n^+ \cdot K_{X_n^+} = n - 1.$$

Thus $X_n^+ \to Y_n$ is the flip of $X_n \to Y_n$ for $n \geq 2$.

Let us note a property of this example. At the isolated fixed point on X we have coordinates (x, v', u) and the curve C is the v'-axis. A typical local μ_n-invariant section of K_X^{-1} is given by $\sigma = (v'^{n-1} - x)(dx \wedge dv' \wedge du)^{-1}$, which has intersection number $(n-1)$ with C. Since this section is invariant, it descends to a local section σ_n of $K_{X_n}^{-1}$. Let $D_n = (\sigma_n = 0)$. By construction $D_n \cong \{(v', u)\text{-plane}\}/\mu_n$ which is a Du Val singularity of type A_{n-1} (cf. section 4.3). Since $C_n \cdot D_n = C_n \cdot K_{X_n}^{-1}$, one can easily see that even globally D_n is a member of $|K_{X_n}^{-1}|$.

The operation that happens in the above example can be formalized as follows:

Definition 2.8. Let $f : X \to Y$ be a proper birational morphism such that the exceptional set $\text{Ex}(f)$ has codimension at least two in X. Assume furthermore that K_X is $\mathbb{Q}$-Cartier and $-K_X$ is f-ample. A variety X^+ together with a proper birational morphism $f^+ : X^+ \to Y$ is called a *flip* of f if

(1) K_{X^+} is $\mathbb{Q}$-Cartier,
(2) K_{X^+} is f^+-ample, and
(3) the exceptional set $\text{Ex}(f^+)$ has codimension at least two in X^+.

By a slight abuse of terminology, the rational map $\phi : X \dashrightarrow X^+$ is also called a flip. We will see that a flip is unique and the main question is its existence. A flip gives the following diagram:

$$\begin{array}{ccc} X & \overset{\phi}{\dashrightarrow} & X^+ \\ & & \\ -K_X \text{ is } f\text{-ample} \searrow & & \swarrow K_{X^+} \text{ is } f^+\text{-ample} \\ & Y & \end{array}$$

If we perform a flip, it is not clear that X^+ is any 'simpler' than X. In the example above this happens since X is singular but X^+ is smooth.

Note 2.9. The terminology in the literature is not uniform. Most works in higher dimensional birational geometry follow (2.8), but in many papers any diagram as above is called a flip if $\text{Ex}(f)$ and $\text{Ex}(f^+)$ have codimension at least two, without assuming anything about the canonical classes. We introduce a more general notion of flip in (3.33).

2.10. Before we formalize the above discussions, it is necessary to settle on a class of singularities where Mori's program actually works. Such a class was already studied in the nineteenth century in the following setting:

Let $Y_d \subset \mathbb{P}^n$ be a singular hypersurface with resolution $f : X \to Y$. The adjunction formula predicts that

$$H^0(X, \mathcal{O}_X(K_X)) \overset{?}{=} H^0(Y, \mathcal{O}_Y(K_Y)) = H^0(Y, \mathcal{O}_Y(d-n-1)) = \binom{d-1}{n}.$$

In general we do not have equality because the singularities of Y pose 'adjunction conditions'. Thus it is of interest to understand which singularities pose no adjunction conditions. In modern terminology these are the canonical singularities:

Definition 2.11. Let Y be a normal variety. We say that Y has *canonical* singularities if

    (1) K_Y is $\mathbb{Q}$-Cartier, and
    (2) $f_* \mathcal{O}_X(mK_X) = \mathcal{O}_Y(mK_Y)$ for every resolution of singularities $f : X \to Y$.

(One can easily prove that it is sufficient to check (2) for one resolution of singularities.)

The following observation leads to an even smaller class. If Y is smooth, $f : X \to Y$ is birational and $\gamma \in H^0(Y, \mathcal{O}_Y(mK_Y))$ then $f^* \gamma$, as a section of $\mathcal{O}_X(mK_X)$, vanishes along the exceptional divisor $\mathrm{Ex}(f)$. This is stronger than (2.11.2), which requires only that $f^* \gamma$ have no poles along $\mathrm{Ex}(f)$. This stronger condition gives us the next definition. One can see that this is the smallest class where Mori's program can work.

Definition 2.12. Let Y be a normal variety. We say that Y has *terminal* singularities if

    (1) K_Y is $\mathbb{Q}$-Cartier, and
    (2) $f_* \mathcal{O}_X(mK_X - E) = \mathcal{O}_Y(mK_Y)$ for every resolution of singularities $f : X \to Y$ where $E \subset X$ is the reduced exceptional divisor.

(One can easily prove that it is sufficient to check (2) for one resolution of singularities.)

In analogy with (1.30) we introduce the concept of higher dimensional minimal models:

Definition 2.13. Let X be a normal and proper variety. We say that X is *minimal* or a *minimal model* if

(1) X has terminal singularities, and
(2) K_X is nef.

If Y is a smooth proper variety birational to X, then X is also called a minimal model of Y. It should be stressed that, unlike for surfaces (1.30), higher dimensional minimal models are not unique.

With these definitions at hand, we can formulate a precise step by step outline of Mori's program.

2.14 (Mori's program or Minimal model program).
Starting with a smooth projective variety X, we perform a sequence of understandable birational modifications, until we arrive at a variety X^*, whose global structure is simpler. The following is a more detailed list of the necessary steps.

Step 0 (Initial datum). We start with a projective variety $X = X_0$ over $\mathbb{C}$ with only $\mathbb{Q}$-factorial and terminal singularities. We set up a recursive procedure which produces intermediate varieties X_i and then stops with a final variety X^*. Assume that we have already constructed X_i.

Step 1 (Preparation). If K_{X_i} is nef, then there is nothing to do and we go directly to Step 3.2. If K_{X_i} is not nef then we establish two results:

(1) (Cone Theorem) $\overline{NE}(X_i) = \overline{NE}(X_i)_{K_{X_i} \geq 0} + \sum \mathbb{R}^+[C_i]$. If K_{X_i} is not nef, we pick a K_{X_i}-negative extremal ray $R_i \subset \overline{NE}(X_i)$. (We try to construct a theory which works with any choice of R_i, though sometimes a careful choice may be very useful.)
(2) (Contraction of an extremal ray) Let $\operatorname{cont}_{R_i} : X_i \to Y_i$ denote the corresponding extremal contraction (2.5).

Step 2 (Birational transformations). By (2.5), we have three possible types of contractions $\operatorname{cont}_{R_i} : X_i \to Y_i$. Two of them are used to produce a new variety X_{i+1} as follows.

(1) (Divisorial contraction) If $\operatorname{cont}_{R_i} : X_i \to Y_i$ is a divisorial contraction as in (2.5.2), then set $X_{i+1} = Y_i$. We will prove that X_{i+1} again has $\mathbb{Q}$-factorial and terminal singularities, so we can go back to Step 0 with X_{i+1} and start anew.
(2) (Flipping contraction) If $\operatorname{cont}_{R_i} : X_i \to Y_i$ is a flipping contraction as in (2.5.3), then set $X_{i+1} = X_i^+$, the flip of $\operatorname{cont}_{R_i} : X_i \to Y_i$. We will prove that X_{i+1} again has $\mathbb{Q}$-factorial and terminal

singularities, so we can go back to Step 0 with X_{i+1} and start anew.

Step 3 (Final outcome). We hope that eventually the procedure stops, and we get one of the following two possibilities:

(1) (Fano fiber space) If $\text{cont}_{R_i} : X_i \to Y_i$ is a Fano contraction as in (2.5.1) then set $X^* := X_i$. The hope is that the investigation of the lower dimensional variety Y_i and of the fibers provides new methods to study X^*.
(2) (Minimal model) If K_{X_i} is nef then again set $X^* := X_i$. We hope to be able to exploit the semi-positivity of the canonical class in further attempts to understand X^*.

The program outlined above is a rather unfinished wishlist. First of all, it is complete only in dimensions 2 and 3. In higher dimensions it is not known if the flip in Step 2.2 can always be performed. It is also not clear that the program eventually stops: a priori we may end up with an infinite sequence of flips. Finally, it is not at all obvious that the end result (Step 3) is of any use to us. We have improved certain global properties of our variety, but at the price of introducing singularities.

It took about a decade to work out the necessary techniques, but by now it is clear that the above program provides a substantial step toward the proof of many results in higher dimensional algebraic geometry.

Example 2.15. Here we give some examples of extremal contractions in higher dimensions:

(2.15.1) If Y is a smooth projective variety and $Y \supset Z$ is a smooth irreducible closed subvariety, then the inverse of the blowing-up $B_Z Y \to Y$ is an extremal contraction.

(2.15.2) Set

$$X = \mathbb{P}_{\mathbb{P}^n}(\underbrace{\mathcal{O}_{\mathbb{P}^n}(1) \oplus \cdots \oplus \mathcal{O}_{\mathbb{P}^n}(1)}_{k\text{-times}} \oplus \mathcal{O}_{\mathbb{P}^n}).$$

The $\mathcal{O}_{\mathbb{P}^n}$ summand gives a natural embedding $\mathbb{P}^n \cong E \subset X$.

If $k \leq n$, then the line $L \subset E \subset X$ generates a K_X-negative extremal ray in $NE(X)$. The corresponding contraction morphism contracts E to a point and is an isomorphism outside E. Thus, if $k \geq 2$, then the exceptional set is not a divisor. This gives such examples for $\dim X \geq 4$.

(2.15.3) Let Y be the space of non-zero linear maps $\mathbb{C}^{n+1} \to \mathbb{C}^n$, modulo scalars. $Y \cong \mathbb{P}^{n(n+1)-1}$, thus Y is smooth. Let X be the set of pairs (g, L) where $g \in Y$, and L is a one-dimensional subspace in the kernel of g. Let $f : X \to Y$ be the natural morphism. f turns out to be an extremal contraction. X has a natural morphism p onto $\mathbb{P}^n$ (= the set of one-dimensional subspaces in $\mathbb{C}^{n+1}$), given by $p(g, L) = L$. The fibers of p are all projective spaces of dimension $n^2 - 1$, thus X is also smooth. Define

$$F = \{g : \operatorname{rank} g \leq n - 1\}, \quad \text{and} \quad E = \{(g, L) : \operatorname{rank} g \leq n - 1\}.$$

The restriction of p to E exhibits E as a fiber bundle over $\mathbb{P}^n$ whose fiber over L is the projectivization of the set of singular maps $\mathbb{C}^{n+1}/L \to \mathbb{C}^n$, thus E is irreducible. If $g \in F$, then $f^{-1}(g)$ is a projective space of dimension $(n - \operatorname{rank} g)$. Thus, for general $g \in F$, it is a $\mathbb{P}^1$. If $n > 2$, then there is a $g \in F$ such that $\operatorname{rank} g = n - 2$, and so $f^{-1}(g) \cong \mathbb{P}^2$.

This shows that f cannot be a smooth blow-up. In fact, one can see that F is singular at g iff $\operatorname{rank} g \leq n - 2$.

(2.15.4) Let C be a smooth projective curve with a distinguished point $p \in C$. Let $S^m C$ denote the mth symmetric product of C and $J(C)$ the Jacobian variety of C. For every m, the choice of p gives a morphism $u_m : S^m C \to J(C)$ which is surjective for $m \geq g(C)$ and birational for $m = g(C)$. One can see that u_m is the contraction of a K-negative extremal ray on $S^m C$ for $m \geq g(C)$ (cf. [Kol96, III.1.6.6]).

This gives many examples of extremal contractions between smooth varieties where the structure of the exceptional set is complicated.

2.2 Extensions of the Minimal Model Program

We discuss four useful extensions of the minimal model program:

(1) relativization
(2) analytic case
(3) varieties with group actions
(4) log varieties

Example 2.16. Relativization

Let $f : X \to Y$ be a projective morphism. Let $N_1(X/Y)$ be the $\mathbb{R}$-vector space generated by irreducible curves $C \subset X$ such that $f(C) =$ point, modulo the relation

$$\sum a_i C_i \sim 0 \Leftrightarrow \sum a_i (C_i \cdot D) = 0 \quad \text{for every Cartier divisor } D \text{ on } X.$$

The elements of $N_1(X/Y)$ are called *relative 1-cycles*. The dimension of $N_1(X/Y)$ is called the *relative Picard number* of X/Y and is denoted by $\rho(X/Y)$. Let $NE(X/Y) \subset N_1(X/Y)$ be the cone generated by the classes of effective curves which are contracted by f. The relative notions $\overline{NE}(X/Y)$, $\overline{NE}(X/Y)_{D\geq 0}$, etc. are defined similarly to the absolute cases.

If X and Y are proper, the intersection number provides a pairing

$$N_1(X/Y) \times (NS(X)/f^*NS(Y)) \otimes_{\mathbb{Z}} \mathbb{R} \to \mathbb{R}.$$

In general this pairing is not perfect since the two sides may even have different dimensions (cf. (1.46)). It is easy to see that $N_1(X/Y)$ is dual to $(NS(X)/f^*NS(Y)) \otimes_{\mathbb{Z}} \mathbb{R}$ if f is birational and Y is $\mathbb{Q}$-factorial.

Let $f : X \to Y$ and $g : Y \to Z$ be proper morphisms. Assume that f is surjective with connected fibers. We get a complex

$$0 \to N_1(X/Y) \to N_1(X/Z) \to N_1(Y/Z) \to 0,$$

which is not exact in general. Exactness holds if

(1) (X, Δ) is klt and $-K_X - \Delta$ is f-ample for some effective $\mathbb{Q}$-divisor Δ [KMM87, 3-2-5],

(2) X and Y have rational singularities and $R^1 f_* \mathcal{O}_X = 0$ [KM92, 12.1.5], or

(3) f is birational and Y is $\mathbb{Q}$-factorial (using the argument of [KM92, 12.1.5]).

The Cone and Contraction Theorems (cf. section 3.1) are just as in the absolute case (with the same proofs). In the technique used to prove the Cone Theorem (1.24), if the starting curve C has $f(C) = $ point, then all curves produced go to the same point in Y.

Assume that X above is a smooth threefold (or has only $\mathbb{Q}$-factorial terminal singularities). If we assume the existence of 3-dimensional flips (which is established in [Mor88]), then successive contractions over Y lead to a morphism $f' : X' \to Y$ where X' has $\mathbb{Q}$-factorial terminal singularities and either

(1) X' is a minimal model over Y, that is $K_{X'}$ is f'-nef, or

(2) X' is a $\mathbb{Q}$-Fano fibration g' over Y, that is, there is a diagram

$$
\begin{array}{ccc}
X' & \xrightarrow{g'} & Z' \\
{\scriptstyle f'}\searrow & & \swarrow{\scriptstyle h'} \\
 & Y &
\end{array}
$$

where g' has connected fibers, $\dim Z' < \dim X'$ and $-K_{X'}$ is g'-ample.

The proofs differ from the global case in technical details only. See [KMM87] for a detailed exposition.

Example 2.17. Analytic case

There are three points of terminology that we have to pay special attention to.

Even for compact analytic manifolds, there may not be a 'canonical divisor' in the sense of (0.4). However we only use their linear equivalence class, which is equivalent to the dualizing sheaf and to the canonical bundle. So this is only abuse of language and does not cause any problems.

In the relative setting we sometimes used the notion of generic fiber. This is again not defined for analytic spaces. One can, however, always think of it as a sufficiently general fiber, without running into any problems.

The last is the notion of $\mathbb{Q}$-factoriality. As defined earlier, an analytic variety X is $\mathbb{Q}$-*factorial* (or *globally analytically $\mathbb{Q}$-factorial*, in contrast with local analytic $\mathbb{Q}$-factoriality) if every Weil divisor on Y is $\mathbb{Q}$-Cartier. X is *locally analytically $\mathbb{Q}$-factorial* if every Euclidean open set of X is $\mathbb{Q}$-factorial.

One has to be extremely careful with these notions. The reason is that algebraic (resp. global analytic) $\mathbb{Q}$-factoriality does not imply global analytic (resp. local analytic) $\mathbb{Q}$-factoriality. (A Cartier divisor on X^{an} may not be algebraic and a Cartier divisor on an open subset may not extend to a Weil divisor on the whole space in the analytic case.) If a projective variety X is $\mathbb{Q}$-factorial then X^{an} is globally analytically $\mathbb{Q}$-factorial because every global divisor of X^{an} is algebraic by Chow's Theorem (cf. [Har77, App.B]).

Keeping these in mind, the situation where a version of the MMP is known is $f : X \to Y$, with Y an analytic space with some mild finiteness assumptions and f projective. The same results hold as in the relative algebraic case, because the required relative vanishing theorems are true in this situation. See [Nak87] for details.

Very little is known if f is not projective. The example of Hironaka reproduced in [Har77, p.443] shows that (1.32) does not generalize to smooth compact complex threefolds. See also (7.80).

Example 2.18. Varieties with group actions

Suppose a projective variety X, smooth or with only $\mathbb{Q}$-factorial terminal singularities, is acted on by a finite group G. Then we have Cone and Contraction Theorems for $NE(X)^G \subset N_1(X)^G$ (where ^{G} denotes G-invariants). The only difference is that the G-orbit of an extremal ray is an extremal face, since K_X is G-invariant. So the Contraction Theorem involves contraction of G-invariant extremal faces.

There are applications in other settings, too. For example, suppose X is a surface defined over a field k. We achieve a minimal model over k by letting $G = \mathrm{Gal}(\bar{k}/k)$, where $\bar{k}$ is an algebraic closure of k. Although this is not a finite group, its action on the Néron–Severi group of $X_{\bar{k}}$ factors through a finite group, so the construction of a G-minimal model proceeds as in the case of an algebraically closed base field.

In case X is a smooth complex projective surface with G-action, G a finite group, we proceed as before with the classification with some minor changes. A G-extremal ray is generated by a 1-cycle of the form $C = \sum C_i$, where the C_i are irreducible rational curves in a G-orbit.

(1) If $(C^2) < 0$, one easily sees that the C_i must be smooth, mutually disjoint, each with self-intersection -1. Thus all the C_i can be blown down to smooth points.

(2) If $(C^2) = 0$, then any connected component of C is either irreducible or is the union of two -1-curves intersecting transversally at a single point. The contraction morphism makes X into a conic bundle over a smooth curve.

(3) If $(C^2) > 0$, then $N_1(X)^G = \mathbb{R}$, and $-K_X$ is ample, so that X is a Del Pezzo surface.

For threefolds, equivariant resolution of singularities followed by the equivariant minimal model program yields the following. ($G\mathbb{Q}$-factorial means that every G-invariant Weil divisor is $\mathbb{Q}$-Cartier.)

Theorem 2.19. *Any proper G-threefold X is G-birational to a terminal and $G\mathbb{Q}$-factorial G-threefold Y such that*

(1) *either, K_Y is nef,*

(2) *or, there is a G-morphism $f : Y \to Z$ such that $-K_Y$ is f-ample and $\dim Z < \dim X$.*

Example 2.20. Log varieties

In the course of the development of the minimal model program it gradually became clear that it is worthwhile to consider 'small perturbations' of the canonical class as well. The first impetus to their study came from Iitaka's approach to open varieties [Iit77].

Instead of concentrating on K_X we consider a divisor of the form $K_X + D$, where X is a normal variety and $D = \sum d_i D_i$ is a formal $\mathbb{Q}$-linear combination of divisors such that the D_i are distinct and $0 \leq d_i \leq 1$. There are at least four reasons to consider these:

(1) Flexibility: By choosing D appropriately, we are able to analyse situations when K_X is small (e.g., $K_X \equiv 0$), or when K_X is not $\mathbb{Q}$-Cartier.

(2) Inductive proofs: In the last few years several procedures were developed to handle some questions of the minimal model program by reducing them to lower dimension. In almost all cases, the reduction is only possible when we work with pairs (X, D) concentrating on $K_X + D$. Such cases are studied in [K$^+$92].

(3) Open varieties: This is the original idea of [Iit77]. Let X be a smooth variety and $X \subset \bar{X}$ a compactification such that $D = \bar{X} - X$ is a divisor with normal crossings. Somewhat surprisingly, cohomology groups of many vector bundles constructed using differential forms with logarithmic poles along D depend only on X, not on the compactification $\bar{X}$. Such examples are

$$H^j(\bar{X}, \Omega^i_{\bar{X}}(\log D)) \quad \text{and} \quad H^0(\bar{X}, \mathcal{O}(m(K_{\bar{X}} + D))) \ (m \geq 0).$$

Thus if we want to study properties of X, it is natural to consider the divisor $K_{\bar{X}} + D$.

(4) Fiber spaces: Consider Kodaira's canonical bundle formula for elliptic surfaces (see e.g. [BPdV84, V.12.1]). Let $f : S \to C$ be a minimal elliptic surface and $m_i F_i = f^*(c_i)$ the multiple fibers. Then

$$\begin{aligned} K_S &= f^* K_C + f^*(f_* K_{S/C}) + \sum (m_i - 1) F_i \\ &\equiv f^* \left[K_C + (f_* K_{S/C}) + \sum \left(1 - \tfrac{1}{m_i}\right) [c_i] \right]. \end{aligned}$$

Thus the study of K_S can be reduced to the study of a divisor of the form $K_C + D$ where D has rational coefficients. The same happens in general for fiber spaces $f : X \to Y$ where the general fiber has trivial canonical class.

The pair (X, D) has to satisfy some technical requirements in order for the proofs to work. In essence, we cannot allow the divisor to be 'too singular'. The precise meaning of this is explored in the next section.

Example 2.21. All of the above

It is also possible to study the cases when several of the above generalizations occur simultaneously. For instance, one can study the relative, G-equivariant, log MMP. All the results generalize to this setting.

2.3 Singularities in the Minimal Model Program

As we saw in (1.32), contractions of extremal rays lead to singular varieties. In order to continue, we have to understand the singularities that occur in the process and we need a new way to 'measure' how singular a variety is. This new measure is called discrepancy.

Definition 2.22. Let X be a normal variety such that mK_X is Cartier for some $m > 0$. Suppose $f : Y \to X$ is a (not necessarily proper) birational morphism from a normal variety Y. Let $E \subset Y$ be an irreducible exceptional divisor, $e \in E$ a general point of E and $\{y_i\}$ a local coordinate system at $e \in Y$ such that $E = (y_1 = 0)$. Then locally near e,

$$f^*(\text{local generator of } \mathcal{O}_X(mK_X)) = y_1^{m \cdot a(E,X)}(\text{unit})(dy_1 \wedge \cdots \wedge dy_n)^{\otimes m}$$

for some rational number $a(E,X)$ such that $m \cdot a(E,X)$ is an integer. $a(E,X)$ is called the *discrepancy* of E with respect to X. $a(E,X)$ is independent of the choice of m.

If $f : Y \to X$ is a proper birational morphism such that K_Y is Cartier (for instance, Y is smooth), then mK_Y is linearly equivalent to

$$f^*(mK_X) + \sum_i (m \cdot a(E_i, X))E_i,$$

where the E_i are the f-exceptional divisors. Using numerical equivalence, we can divide by m and write

$$K_Y \equiv f^*K_X + \sum a(E_i, X)E_i.$$

Remark 2.23. Let $k(X)$ denote the field of rational functions on X. The local ring $\mathcal{O}_{E,Y} \subset k(X)$ (that is, the local ring of the generic point of E) is a discrete valuation ring which corresponds to a valuation $v(E,Y)$ of $k(X)$. Such valuations of $k(X)$ are called *algebraic valuations*. (An abstract characterization of algebraic valuations is given in (2.45).)

Let $f' : Y' \to X$ be another birational morphism and $E' \subset Y'$ an irreducible divisor such that $v(E,Y) = v(E',Y')$. This holds iff the rational map $Y \to X \dashrightarrow Y'$ is an isomorphism at the generic points $e \in E$ and $e' \in E'$. Then $a(E,X) = a(E',X)$, as one can see from the definition.

Thus $a(E,X)$ depends only on the valuation $v(E,Y)$ but not on the particular choice of f and Y. This is why f and Y are suppressed in the notation.

Definition 2.24. Let X be a variety, $f : Y \to X$ a (not necessarily proper) birational morphism from a normal variety Y and $E \subset Y$ an irreducible divisor. Any such E is called a *divisor over X*. The closure of $f(E) \subset Y$ is called the *center* of E on X. It is denoted by $\text{center}_X E$. As above, the center depends only on the valuation $v(E, Y)$.

The definition (2.22) can be generalized to pairs (X, Δ) such that $K_X + \Delta$ is $\mathbb{Q}$-Cartier:

Definition 2.25. Let (X, Δ) be a pair where X is a normal variety and $\Delta = \sum a_i D_i$ is a sum of distinct prime divisors. (We allow the a_i to be arbitrary rational numbers.) Assume that $m(K_X + \Delta)$ is Cartier for some $m > 0$. Suppose $f : Y \to X$ is a birational morphism from a normal variety Y. Let $E \subset Y$ denote the exceptional locus of f and $E_i \subset E$ the irreducible exceptional divisors. Let

$$f_*^{-1}\Delta := \sum a_i f_*^{-1} D_i$$

denote the birational transform of Δ. The two line bundles

$$\mathcal{O}_Y(m(K_Y + f_*^{-1}\Delta))|_{Y-E} \quad \text{and} \quad f^*\mathcal{O}_X(m(K_X + \Delta))|_{Y-E}$$

are naturally isomorphic. Thus there are rational numbers $a(E_i, X, \Delta)$ such that $m \cdot a(E_i, X, \Delta)$ are integers, and

$$\mathcal{O}_Y(m(K_Y + f_*^{-1}\Delta)) \cong f^*\mathcal{O}_X(m(K_X + \Delta))(\sum_i (m \cdot a(E_i, X, \Delta))E_i).$$

By definition $a(D_i, X, \Delta) = -a_i$ and $a(D, X, \Delta) = 0$ for any divisor $D \subset X$ which is different from the D_i. $a(E, X, \Delta)$ is called the *discrepancy* of E with respect to (X, Δ). We frequently write $a(E)$ if no confusion is likely.

As in the $\Delta = 0$ case, $a(E_i, X, \Delta)$ depends only on E_i but not on f.

Notation 2.26. Using numerical equivalence, we can divide by m and write

$$K_Y + f_*^{-1}\Delta \equiv f^*(K_X + \Delta) + \sum_{E_i:\text{exceptional}} a(E_i, X, \Delta)E_i, \quad \text{or}$$

$$K_Y \equiv f^*(K_X + \Delta) + \sum_{E_i:\text{arbitrary}} a(E_i, X, \Delta)E_i.$$

We frequently refer to these formulas by saying: 'write $K_Y \equiv f^*(K_X + \Delta) + A$'. In this case it is understood that A is chosen as above. That is, we have to make sure that the coefficients of the non-exceptional divisors are as expected. Equivalently, $f_* A = -\Delta$.

The discrepancies have the following obvious monotonicity property.

Lemma 2.27. *Notation as above. Assume that Δ' is effective and $\mathbb{Q}$-Cartier. Then $a(E, X, \Delta) \geq a(E, X, \Delta + \Delta')$ for every divisor E over X, and strict inequality holds iff* center$_X E \subset \operatorname{Supp} \Delta'$.   □

For us the most important values are the minima of $a(E, X, \Delta)$ as E runs through various sets of divisors. We use several versions:

Definition 2.28. The *discrepancy* of (X, Δ) is given by

$$\operatorname{discrep}(X, \Delta) := \inf_E \{a(E, X, \Delta) : E \text{ is an exceptional divisor over } X \}.$$

(That is, E runs through all the irreducible exceptional divisors of all birational morphisms $f : Y \to X$.) Equivalently, $\operatorname{discrep}(X, \Delta) := \inf_v \{a(v, X, \Delta)\}$ where v runs through all algebraic valuations of $k(X)$ such that center$_X v$ is non-empty and has codimension at least 2 in X.

The *total discrepancy* of (X, Δ) is defined as

$$\operatorname{totaldiscrep}(X, \Delta) := \inf_E \{a(E, X, \Delta) : E \text{ is a divisor over } X \}.$$

(That is, $E \subset Y$ runs through all the irreducible exceptional divisors for all birational morphisms $f : Y \to X$ and through all the irreducible divisors of X.) Equivalently, $\operatorname{totaldiscrep}(X, \Delta) := \inf_v \{a(v, X, \Delta)\}$ where v runs through all algebraic valuations of $k(X)$ such that center$_X v$ is non-empty.

We usually write $\operatorname{discrep}(X)$ instead of $\operatorname{discrep}(X, 0)$ and similarly for $\operatorname{totaldiscrep}(X)$.

The following two lemmas make it possible to compute discrepancies in many cases. The proofs are straightforward. The first one shows what happens under one blow up, and the second shows how to use this step inductively. The fact that all divisors are covered by this method follows from (2.45).

Lemma 2.29. *Let X be a smooth variety and $\Delta = \sum a_i D_i$ a sum of distinct prime divisors. Let $Z \subset X$ be a closed subvariety of codimension k. Let $p : B_Z X \to X$ be the blow up of Z and $E \subset B_Z X$ the irreducible* component of the exceptional divisor which dominates Z. (If Z is smooth, then this is the only component.) Then,

$$a(E, X, \Delta) = k - 1 - \sum_i a_i \cdot \operatorname{mult}_Z D_i. \quad \square$$

Lemma 2.30. *Let $f : Y \to X$ be a proper birational morphism between normal varieties. Let Δ_Y resp. Δ_X be $\mathbb{Q}$-divisors on Y resp. X such that*

$$K_Y + \Delta_Y \equiv f^*(K_X + \Delta_X) \quad \text{and} \quad f_*\Delta_Y = \Delta_X.$$

Then, for any divisor F over X,

$$a(F, Y, \Delta_Y) = a(F, X, \Delta_X). \quad \square$$

The following is the first example computing discrepancies.

Corollary 2.31. *Assume that $\Delta = \sum a_i D_i$ is a $\mathbb{Q}$-divisor.*

(1) *Either $\operatorname{discrep}(X, \Delta) = -\infty$, or*
   $-1 \le \operatorname{totaldiscrep}(X, \Delta) \le \operatorname{discrep}(X, \Delta) \le 1$.
(2) *If X is smooth then $\operatorname{discrep}(X, 0) = 1$.*
(3) *Assume that X is smooth, $\sum D_i$ is an snc divisor and $a_i \le 1$ for every i. Then*

$$\operatorname{discrep}(X, \Delta) = \min \left\{ \min_{i \ne j, D_i \cap D_j \ne \emptyset} \{1 - a_i - a_j\}, \min_i \{1 - a_i\}, 1 \right\}.$$

Proof: Blowing up a locus of codimension two which intersects the set of smooth points of X, one sees that $\operatorname{discrep}(X, \Delta) \le 1$.

Assume that E is a divisor over X such that $a(E, X, \Delta) = -1 - c$ with $c > 0$. Take a birational morphism $f : Y \to X$ such that $\operatorname{center}_Y E$ is a divisor on Y and let $K_Y + \Delta_Y \equiv f^*(K_X + \Delta)$ as in (2.26).

Let Z_0 be a codimension 2 locus contained in E but not in any other exceptional divisor of f or in $f_*^{-1}\Delta$. Assume that Y is smooth at the generic point of Z_0. Let $g_1 : Y_1 = B_{Z_0}Y \to Y$ with exceptional divisor $E_1 \subset Y_1$. Then

$$a(E_1, X, \Delta) = a(E_1, Y, \Delta_Y) = -c.$$

Let $Z_1 \subset Y_1$ be the intersection of E_1 and of the birational transform of E. Let $g_2 : Y_2 = B_{Z_1}Y_1 \to Y_1$ with exceptional divisor $E_2 \subset Y_2$. Then

$$a(E_2, X, \Delta) = a(E_2, Y, \Delta_Y) = -2c.$$

Repeat the blowing-up, this time along the intersection of the birational transform of E and E_2 to get a divisor with discrepancy $-3c$, etc. This shows the first part.

The following picture illustrates the above argument.

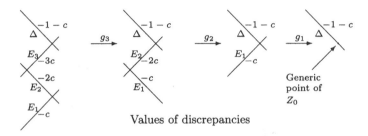

Values of discrepancies

The assertion (2) is a special case of (3).

Let $r(X, \Delta)$ be the right hand side of the equality in (3). Blowing up $D_i \cap D_j$ shows $\mathrm{discrep}(X, \Delta) \leq r(X, \Delta)$. Let D be an exceptional divisor for some birational morphism $f : Y \to X$. We need to prove $a(D, X, \Delta) \geq r(X, \Delta)$. By (2.45), we can assume that E is obtained by a succession of (say t) blow ups along smooth centers (followed by shrinking). We prove the inequality by induction on t. We note that $r(X, \Delta)$ does not decrease if we shrink X. Therefore we can assume that $f(D)$ is a smooth closed subvariety of X. Let $g_1 : X_1 \to X$ be the blow up along $f(D)$ and $E_1 \subset X_1$ the exceptional divisor. By shrinking X around a general point of $f(D)$, we may assume that $E_1 \cup (g_1)_*^{-1}\Delta$ is snc. By shrinking Y around a general point of D, we may assume that $f_1 : Y \to X_1$ is a morphism. After renumbering the D_i, we may assume that $\mathrm{codim}\, f(D) = k \geq 2$ and $f(D) \subset D_i$ iff $i \leq b$ for some $b \leq k$. The blow up formula says that $a(E_1, X, \Delta) = k - 1 - \sum_{l \leq b} a_l$. We treat three cases: $b \leq 0, b = 1$ and $b \geq 2$.

If $b \leq 0$, then $a(E_1, X, \Delta) \geq 1 \geq r(X, \Delta)$.

If $b = 1$ then $a(E_1, X, \Delta) \geq 1 - a_1 \geq r(X, \Delta)$.

If $b \geq 2$, then one has

$$
\begin{aligned}
a(E_1, X, \Delta) &\geq (k - b - 1) + \sum_{1 \leq l \leq b} (1 - a_l) \\
&\geq -1 + (1 - a_1) + (1 - a_2) \geq r(X, \Delta).
\end{aligned}
$$

Thus the case $t = 1$ is settled. On the other hand, if we define Δ_1 on X_1 by $K_{X_1} + \Delta_1 \equiv g_1^*(K_X + \Delta)$, then

$$
\begin{aligned}
r(X_1, \Delta_1) &\geq \min\{r(X, \Delta), 1 + a(E_1, X, \Delta) - \max_{D_i \cap f(D) \neq \emptyset} a_i\} \\
&\geq \min\{r(X, \Delta), a(E_1, X, \Delta)\} \geq r(X, \Delta).
\end{aligned}
$$

Since $\mathrm{Supp}\, \Delta_1$ is a normal crossing divisor and the coefficient of E_1 in

Δ_1 is $-a(E_1, X, \Delta) \leq 1$, one has $a(D, X, \Delta) \geq r(X_1, \Delta_1) \geq r(X, \Delta)$ by the induction hypothesis on f_1. $\qquad\square$

Under some conditions, the discrepancy can be computed from the exceptional divisors occurring on a given resolution:

Corollary 2.32. *Given X, let $f : Y \to X$ be any resolution of singularities with $E \subset Y$ the exceptional set and $E_i \subset E$ all the irreducible exceptional divisors.*

(1) *Assume that $1 \geq \min_i\{a(E_i, X)\} \geq 0$. Then*

$$\mathrm{discrep}(X) = \min_i\{a(E_i, X)\}.$$

(2) *Let $\Delta = \sum a_j D_j$, $a_j \leq 1$. Then there is a log resolution f for (X, Δ) such that $\sum f_*^{-1} D_j$ is smooth. Let f be any such. If $a(E_i, X, \Delta) \geq -1$ for every i, then*

$$\mathrm{discrep}(X, \Delta) = \min\left\{\min_i\{a(E_i, X, \Delta)\}, \min_j\{1 - a_j\}, 1\right\}.$$

Proof: For (1), let Δ_Y be the $\mathbb{Q}$-divisor on Y such that $K_Y + \Delta_Y \equiv f^* K_X$. Then $\Delta_Y = -\sum_i a(E_i, X) E_i \leq 0$ and therefore $\mathrm{discrep}(Y, \Delta_Y) \geq \mathrm{discrep}(Y, 0) = 1$ by (2.31). Thus

$$
\begin{aligned}
\mathrm{discrep}(X) &= \min\{\mathrm{discrep}(Y, \Delta_Y), \min_i\{a(E_i, X)\}\} \\
&= \min_i\{a(E_i, X)\}.
\end{aligned}
$$

For the existence in (2), take any log resolution $g : Z \to X$ for (X, Δ). Then $\mathrm{Sing}(\sum g_*^{-1} D_i)$ is a union of finitely many (say $k(g)$) non-singular subvarieties S of codimension 2. If $h : Z' \to Z \to X$ is the blow up of Z along one irreducible component S, then $k(h) = k(g) - 1$. So we obtain f with $k(f) = 0$, that is, $\sum f_*^{-1} D_i$ smooth by induction on $k(g)$. Let Δ_Y be the $\mathbb{Q}$-divisor on Y such that $K_Y + \Delta_Y \equiv f^*(K_X + \Delta)$. Set $b_i = -a(E_i, X, \Delta)$. By (2.31), $\mathrm{discrep}(Y, \Delta_Y)$ is a minimum of certain numbers of the form $1 - b_i, 1 - b_i - b_{i'}, 1 - a_j, 1 - b_i - a_j$ and of 1. (We do not have to consider $1 - a_j - a_{j'}$ since $f_*^{-1} D_j$ and $f_*^{-1} D_{j'}$ are disjoint.) Furthermore,

$$
\begin{aligned}
\mathrm{discrep}(X, \Delta) &= \min\{\mathrm{discrep}(Y, \Delta_Y), \min_i\{a(E_i, X, \Delta)\}\} \\
&= \min\{\mathrm{discrep}(Y, \Delta_Y), \min_i\{-b_i\}\}.
\end{aligned}
$$

Notice that $-b_i \leq 1 - b_i - b_{i'}$ and $-b_i \leq 1 - b_i - a_j$, thus

$$\mathrm{discrep}(X, \Delta) = \min\{\min_j\{1 - a_j\}, \min_i\{-b_i\}\}. \quad\square$$

The following is the first example showing that the discrepancy is a lower semi-continuous function. A much more sophisticated manifestation of this principle is discussed in section 5.4.

Corollary 2.33. *Let (X, Δ) be a pair and $|L|$ a linear system on X. Let $L_0 \in |L|$ be a member and $L_g \in |L|$ a general member. Then* $\mathrm{discrep}(X, \Delta + cL_0) \leq \mathrm{discrep}(X, \Delta + cL_g)$.

Proof. Choose a resolution of singularities $f : Y \to X$ such that

  (1) $f^*|L| = B + |F|$, where $|F|$ is free,
  (2) $f_*^{-1} \operatorname{Supp} \Delta$ is smooth (cf. (2.32)),
  (3) $B + f_*^{-1} \operatorname{Supp} \Delta + \operatorname{Ex}(f)$ is a snc divisor.

For any $L_\lambda \in |L|$ we can write

$$f^*(K_X + \Delta + cL_\lambda) \equiv K_Y + \Delta_Y + cF_\lambda$$

where $F_\lambda \in |F|$. Let $C(\lambda)$ denote the largest coefficient of an f-exceptional divisor in $\Delta_Y + cF_\lambda$.

$|F|$ is a free linear system, thus $\Delta_Y + cF_g$ is a snc divisor for general $F_g \in |F|$. By (2.32.2) we obtain that $\mathrm{discrep}(X, \Delta + cL_g) = -C(g)$.

$C(0) \geq C(g)$ for any $F_0 \in |F|$. By definition, we have an inequality $\mathrm{discrep}(X, \Delta + cL_0) \leq -C(0)$. $\qquad\square$

Next we define five of the six classes of singularities that are most important for the minimal model program.

Definition 2.34. Let (X, Δ) be a pair where X is a normal variety and $\Delta = \sum a_i D_i$ is a sum of distinct prime divisors. (We allow the a_i to be arbitrary rational numbers.) Assume that $m(K_X + \Delta)$ is Cartier for some $m > 0$. We say that (X, Δ) is

$$
\left.
\begin{array}{r}
terminal \\
canonical \\
klt \\
plt \\
lc
\end{array}
\right\}
\quad \text{if } \mathrm{discrep}(X, \Delta)
\left\{
\begin{array}{l}
> 0, \\
\geq 0, \\
> -1 \quad \text{and } \lfloor \Delta \rfloor \leq 0, \\
> -1, \\
\geq -1.
\end{array}
\right.
$$

Here klt is short for '*Kawamata log terminal*', plt for '*purely log terminal*' and lc for '*log canonical*'. (The frequently used phrase '(X, Δ) has terminal, etc., singularities' may be confusing since it could refer to the singularities of $(X, 0)$ instead.)

If $(X, \Delta = \sum a_i D_i)$ is lc (and the D_i are distinct) then $a_i \leq 1$ for every i by (2.31.1).

Each of these five notions has an important place in the theory of minimal models:

(1) Terminal: Assuming $\Delta = 0$, this is the smallest class that is necessary to run the minimal model program for smooth varieties. The $\Delta \neq 0$ case appears only infrequently.

(2) Canonical: Assuming $\Delta = 0$, these are precisely the singularities that appear on the canonical models of varieties of general type. This class is especially important for moduli problems.

(3) Kawamata log terminal: The proofs of the vanishing theorems seem to run naturally in this class. In general, proofs that work with canonical singularities frequently work with klt; see Chapter 3. This class does not contain the case of open varieties and is also not suitable for inductive proofs.

   If $\Delta = 0$ then the notions klt, plt and dlt (2.37) coincide and in this case we say that X has *log terminal* (abbreviated to *lt*) singularities.

(4) Purely log terminal: This class was invented for inductive purposes. We do not use it much.

(5) Log canonical: This is the largest class where discrepancy still makes sense. It contains many cases that are rather complicated from the cohomological point of view. Therefore it is very hard to work with. A sixth class is introduced in (2.37) to overcome some of these problems.

(2.27) and (2.32) imply the following continuity properties of these notions.

Corollary 2.35. *Let (X, Δ) be a pair and Δ' an effective $\mathbb{Q}$-Cartier divisor. Then*

(1) *If $(X, \Delta + \Delta')$ is terminal (resp. canonical, klt, plt, lc) then (X, Δ) is also terminal (resp. canonical, klt, plt, lc).*

(2) *If (X, Δ) is terminal (resp. klt) then $(X, \Delta + \epsilon \Delta')$ is also terminal (resp. klt) for $0 \leq \epsilon \ll 1$.*

(3) *If (X, Δ) is plt then $(X, \Delta + \epsilon \Delta')$ is also plt for $0 \leq \epsilon \ll 1$, assuming that Δ and Δ' have no common irreducible components.*

(4) *If (X, Δ) is terminal then $(X, \Delta + \Delta')$ is canonical iff $(X, \Delta + c\Delta')$ is terminal for every $c < 1$.*

(5) *If (X, Δ) is klt (resp. plt) then $(X, \Delta + \Delta')$ is lc iff $(X, \Delta + c\Delta')$ is klt (resp. plt) for every $c < 1$.* ☐

Proposition 2.36. *Let (X, Δ) be a klt pair.*

(1) *There exists a log resolution $f : X' \to X$ such that if we write $f^*(K_X + \Delta) \equiv K_{X'} + A_{X'} - B_{X'}$ using effective divisors $A_{X'}, B_{X'}$ without common components, then $\operatorname{Supp} A_{X'}$ is smooth.*
(2) *If $a(E, X, \Delta) < 1 + \operatorname{totaldiscrep}(X, \Delta)$, then $\operatorname{center}_{X'} E$ is a divisor. In particular, there are only finitely many exceptional divisors E over X such that $a(E, X, \Delta) < 1 + \operatorname{totaldiscrep}(X, \Delta)$.*

Proof. For (1), let us start with an arbitrary log resolution $f : Y \to X$ and $A_Y = \sum_{i \in I}(1 - \alpha_i)A_i$ $(0 < \alpha_i < 1)$ using distinct prime components. Let $S(Y) = \sum \mathbb{Z}_{\geq 0}\alpha_i \subset \mathbb{Z}_{\geq 0}$. We also set

$$w(J) = \begin{cases} \sum_{i \in J} \alpha_i & \text{if } \cap_{i \in J} A_i \neq \emptyset \text{ and } \#J \geq 2 \\ \infty & \text{otherwise} \end{cases} \quad (J \subset I),$$
$$M(Y) = \min\{w(J) \mid J \subset I\},$$
$$r(Y) = \#\{J \subset I \mid w(J) = M(Y)\}.$$

We note that $M(Y) \in (0, \dim X) \cap S(Y)$ or $M(Y) = \infty$ and that $M(Y) = \infty$ iff $\operatorname{Supp} A_Y$ is smooth.

Assume $M(Y) < \infty$. Let $J \subset I$ be such that $w(J) = M(Y)$. Set $Z = \cap_{i \in J}A_i$. Let $p : Y' \to Y$ be the blow up along Z and E the exceptional divisor, then the coefficient of E in $(f \circ p)^*(K_X + \Delta)$ is equal to $1 - \sum_{i \in J} \alpha_i$. So $S(Y') = S(Y)$, and it is easy to see that $M(Y') > M(Y)$ or $M(Y') = M(Y)$ and $r(Y') < r(Y)$. Since $M(Y)$ can take only a finite number of values, we cannot keep blowing up indefinitely. Hence we get $M(Y) = \infty$ after a finite number of blowups.

By (2.31), if E is exceptional over X', then

$$a(E, X, \Delta) \geq a(E, X', A_{X'}) \geq \min\{1, \min_i\{\alpha_i\}\},$$

where $A_{X'} = \sum_{i \in I}(1 - \alpha_i)A_i$. Thus (2) follows from (1). ☐

Definition 2.37. Let (X, Δ) be a pair where X is a normal variety and $\Delta = \sum a_i D_i$ is a sum of distinct prime divisors, $0 \leq a_i \leq 1$. Assume that $m(K_X + \Delta)$ is Cartier for some $m > 0$. We say that (X, Δ) is *dlt* or *divisorial log terminal* iff there is a closed subset $Z \subset X$ such that

(1) $X \setminus Z$ is smooth and $\Delta|_{X \setminus Z}$ is a snc divisor.
(2) If $f : Y \to X$ is birational and $E \subset Y$ is an irreducible divisor such that $\operatorname{center}_X E \subset Z$ then $a(E, X, \Delta) > -1$.

Remark 2.38. This definition is quite delicate. It is crucial that in (1) $\Delta|_{X \setminus Z}$ be a *simple* nc divisor. Many of the proofs break down if we allow self-intersections in Δ, though the resulting more general class may behave quite well.

Also, (2.37) is not the usual definition (cf. [K⁺92, 2.13])), though the two versions are equivalent by (2.44). The point of the new definition is that it allows us to prove all the necessary results without using (2.44), which requires a quite delicate use of the methods of [Hir64].

The following continuity properties of dlt pairs correspond to (2.35) and follow from (2.27) and (2.32) similarly.

Corollary 2.39. *Let* (X, Δ) *be a pair with* Δ *effective, and* Δ' *an effective* $\mathbb{Q}$*-Cartier divisor. Then*

(1) *If* $(X, \Delta + \Delta')$ *is dlt then* (X, Δ) *is also dlt.*
(2) *If* (X, Δ) *is dlt then* $(X, \Delta + \epsilon\Delta')$ *is also dlt for* $0 \leq \epsilon \ll 1$, *assuming that* $\mathrm{Supp}\,\Delta' \subset \mathrm{Supp}(\Delta - \lfloor\Delta\rfloor)$. $\qquad\square$

Proposition 2.40. *Let* (X, Δ) *be a pair where* X *is a normal variety and* $\Delta = \sum a_i D_i$ *is a sum of distinct prime divisors,* $0 \leq a_i \leq 1$. *Assume that* $m(K_X + \Delta)$ *is Cartier for some* $m > 0$. *Then* (X, Δ) *is dlt iff there is a closed subset* $Z \subset X$ *such that*

(1) $X \setminus Z$ *is smooth and* $\Delta|_{X \setminus Z}$ *is a snc divisor.*
(2) *There is a log resolution* $f : Y \to X$ *of* (X, Δ) *such that* $f^{-1}(Z) \subset Y$ *has pure codimension 1 and* $a(E, X, \Delta) > -1$ *for every irreducible divisor* $E \subset f^{-1}(Z)$.

Proof. If (X, Δ) is dlt then any log resolution such that $f^{-1}(Z) \subset Y$ has pure codimension 1 has the above properties.

Conversely, assume that $f : Y \to X$ exists as above. Write $K_Y + \Delta_Y \equiv f^*(K_X + \Delta)$ and let Δ' be an effective divisor whose support equals $\mathrm{Supp}\, f^{-1}(Z)$. By (2.40.2) every irreducible component of Δ' has coefficient < 1 in Δ_Y. Thus every irreducible component of $\Delta_Y + \epsilon\Delta'$ has coefficient ≤ 1 for $0 < \epsilon \ll 1$, hence $(Y, \Delta_Y + \epsilon\Delta')$ is lc by (2.31.3).

If E is any divisor over X whose center is contained in Z, then $\mathrm{center}_Y\, E \subset f^{-1}(Z)$, thus by (2.27)

$$a(E, X, \Delta) = a(E, Y, \Delta_Y) > a(E, Y, \Delta_Y + \epsilon\Delta') \geq -1.$$

This shows (2.37.2). $\qquad\square$

Proposition 2.41. *A dlt pair (X, Δ) is klt iff $\lfloor \Delta \rfloor = 0$.*

Proof. If (X, Δ) is klt then it is dlt with $Z = X$. Assume that (X, Δ) is dlt and $\lfloor \Delta \rfloor = 0$. Let $f : Y \to X$ be a birational morphism and $E \subset Y$ an exceptional divisor. If center$_X E \subset Z$ then $a(E, X, \Delta) > -1$ by the definition of dlt. If center$_X E \not\subset Z$ then $a(E, X, \Delta) > -1$ by (2.31.3). Thus (X, Δ) is klt. □

Proposition 2.42. *Let $(x \in X, \Delta)$ be a surface dlt pair. Then, in a neighbourhood of x, either X is smooth and $\Delta = \lfloor \Delta \rfloor$ has two irreducible components intersecting transversally, or (X, Δ) is plt.*

Proof. If $x \notin Z$ we get the first case. If $x \in Z$, let $E \subset Y$ be an exceptional divisor of $f : Y \to X$. If $f(E) \in Z$ then $a(E, X, \Delta) > -1$ by the definition of dlt. If $f(E) \not\subset Z$ then $a(E, X, \Delta) \geq 0$ by (2.31.3) since two irreducible components of Δ do not intersect in a punctured neighbourhood of x. □

A dlt pair is always a limit of klt pairs:

Proposition 2.43. *Assume that (X, Δ) is dlt and X is quasi projective with ample divisor H. Let Δ_1 be an effective $\mathbb{Q}$-divisor (not necessarily $\mathbb{Q}$-Cartier) such that $\Delta - \Delta_1$ is effective. Then there exist a rational number $c > 0$ and an effective $\mathbb{Q}$-divisor $D \equiv \Delta_1 + cH$ such that $(X, \Delta - \epsilon\Delta_1 + \epsilon D)$ is dlt for all rational numbers $0 < \epsilon \ll 1$.*

If $\operatorname{Supp} \Delta_1 = \operatorname{Supp} \Delta$, then $(X, \Delta - \epsilon\Delta_1 + \epsilon D)$ is klt for all sufficiently small rational numbers $\epsilon > 0$.

Proof. Choose $m, m' \in \mathbb{Z}_{>0}$ such that $m\Delta_1$ is an integral divisor and the sheaf $\mathcal{O}_X(m\Delta_1 + m'H)$ is generated by global sections. Let $D' \in |m\Delta_1 + m'H|$ be a general member and set $D = (1/m)D'$. Since $-m\Delta_1 + mD \sim m'H$ is Cartier, $K_X + \Delta - \epsilon\Delta_1 + \epsilon D$ is $\mathbb{Q}$-Cartier.

Let $Z \subset X$ be as in the definition of dlt. Then $m\Delta_1$ is Cartier on $X \setminus Z$, thus $|m\Delta_1 + m'H|$ is basepoint free on $X \setminus Z$. Since D' is a general member, $\Delta + D$ is an snc divisor on $X \setminus Z$.

Let $f : Y \to X$ be any log resolution of $(X, \Delta + D)$. If $E \subset f^{-1}(Z)$ is any irreducible divisor then

$$a(E, X, \Delta - \epsilon\Delta_1 + \epsilon D) \to a(E, X, \Delta) > -1 \quad \text{as} \quad \epsilon \to 0.$$

Thus $(X, \Delta - \epsilon\Delta_1 + \epsilon D)$ is dlt for $0 < \epsilon \ll 1$ by (2.40).

The last statement follows from (2.41). □

Theorem 2.44. *[Sza95] Let $(X, \Delta = \sum d_i D_i)$ be a pair, $0 \leq d_i \leq 1$. The following are equivalent:*

(1) (X, Δ) *is dlt.*

(2) *There exists a log resolution* $f : X' \to X$ *of* (X, Δ) *such that* $a(E_i, X, \Delta) > -1$ *for every exceptional divisor* $E_i \subset X'$. □

Finally we recall a result of Zariski which shows that every divisor can be reached by a sequence of blow-ups. In many cases this can be used as a substitute for resolution of singularities. The proof is taken from [Art86].

Lemma 2.45. *Let* X *be an algebraic variety over a field* k. *Let* (R, m_R) *be a DVR of the quotient field* $k(X)$ *of* X *with* $\operatorname{trdeg}(R/m_R : k) = \dim X - 1$. *Let* $Y = \operatorname{Spec} R$, $y \in Y$ *the closed point and* $f : Y \to X$ *the induced birational morphism. We define a sequence of varieties and maps as follows:*

$X_0 = X, f_0 = f.$

If $f_i : Y \to X_i$ *is already defined, then let* $Z_i \subset X_i$ *be the closure of* $x_i = f_i(y)$. *Let* $X_{i+1} = B_{Z_i} X_i$ *and* $f_{i+1} : Y \to X_{i+1}$ *the induced map.*

Then $f_n : Y \to X_n$ *induces an isomorphism* $\mathcal{O}_{x_n, X_n} \cong R$ *for some* $n \geq 0$.

Proof. We recall that R defines a valuation v_R by setting

$$v_R(g) := \max\{s \in \mathbb{Z} : g \in m_R^s\} \quad \text{for } 0 \neq g \in k(X).$$

The following is the ring theoretic version of the sequence of blow ups. Set $\mathcal{O}_0 := \mathcal{O}_{x_0, X_0}$ with maximal ideal m_0. Assume that $\mathcal{O}_n$ and m_n are already defined. Let $z_1, \ldots, z_r$ be generators of m_n such that $v_R(z_1) \leq \cdots \leq v_R(z_r)$. Let $\mathcal{O}'_n := \mathcal{O}_n[z_2/z_1, \cdots, z_r/z_1]$, $\mathcal{O}_{n+1}$ the localization of $\mathcal{O}'_n$ at $\mathcal{O}'_n \cap m_R$ and m_{n+1} the maximal ideal of $\mathcal{O}_{n+1}$.

We claim that $R = \cup_n \mathcal{O}_n$. To see this, take an arbitrary $u \in R \setminus \mathcal{O}_n$. Write $u = y_1/y_2$ where $y_i \in \mathcal{O}_n$ are chosen so that $v_R(y_2) \geq 0$ is the smallest possible. One sees $y_i \in m_n$ by

$$u \notin \mathcal{O}_n \Leftrightarrow y_2 \in m_n \Leftrightarrow v_R(y_2) > 0 \Rightarrow y_1 \in m_n.$$

We can write $y_i = \sum_j z_j y_{ij}$ where $y_{ij} \in \mathcal{O}_n$ and the z_j are generators of m_n as above. Let $y'_i := \sum_j (z_j/z_1) y_{ij} \in \mathcal{O}_{n+1}$. Then $y_i = z_1 y'_i$ and $u = y'_1/y'_2$. By construction $v_R(y'_2) < v_R(y_2)$. Iterating this procedure eventually we obtain that $u = y_1^{(s)}/y_2^{(s)}$ and $v_R(y_2^{(s)}) = 0$. This implies that $u \in \mathcal{O}_{n+s}$ for some s. Hence $R = \cup_n \mathcal{O}_n$.

Pick $u_1, \ldots, u_r \in k(X)$ which give a transcendence basis of R/m_R over k. Then $u_1, \ldots, u_r \in \mathcal{O}_n$ for $n \gg 1$, hence

$$\operatorname{trdeg}(k(x_n) : k) = \operatorname{trdeg}(R/m_R : k) = \dim X - 1 \quad \text{for } n \gg 1.$$

Since $\mathcal{O}_n$ is the localization of a k-algebra of finite type by construction, one has

$$\mathrm{trdeg}(k(X) : k) - \mathrm{trdeg}(k(x_n) : k) = \dim \mathcal{O}_n.$$

This implies that

$$\mathrm{trdeg}(R/m_R : k(x_n)) = \dim \mathcal{O}_n - 1 = 0 \quad \text{for } n \gg 1.$$

Thus R/m_R is an algebraic extension of $k(x_n)$ and $\dim \mathcal{O}_n = 1$. Then R is a localization of the normalization R' of $\mathcal{O}_n$, and R' is a finite $\mathcal{O}_n$-module with generators $v_1, \ldots, v_s$. For $p \gg n$, one has $v_1, \ldots, v_s \in \mathcal{O}_p$. Thus $\mathcal{O}_p = R$ for $p \gg 1$. $\square$

2.4 The Kodaira Vanishing Theorem

In this section we prove the Kodaira Vanishing Theorem. In the next section we discuss various refinements; these more general vanishing results are crucial in Chapter 3. The proof is based on the following:

Principle 2.46. *If the cohomology of a sheaf F comes from topological cohomology, then there is a Kodaira-type vanishing theorem.*

A detailed explanation of this principle can be found in [Kol86b, Section 5]. Here we illustrate this principle by using it to prove the classical case. The proof is taken from [Kol95].

Theorem 2.47 (Kodaira Vanishing Theorem). *Let X be a smooth projective variety and L an ample line bundle on X. Then*

$$H^i(X, L^{-1}) = 0, \quad for \quad i < \dim X.$$

Step 1 (GAGA principle).
    Let X be a scheme of finite type over $\mathbb{C}$. X can be viewed as a complex analytic space, denoted by X^{an}. If F is a coherent sheaf on X then F^{an} denotes the corresponding coherent analytic sheaf. At the level of stalks this is obtained as

$$F^{an}_{x,X^{an}} = F_{x,X} \otimes_{\mathcal{O}_{x,X}} \mathcal{O}^{an}_{x,X^{an}}.$$

The so-called GAGA principle (acronym of the title of [Ser56]) asserts that in many cases algebraic and analytic objects behave the same. The following special case is formulated for higher direct images, although for now we need only the cohomological version.

Theorem 2.48. *Let X, Y be separated schemes of finite type over $\mathbb{C}$, $f : X \to Y$ a proper morphism and F a coherent sheaf on X. Then*

$$(R^i f_* F)^{an} \cong R^i (f^{an})_* (F^{an}) \quad \text{for every } i.$$

If X is proper, then $H^i(X, F) \cong H^i(X^{an}, F^{an})$ for every i. □

Because of this, for the rest of the section, we can compute the cohomologies in the complex analytic setting. To simplify notation, we will not use the superscript *an*.

Step 2. A coherent sheaf F is also a sheaf of abelian groups. Find a topologically constructible sheaf **F** and a natural map **F** $\to F$ such that the induced map on cohomologies is surjective. (The sheaf cohomology of a coherent analytic sheaf is the same as its cohomology as a sheaf of abelian groups.) For $F = \mathcal{O}_X$ the constant sheaf $\mathbb{C}_X$ suffices, since Hodge theory (see, e.g. [GH78, p.116]) tells us that the natural mapping $H^i(X, \mathbb{C}_X) \to H^i(X, \mathcal{O}_X)$, induced by inclusion of sheaves, is surjective.

Step 3. Cyclic covers.

The construction of cyclic covers is useful in many different contexts, so we give the general definition in three steps:

Definition 2.49 (Unramified cyclic covers). Let X be a normal variety (or complex analytic space) and L a line bundle on X such that $L^m \cong \mathcal{O}_X$ for some $m > 0$. The corresponding cyclic cover is a degree m finite étale morphism $p : X_{m,L} \to X$ defined in three equivalent ways:

(1) Topological. $c_1(L) \in H^2(X, \mathbb{Z})$ is m-torsion by assumption, so it corresponds to a quotient $\pi_1(X) \to H_1(X, \mathbb{Z}) \to \mathbb{Z}_m$. This in turn defines an m-sheeted cover of X.

(2) Geometric. We can view L as a $\mathbb{C}^1$-bundle $L \to X$. Let $h : L \to X \times \mathbb{C}$ be the morphism $(x, l) \mapsto (x, l^{\otimes m})$. Set $X_{m,L} := h^{-1}(X \times \{1\})$.

(3) Algebraic. $\oplus_{i=0}^{m-1} L^{-i}$ is a sheaf of algebras on X where we use the multiplication

$$L^{-i} \otimes L^{-j} \cong L^{-i-j} \cong L^{-i-j+m} \quad \text{for } i + j \geq m.$$

Let $X_{m,L} := \operatorname{Spec}_X \oplus_{i=0}^{m-1} L^{-i}$. (The negative exponents are a matter of preference. If $L^m \cong \mathcal{O}_X$ then also $(L^{-1})^m \cong \mathcal{O}_X$ and they give isomorphic covers. This choice fits better with the ramified covers to be studied next.)

From all three descriptions one can see that p^*L is the trivial line bundle on $X_{m,L}$. For instance, working algebraically we obtain that

$$p_*(p^*L) \cong L \otimes \oplus_{i=0}^{m-1} L^{-i} \cong \oplus_{i=0}^{m-1} L^{1-i}.$$

For $i = 1$ we obtain the summand $L^0 \cong \mathcal{O}_X$. The constant sections in $\mathcal{O}_X$ give nowhere zero sections of p^*L.

Definition 2.50 (Ramified cyclic covers, line bundle case). Let X be a normal variety (or complex analytic space), L a line bundle on X and $s \in H^0(X, L^m)$ a section with zero divisor $D = (s = 0)$.

$L|_{X \setminus D}$ is an invertible sheaf such that $(L|_{X \setminus D})^m \cong \mathcal{O}_{X \setminus D}$. (2.49) gives an m-sheeted cover $p' : Z' \to X \setminus D$ which can be extended to a ramified cover $p : Z \to X$.

The algebraic description of Z' gives the fastest way to obtain Z directly:

The section s can be viewed as a map of sheaves $s : \mathcal{O}_X \to L^m$. Therefore $\oplus_{i=0}^{m-1} L^{-i}$ is a sheaf of algebras on X where, for $i + j \geq m$, we use the multiplication

$$L^{-i} \otimes L^{-j} \cong L^{-i-j} \otimes \mathcal{O}_X \xrightarrow{id \otimes s} L^{-i-j} \otimes L^m \cong L^{-i-j+m}.$$

Let $Z = \mathrm{Spec}_X \oplus_{i=0}^{m-1} L^{-i}$ with projection $p : Z \to X$.

It is easy to get a local description of Z. Pick $x \in X$, local coordinates x_i and let $z \in \Gamma(L)$ be a local generator at x. The image of $z^m \in \Gamma(L^m)$ in $\mathcal{O}_X$ is a function $s(x_1, \ldots, x_n)$ which is a local equation of D at x. Then, locally near x, $\oplus_{i=0}^{m-1} L^{-i}$ is generated by $\mathcal{O}_X$ and z, subject to the relation $z^m = s(x_1, \ldots, x_n)$. In particular, we conclude that:

Lemma 2.51. Z is smooth if and only if X and D are smooth.　　□

Later we use the following generalization of ramified cyclic covers:

Definition 2.52 (Ramified cyclic covers, general case).

Let X be a normal variety (or complex analytic space) and L a rank 1 torsion free sheaf on X. Let $X^0 \subset X$ be the largest open set such that $L|_{X^0}$ is locally free. Then $X \setminus X^0$ has codimension at least 2. The sheaf $L^{\otimes i}$ can even have torsion supported in $X \setminus X^0$. To remedy this, we let $L^{[i]}$ denote the double dual of $L^{\otimes i}$. $L^{[i]}$ is the unique reflexive sheaf on X such that $L^{[i]}|_{X^0} \cong (L|_{X^0})^{\otimes i}$. In practice this means that whatever we do with the sheaves $(L|_{X^0})^{\otimes i}$ automatically extends to the sheaves $L^{[i]}$. For cyclic covers we obtain the following construction.

Let $s \in H^0(X, L^{[m]})$ be a section with zero divisor $D = (s = 0)$. s can be viewed as a map of sheaves $s : \mathcal{O}_X \to L^{[m]}$. Therefore $\oplus_{i=0}^{m-1} L^{[-i]}$ is a sheaf of algebras on X where, for $i + j \geq m$, we use the multiplication

$$L^{[-i]} \otimes L^{[-j]} \cong L^{[-i-j]} \otimes \mathcal{O}_X \xrightarrow{id \otimes s} L^{[-i-j]} \otimes L^{[m]} \cong L^{[-i-j+m]}.$$

Let $Z = \operatorname{Spec}_X \oplus_{i=0}^{m-1} L^{[-i]}$ with projection $p : Z \to X$.

Lemma 2.53. *With the notation as above, assume in addition that* $L^{[m]} \cong \mathcal{O}_X$. *Then* $(p^*L)^{[1]} \cong \mathcal{O}_Z$.

Proof. p^*L need not be reflexive, and $(p^*L)^{[1]}$ denotes the double dual. We noted in (2.49) that $p^*(L|_{X^0}) \cong \mathcal{O}_{p^{-1}(X^0)}$. This isomorphism extends automatically to Z. □

Step 4. Going back to the original setup, assume that X is smooth and L ample. Choose m such that L^m is very ample and let s be a general section. Then $D = (s = 0)$ is a smooth, very ample divisor. Thus Z is smooth and $p : Z \to X$ is étale over $X \setminus D$.

By Hodge theory, the map $\tau : H^i(Z, \mathbb{C}_Z) \to H^i(Z, \mathcal{O}_Z)$ is surjective. Since the fibers of p are zero-dimensional, there are no higher direct-image sheaves, and

$$p_*\tau : H^i(X, p_*\mathbb{C}_Z) \to H^i(X, p_*\mathcal{O}_Z) \quad \text{is surjective.}$$

The action of $\mathbb{Z}/m\mathbb{Z}$ on Z decomposes this last map into a direct sum of maps of eigensheaves as follows.

Set $\xi = e^{2\pi i/m}$. Since the action of $\mathbb{Z}/m\mathbb{Z}$ on Z is continuous, we can decompose

$$p_*\mathbb{C}_Z \cong \oplus_{r=0}^{m-1} \mathbb{C}[\xi^r], \quad \text{and} \quad p_*\mathcal{O}_Z \cong \oplus_{r=0}^{m-1} L^{-r},$$

where $\mathbb{C}[\xi^r]$ denotes the local system that has monodromy ξ^r if we go around the divisor D once. For every r we have natural inclusions $\mathbb{C}[\xi^r] \to L^{-r}$. $p_*\tau$ restricts to surjections between the corresponding eigenspaces, hence

$$H^i(X, \mathbb{C}[\xi^r]) \to H^i(X, L^{-r}) \quad \text{is surjective for every } i, r.$$

Step 5. The sheaves $\mathbb{C}[\xi^r]$ for $0 < r < m$ have non-trivial monodromy around D, thus they have the following simple property:

Corollary 2.54. *Let* $U \subset X$ *be a connected open set such that* $U \cap D \neq \emptyset$. *Then* $H^0(U, \mathbb{C}[\xi^r]|_U) = 0$. □

This property is utilized via the following obvious fact.

Lemma 2.55. *Let F be a sheaf of Abelian groups on a topological space X and $F_1, F_2 \subset F$ subsheaves. Let $D \subset X$ be a closed subset. Assume that:*

(1) $F_2|_{X \setminus D} = F|_{X \setminus D}$, *and*

(2) *if U is connected, open and $U \cap D \neq \emptyset$, then $H^0(U, F_1|_U) = 0$.*

Then F_1 is a subsheaf of F_2. $\square$

As a corollary we obtain:

Corollary 2.56. *Let $0 < r < m$ and $M \subset L^{-r}$ be a subsheaf such that $M|_{X \setminus D} = L^{-r}|_{X \setminus D}$. Then the injection*

$$\mathbb{C}[\xi^r] \to L^{-r} \quad \text{factors as} \quad \mathbb{C}[\xi^r] \to M \to L^{-r}.$$

Therefore $H^i(X, M) \to H^i(X, L^{-r})$ is surjective for every i and $0 < r < m$.

Proof. The first part is clear from (2.54) and (2.55). This implies that we have maps

$$H^i(X, \mathbb{C}[\xi^r]) \to H^i(X, M) \to H^i(X, L^{-r}).$$

As we saw above, the composite is surjective. Hence so is the map on the right. $\square$

Step 6. We are ready to finish the argument. Set $M = L^{-r}(-kD) \cong L^{-(r+mk)}$. By (2.56) we see that

$$H^i(X, L^{-(r+mk)}) \to H^i(X, L^{-r}) \quad \text{is surjective for every } i, k \geq 0.$$

Choose $r = 1$ and $k \gg 1$. The left hand side vanishes for $i < n$ by duality and Serre vanishing. Thus $H^i(X, L^{-1}) = 0$ for $i < n$. This completes the proof of (2.47). $\square$

As an application of Kodaira vanishing, we prove that the Euler characteristic of a divisor D depends only on the numerical equivalence class of D. This is a consequence of the general Riemann–Roch theorem (cf. [Har77, App. A]). Here we give a direct proof for smooth varieties when the base field is $\mathbb{C}$. This is the only case that we need later.

Proposition 2.57. *Let X be a smooth projective variety over $\mathbb{C}$ and D, D' Cartier divisors on X such that $D \equiv D'$. Then $\chi(X, \mathcal{O}_X(D)) = \chi(X, \mathcal{O}_X(D'))$.*

Proof. Let H be an ample divisor on X. Set $P(v) := \chi(X, \mathcal{O}_X(D + vH))$ and $P'(u, v) := \chi(X, \mathcal{O}_X(D + u(D' - D) + vH))$. P and P' are polynomials in u, v by (1.36). Fix any $v \gg 1$ such that $D + vH - K_X$ is ample. Then $D + u(D' - D) + vH - K_X$ is also ample by (1.18) since $D' - D \equiv 0$. $H^i(X, \mathcal{O}_X(D+vH)) = 0 = H^i(X, \mathcal{O}_X(D+u(D'-D)+vH))$ for all u and $i > 0$ by (2.47). Hence $P'(u, v) = h^0(X, \mathcal{O}_X(D + u(D' - D) + vH))$ for every u. We would like to conclude from this that for every fixed $v \gg 1$, $P'(u, v)$ is bounded as a function of u. One can prove this by induction on $n = \dim X$ as follows. Pick $m \gg 1$ such that $(D + vH - mH) \cdot H^{n-1} < 0$ and let $Y \in |mD|$ be a smooth divisor. Then $H^0(X, \mathcal{O}_X(D + u(D' - D) + vH - mH)) = 0$, hence

$$H^0(X, \mathcal{O}_X(D+u(D'-D)+vH)) \leq H^0(Y, \mathcal{O}_Y((D+u(D'-D)+vH)|_Y)).$$

A bounded polynomial is constant, thus $P'(u, v) = P(v)$ for $v \gg 1$, hence for every v. Therefore

$$\chi(X, \mathcal{O}_X(D)) = P(0) = P'(1, 0) = \chi(X, \mathcal{O}_X(D')). \quad \square$$

2.5 Generalizations of the Kodaira Vanishing Theorem

In this section we show that the Vanishing Theorem (2.47) still holds if L is only 'close to ample'. The precise meaning of 'close to ample' is not at all obvious.

Lemma 2.58. *Let X be a projective scheme of dimension n over a field and B a Cartier divisor on X. Then $h^0(X, \mathcal{O}_X(kB)) \leq C \cdot k^n$ for some $C > 0$ and every $k > 0$.*

Proof. Let H be very ample on X such that $H - B$ is linearly equivalent to an effective divisor. Then $h^0(X, \mathcal{O}_X(kB)) \leq h^0(X, \mathcal{O}(kH))$ and the growth of the latter is given by its Hilbert polynomial. (The lemma still holds if X is proper. The proof can be reduced to the projective case by the Chow Lemma.) $\quad \square$

Definition 2.59. *Let X be a proper variety of dimension n. A Cartier divisor D is called big if $h^0(X, \mathcal{O}(kD)) > c \cdot k^n$ for some $c > 0$ and $k \gg 1$.*

Being big is essentially the birational version of being ample. If $f : Y \to X$ is birational and D is a Cartier divisor on X then D is big iff $f^* D$ is big.

Lemma 2.60. *Let X be a projective variety of dimension n and D a Cartier divisor. Then the following are equivalent:*

(1) D is big,

(2) $mD \sim A + E$ where A is ample and E is effective for some $m > 0$,

(3) for some $m > 0$, the rational map $\phi_{|mD|}$ associated to the linear system $|mD|$ is birational, and

(4) the image of $\phi_{|mD|}$ has dimension n for some $m > 0$.

Proof. Note that (2) $\Rightarrow$ (3) $\Rightarrow$ (4) is obvious. To prove (4) $\Rightarrow$ (1), assume that $Y := \phi_{|D|}(X) \subset \mathbb{P}^N$ has dimension n.

By [Har77, I.7.5] the Hilbert polynomial of Y is

$$h^0(Y, \mathcal{O}_Y(k)) = (\deg Y / n!) k^n + (\text{lower order terms}).$$

$\phi_{|D|}^*$ induces an injection $H^0(Y, \mathcal{O}_Y(k)) \subset H^0(X, \mathcal{O}(kD))$, and this proves (4) $\Rightarrow$ (1).

To prove (1) $\Rightarrow$ (2), let D be big and A ample and effective. We have an exact sequence

$$0 \to H^0(X, \mathcal{O}(kD - A)) \to H^0(X, \mathcal{O}(kD)) \to H^0(A, \mathcal{O}(kD|_A)).$$

$\dim A = n - 1$, thus $h^0(A, \mathcal{O}(kD|_A))$ grows at most like const $\cdot \, k^{n-1}$ by (2.58). Thus $H^0(X, \mathcal{O}(kD - A)) \neq 0$ for $k \gg 1$. Let E be any effective divisor in $|kD - A|$. This proves (1) $\Rightarrow$ (2).   □

Proposition 2.61. *Let X be a projective variety of dimension n and D a Cartier divisor. The following are equivalent:*

(1) D *is nef and big.*

(2) D *is nef and $(D^n) > 0$.*

(3) *There is an effective divisor E and ample $\mathbb{Q}$-divisors A_k such that $D \equiv A_k + (1/k)E$ for $k \gg 1$.*

If X has characteristic zero, then the above are further equivalent to

(4) *For any divisor $\Delta \subset X$, there is a log resolution $f : Y \to X$ of (X, Δ), an effective snc divisor E' and ample $\mathbb{Q}$-divisors A_k' on Y such that $f^*D \equiv A_k' + (1/k)E'$ for $k \gg 1$.*

Proof. If D is nef and big, then by (2.60) $mD \sim A + E$, where A is very ample and E is effective. We prove by induction on n that $m^n(D^n) \geq (A^n)$.

$$m^n(D^n) = (A \cdot (mD)^{n-1}) + m^{n-1}(E \cdot D^{n-1})$$
$$= m^{n-1}[((D|_A)^{n-1}) + ((D|_E)^{n-1})].$$

$((D|_E)^{n-1})$ is non-negative by (1.38). By (2.60), $D|_A$ is nef and big. By induction the first term satisfies the inequality

$$m^{n-1}((D|_A)^{n-1}) \geq ((A|_A)^{n-1}) = (A^n).$$

This shows that (1) implies (2). (1) implies (3) by the formula

$$D = \frac{1}{k}(A + (k-m)D) + \frac{1}{k}E,$$

where $A + (k-m)D$ is ample for $k \geq m$ by (1.18).

Assume next that D is nef and $(D^n) > 0$. We prove that D is big, assuming that we are over $\mathbb{C}$. The general case needs rather different arguments, see [Fuj83] and [Kol96, VI.2.15].

It is sufficient to do this after pulling back everything by a birational morphism $f : X' \to X$, thus we may assume that X is smooth. Let $B \subset X$ be an effective ample divisor such that $B - K_X$ is also ample. From Riemann–Roch and (2.47) we get that

$$h^0(X, \mathcal{O}(mD + B)) = \chi(X, \mathcal{O}(mD + B)) = \frac{(D^n)}{n!}m^n + O(m^{n-1}).$$

From the sequence

$$0 \to H^0(X, \mathcal{O}(mD)) \to H^0(X, \mathcal{O}(mD + B)) \to H^0(B, \mathcal{O}(mD + B|_B))$$

we conclude that

$$\begin{aligned} h^0(X, \mathcal{O}(mD)) &\geq h^0(X, \mathcal{O}(mD + B)) - h^0(B, \mathcal{O}(mD + B|_B)) \\ &= \frac{(D^n)}{n!}m^n + O(m^{n-1}). \end{aligned}$$

Finally assume (3). Then $D = \lim A_k$ is nef, hence it is also big by (2.60). Similarly, (4) implies (1).

In order to show that the first three properties imply the last, we can start with any resolution $g : X' \to X$ such that $g^{-1}(\Delta)$ is a divisor. If (4) holds for X', g^*D and $g^{-1}(\Delta)$ then it also holds for X, D and Δ. Thus we may assume that X itself is smooth.

By (2.60) we can write $D \equiv A + E$ where A is ample and E is an effective $\mathbb{Q}$-divisor. Let $f : Y \to X$ be a log resolution of $(X, E + \Delta)$. Then $f^*D \equiv f^*A + f^*E$ and f^*E is a snc divisor, but f^*A is not ample. By (2.62) there is an effective f-exceptional divisor F such that $-F$ is f-ample. Then $A' := f^*A - \epsilon F$ is ample for $0 < \epsilon \ll 1$ (1.45) and $E' := f^*E + \epsilon F$ is an effective snc divisor. As before, set $A'_k := (1/k)(A' + (k-1)f^*D)$. $\qquad\square$

Lemma 2.62. *Let* $f : Y \to X$ *be a birational morphism. Assume that* Y *is projective and* X *is* $\mathbb{Q}$*-factorial. Then there is an effective* f*-exceptional divisor* F *such that* $-F$ *is* f*-ample.*

Proof. Let H be an effective very ample divisor on Y. Such an H exists by the extra assumption on X. f_*H is $\mathbb{Q}$-Cartier and so $f^*(f_*H)$ is defined. $f^*(f_*H) = H + F$ for some effective f-exceptional divisor F and $-F \equiv_f H$. □

Corollary 2.63. *Notation and assumptions as in (2.62). Then the exceptional set* $\mathrm{Ex}(f)$ *of* f *is of pure codimension one.*

Proof. Let H, F be as in the proof of (2.62). $H = f^*(f_*H) - F$ is very ample on Y. $F = f^{-1}f(F)$ as sets since $f^*f_*\mathcal{O}_Y(-F) \twoheadrightarrow \mathcal{O}_Y(-F)$ is surjective. Thus on $Y \setminus \mathrm{Supp}\, F = f^{-1}(X \setminus f(\mathrm{Supp}\, F))$, $f^*(f_*H) = H$ is very ample and f is an isomorphism. Hence $\mathrm{Ex}(f) = \mathrm{Supp}\, F$.

The arguments of [Sha94, II.4.4] can also be generalized to this situation. □

We are now ready to formulate and prove the promised generalization of (2.47).

Theorem 2.64 (Kodaira Vanishing Theorem II). *[Kaw82, Vie82] Let* X *be a smooth complex projective variety. Let* L *be a line bundle on* X *such that* $L \equiv M + \sum a_i D_i$, *where*

(1) M *is a nef and big* $\mathbb{Q}$*-divisor,*

(2) $\sum D_i$ *is a snc divisor,*

(3) $0 \le a_i < 1$, *and* $a_i \in \mathbb{Q}$ *for all* i.

Then $H^i(X, L^{-1}) = 0$ *for* $i < \dim X$.

Proof: We reduce (2.64) to the Kodaira Vanishing Theorem in several steps.

Step 1. Let $f : Y \to X$ be a finite morphism between n-dimensional normal varieties. If the characteristic is zero, then $\frac{1}{n}\mathrm{Trace}_{Y/X}$ (cf. (5.6)) splits the inclusion $\mathcal{O}_X \to f_*\mathcal{O}_Y$. Thus if F is a coherent sheaf on X, then F is a direct summand of $f_*(f^*F)$, hence $H^i(X, F)$ is a direct summand of $H^i(Y, f^*F)$. We use this repeatedly for $F = L^{-1}$.

Step 2. In this step we reduce (2.64) to the case when $L = M$, by induction on the number of divisors D_i. Since the D_i appear with rational coefficients, this is not straightforward since $L - a_1 D_1$ is not a Cartier divisor. We use Step 1 and (2.67) to reduce to the case when $a_1 D_1$ is at least numerically equivalent to a Cartier divisor. To facilitate induction, we prove the following more general result.

Claim 2.65. Let X be a smooth complex projective variety and L a line bundle on X such that $L \equiv M + \sum a_i D_i$, where each D_i is a reduced and smooth (possibly disconnected) divisor, $\sum D_i$ is a snc divisor and $0 \leq a_i < 1$ are rational.

Then there is a finite and surjective morphism $p : Z \to X$ from a smooth projective variety Z and a line bundle M_Z on Z such that $M_Z \equiv p^* M$ and $H^i(X, L^{-1})$ is a direct summand of $H^i(Z, M_Z^{-1})$.

Proof. We use induction on the number of divisors D_i. Write $a_1 = b/m$ where $m \in \mathbb{Z}_{>0}$. By (2.67) there is a finite and surjective morphism $p_1 : X_1 \to X$ such that $p_1^* D_1 \sim mD$ for some Cartier divisor D on X_1. We can also assume that each $p_1^*(D_i)$ is smooth and $\sum p_1^*(D_i)$ is a snc divisor. Also, by Step 1, $H^i(X, L^{-1})$ is a direct summand of $H^i(X_1, p_1^* L^{-1})$. D_1 corresponds to a section of $\mathcal{O}_X(mD)$; let $p_2 : X_2 \to X_1$ be the corresponding cyclic cover. X_2 is smooth by (2.51) and similarly we see that each $p_2^*(D_i)$ is smooth and $p_2^*(\sum_{i>1} D_i)$ is a snc divisor. $(p_2)_* \mathcal{O}_{X_2} = \oplus_{j=0}^{m-1} \mathcal{O}_{X_1}(-jD)$, thus

$$H^i(X_2, p_2^* p_1^* L^{-1}(bD)) = \oplus_{j=0}^{m-1} H^i(X_1, p_1^* L^{-1}((b - j)D)).$$

The $j = b$ case shows that $H^i(X_1, p_1^* L^{-1})$ is a direct summand of $H^i(X_2, p_2^* p_1^* L^{-1}(bD))$.

$p_2^* p_1^* L(-bD) \equiv p_2^* p_1^* M + \sum_{i>1} a_i p_2^* p_1^*(D_i)$ satisfies the assumptions of (2.65). By induction on the number of summands in $\sum a_i D_i$ we obtain $Z \to X_2$. The composite $Z \to X_2 \to X$ satisfies the requirements of (2.65). $\quad\square$

Step 3. Assume that M is ample in (2.64). Then the previous step reduces (2.64) to (2.47). Thus in this case the proof is already complete.

Step 4. Assume now that L is a nef and big line bundle. We need to show that $H^i(X, L^{-1}) = 0$ for $i < \dim X$. By Serre duality this is equivalent to $H^i(X, \omega_X \otimes L) = 0$ for $i > 0$.

Pick an arbitrary ample divisor H on X. By (2.60.4) there is a smooth projective variety Y and a birational morphism $f : Y \to X$ such that $f^* L \equiv A + E$ where A is ample and $E = \sum e_j E_j$ is a snc divisor with $0 \leq e_j < 1$ for every j. There is a Leray spectral sequence

$$H^i(X, L(rH) \otimes R^j f_* \omega_Y) \Rightarrow H^{i+j}(Y, \omega_Y \otimes f^* L(rH)).$$

$f^* L(rH) \equiv (A + rf^* H) + E$ and $A + rf^* H$ is ample. Thus, as we established in the previous step, $H^k(Y, f^* L(rH) \otimes \omega_Y) = 0$ for $k > 0$.

First choose $r \gg 1$. Then $H^i(X, L(rH) \otimes R^j f_* \omega_Y) = 0$ for $i > 0$, thus the spectral sequence gives that

$$H^0(X, L(rH) \otimes R^j f_* \omega_Y) = H^j(Y, \omega_Y \otimes f^* L(rH)) = 0.$$

Since $r \gg 1$, this implies that $R^j f_* \omega_Y = 0$ for $j > 0$.

Finally set $r = 0$. Then we obtain that

$$H^i(X, L \otimes f_* \omega_Y) = H^i(Y, f^* L \otimes \omega_Y) = 0.$$

This completes the proof of (2.64).   □

Remark 2.66. By Serre duality , the vanishing of $H^i(X, L^{-1})$ is equivalent to the vanishing of $H^{n-i}(X, \omega_X \otimes L)$, but during the proof of (2.64) the two versions behave rather differently. Steps 1–3 work with the $H^i(X, L^{-1})$ version. It is possible to do the same steps with the dual vanishing, but it would require a careful comparison of ω_Z and of ω_X which is not straightforward.

In the last step we used the $H^{n-i}(X, \omega_X \otimes L)$ version. Here it would be much harder to work with the groups $H^i(X, L^{-1})$.

During the proof we have used the following lemma of [BG71], which is useful in may situations.

Proposition 2.67. *Let X be a quasi projective variety, D a Cartier divisor on X and m a natural number. Then there is a normal variety Y, a finite and surjective morphism $g : Y \to X$ and a Cartier divisor D' on Y such that $g^* D \sim m D'$.*

If X is smooth and $\sum F_j$ is a snc divisor on X then we can choose Y to be smooth such that the $g^* F_j$ are smooth and $\sum g^* F_j$ is a snc divisor.

Proof. Let $\pi : \mathbb{P}^n \to \mathbb{P}^n$ be the morphism $(x_0 : \cdots : x_n) \mapsto (x_0^m : \cdots : x_n^m)$, where $n = \dim X$. Then $\pi^* \mathcal{O}_{\mathbb{P}^n}(1) \cong \mathcal{O}_{\mathbb{P}^n}(m)$.

Let L be a very ample divisor X. Then there is a morphism $h : X \to \mathbb{P}^n$ such that $L \cong h^* \mathcal{O}_{\mathbb{P}^n}(1)$. Let Y be the normalization of the fiber product sitting in the diagram

$$
\begin{array}{ccc}
Y & \overset{h_Y}{\to} & \mathbb{P}^n \\
g \downarrow & & \downarrow \pi \\
X & \overset{h}{\to} & \mathbb{P}^n.
\end{array}
$$

Then $g^* \mathcal{O}_X(D) \cong h_Y^*(\pi^* \mathcal{O}_{\mathbb{P}^n}(1)) \cong h_Y^* \mathcal{O}_{\mathbb{P}^n}(m)$. We are done with the first part if D is very ample.

If X is smooth then instead of π we consider $\pi' : \mathbb{P}^n \to \mathbb{P}^n$ which is the composition of π with a general automorphism of the target $\mathbb{P}^n$.

Kleiman's Bertini-type theorem (cf. [Har77, III.10.8]) shows that Y is smooth and g^*F is a snc divisor.

In general, write $D \sim L_1 - L_2$. Using the above argument twice, we obtain $g : Y \to X$ such that $g^*L_i \sim mL'_i$ for some Cartier divisors L'_i. Set $D' := L'_1 - L'_2$. □

A special case of (2.64), first proved in [GR70] for $L = 0$, is worth mentioning:

Corollary 2.68. *Let $f : Y \to X$ be a birational morphism between projective varieties, Y smooth. Let L be a line bundle on Y and assume that $L \equiv M + \sum a_i D_i$ where*

(1) *M is nef,*
(2) *$\sum D_i$ is a snc divisor,*
(3) *$0 \le a_i < 1$, and $a_i \in \mathbb{Q}$ for all i.*

Then $R^i f_*(\omega_Y \otimes M) = 0$ for $i > 0$. In particular, $R^i f_* \omega_Y = 0$ for $i > 0$.

Proof: Let H be an ample divisor on X. Apply (2.64) to $L(rf^*H)$ on Y and then use (2.69). □

Proposition 2.69. *Let $f : Y \to X$ be a proper morphism and F a coherent sheaf on Y. The following are equivalent:*

(1) *$H^j(Y, F \otimes f^*H) = 0$ for H sufficiently ample,*
(2) *$R^j f_* F = 0$.*

Proof: This was essentially done during Step 4 above. Choose H such that $H^i(X, H \otimes R^k f_* F) = 0$ for all $i > 0$ and k. Then the Leray spectral sequence degenerates at E_2. Thus $H^j(Y, F \otimes f^*H) = H^0(X, H \otimes R^j f_* F)$. □

(2.64) can be generalized even further, to obtain the following:

Theorem 2.70 (General Kodaira Vanishing Theorem).
Let (X, Δ) be a proper klt pair. Let N be a $\mathbb{Q}$-Cartier Weil divisor on X such that $N \equiv M + \Delta$, where M is a nef and big $\mathbb{Q}$-Cartier $\mathbb{Q}$-divisor. Then $H^i(X, \mathcal{O}_X(-N)) = 0$ for $i < \dim X$.

It is not hard to reduce (2.70) to (2.64). This is the approach taken in [Kol95, Chs. 9–11]. Another proof, using positive characteristic techniques, can be found in [EV92]. See also [Kol97, sec. 2] for a recent summary of further generalizations.

3

Cone Theorems

In Chapter 1 we proved the Cone Theorem for smooth projective varieties, and we noted that the proof given there did not work for singular varieties. For the minimal model program certain singularities are unavoidable and it is essential to have the Cone Theorem for pairs (X, Δ). Technically and historically this is a rather involved proof, developed by several authors. The main contributions are [Kaw84a, Rei83c, Sho85].

Section 1 states the four main steps of the proof and explains the basic ideas behind it. There is a common thread running through all four parts, called the basepoint-freeness method. This technique appears transparently in the proof of the Basepoint-free Theorem. For this reason in section 2 we present the proof of the Basepoint-free Theorem, though logically this should be the second step of the proof.

The basepoint-freeness method has found applications in many different contexts as well, some of which are explained in [Laz96] and [Kol97, Sec.5].

The remaining three steps are treated in the next three sections, the proof of the Rationality Theorem being the most involved.

In section 6 we state and explain the relative versions of the Basepoint-free Theorem and the Cone Theorem.

With these results at our disposal, we are ready to formulate in a precise way the log minimal model program. This is done in section 7. In dimension two the program does not involve flips, and so we are able to treat this case completely.

In section 7 we study minimal models of pairs. It turns out that this concept is not a straightforward generalization of the minimal models of smooth varieties (2.13). The definitions are given in (3.50) and their basic properties are described in (3.52).

3.1 Introduction to the Proof of the Cone Theorem

In section 1.3, we proved the Cone Theorem for smooth varieties. We now begin a sequence of theorems leading to the proof of the Cone Theorem in the general case. This proof is built on a very different set of ideas. Applied even in the smooth case, it gives results not accessible by the previous method; namely it proves that extremal rays can always be contracted. On the other hand, it gives little information about the curves that span an extremal ray. Also, this proof works only in characteristic 0. Before proceeding, we reformulate slightly the Vanishing Theorem (2.64):

Theorem 3.1. *Let Y be a smooth complex projective variety, $\sum d_i D_i$ a $\mathbb{Q}$-divisor (written as a sum of distinct prime divisors) and let L be a line bundle (or Cartier divisor). Assume that $D := L + \sum d_i D_i$ is nef and big and that $\sum D_i$ has only simple normal crossings. Then*

$$H^i(Y, \mathcal{O}_Y(K_Y + \lceil D \rceil)) = 0 \quad for \quad i > 0. \qquad \square$$

3.2. We prove four basic theorems finishing with the Cone Theorem. The proofs of these four theorems are fairly interwoven in history. For smooth threefolds [Mor82] obtained some special cases. The first general result for threefolds was obtained by [Kaw84b], and completed by [Ben83] and [Rei83c]. Non-vanishing was done by [Sho85]. The Cone Theorem appears in [Kaw84a] and is completed in [Kol84]. See [KMM87] for a detailed treatment and for generalizations to the relative case.

Theorem 3.3 (Basepoint-free Theorem). *Let (X, Δ) be a proper klt pair with Δ effective. Let D be a nef Cartier divisor such that $aD - K_X - \Delta$ is nef and big for some $a > 0$. Then $|bD|$ has no basepoints for all $b \gg 0$.*

Theorem 3.4 (Non-vanishing Theorem). *Let X be a proper variety, D a nef Cartier divisor and G a $\mathbb{Q}$-divisor. Suppose*

(1) $aD + G - K_X$ is $\mathbb{Q}$-Cartier, nef and big for some $a > 0$, and
(2) $(X, -G)$ is klt.

Then, for all $m \gg 0$, $H^0(X, mD + \lceil G \rceil) \neq 0$.

Theorem 3.5 (Rationality Theorem). *Let (X, Δ) be a proper klt pair with Δ effective such that $K_X + \Delta$ is not nef. Let $a(X) > 0$ be an integer* *such that $a(X) \cdot (K_X + \Delta)$ is Cartier. Let H be a nef and big Cartier divisor, and define*

$$r = r(H) := \max\{t \in \mathbb{R} : H + t(K_X + \Delta) \text{ is nef}\}.$$

Then r is a rational number of the form u/v ($u, v \in \mathbb{Z}$) where

$$0 < v \le a(X) \cdot (\dim X + 1).$$

Complement 3.6. *Notation as above. Then there is an extremal ray R such that $R \cdot (K_X + \Delta) < 0$ and $R \cdot (H + r(K_X + \Delta)) = 0$.*

Theorem 3.7 (Cone Theorem). *Let (X, Δ) be a projective klt pair with Δ effective. Then:*

(1) *There are (countably many) rational curves $C_j \subset X$ such that $0 < -(K_X + \Delta) \cdot C_j \le 2 \dim X$, and*

$$\overline{NE}(X) = \overline{NE}(X)_{(K_X+\Delta)\ge 0} + \sum \mathbb{R}_{\ge 0}[C_j].$$

(2) *For any $\epsilon > 0$ and ample $\mathbb{Q}$-divisor H,*

$$\overline{NE}(X) = \overline{NE}(X)_{(K_X+\Delta+\epsilon H)\ge 0} + \sum_{\text{finite}} \mathbb{R}_{\ge 0}[C_j].$$

(3) *Let $F \subset \overline{NE}(X)$ be a $(K_X + \Delta)$-negative extremal face. Then there is a unique morphism $\mathrm{cont}_F : X \to Z$ to a projective variety such that $(\mathrm{cont}_F)_* \mathcal{O}_X = \mathcal{O}_Z$ and an irreducible curve $C \subset X$ is mapped to a point by cont_F iff $[C] \in F$. cont_F is called the contraction of F.*

(4) *Let F and $\mathrm{cont}_F : X \to Z$ be as in (3). Let L be a line bundle on X such that $(L \cdot C) = 0$ for every curve C with $[C] \in F$. Then there is a line bundle L_Z on Z such that $L \cong \mathrm{cont}_F^* L_Z$.*

Note. Part (3) is frequently called the Contraction Theorem.

3.8. The logical order of proof of these theorems is the following:

Non-vanishing $\Rightarrow$ Basepoint-free $\Rightarrow$ Rationality $\Rightarrow$ Cone.

However, for better understanding we prove first basepoint freeness, because its proof shows the basic underlying ideas the best.

3.9 (Basic strategy). The main idea for proving the Basepoint-free Theorem (as well as the Non-vanishing and Rationality Theorems) is the following.

Assume for simplicity that X is smooth. Let M be an ample Cartier divisor and $F \subset X$ an irreducible divisor. We have an exact sequence

$$0 \to \mathcal{O}_X(K_X + M) \to \mathcal{O}_X(K_X + M + F) \to \mathcal{O}_F(K_F + M|_F) \to 0,$$

which gives a surjection by (2.64)

$$H^0(X, \mathcal{O}_X(K_X + M + F)) \twoheadrightarrow H^0(F, \mathcal{O}_F(K_F + M|_F)).$$

Notice that we do not need M to be ample, we are in good shape if our vanishing theorem applies to $K_X + M$. Thus if we can write $bD - K_X = M + F$ as above, then we can hope to get sections by induction on the dimension. In general of course this cannot be done.

We have a little more room if we make a birational modification $f : Y \to X$. $H^0(X, \mathcal{O}_X(bD)) = H^0(Y, \mathcal{O}_Y(bf^*D + A))$ if $A \subset Y$ is an effective exceptional divisor. Thus the above sequence can be replaced by

$$0 \to \mathcal{O}_Y(K_Y + M) \to \mathcal{O}_Y(K_Y + M + F) \to \mathcal{O}_F(K_F + M|_F) \to 0,$$

where $M \equiv bf^*D - K_Y + A - F$ and vanishing applies to M. In practice this means that

$$bf^*D - K_Y \equiv (\text{nef and big}) + \Delta + F - A.$$

How can we get b and F? We should use some linear combination of divisors given by the problem. That is, we have at our disposal $aD - K_X$ and D. Moreover, if $|mD|$ is not basepoint free, then we can assume that $f^*|mD| = |L| + \sum r_j F_j$ where $|L|$ is basepoint free and $\sum r_j F_j$ is the fixed part. Let $K_Y = f^*K_X + \sum a_j F_j$. (In $\sum a_j F_j$ it is natural to let F_j run through the exceptional divisors only, while $\sum r_j F_j$ may involve non-exceptional divisors as well. We have to pay attention to this.)

Thus we can write

$$
\begin{aligned}
bf^*D - K_Y &= (b - cm - a)f^*D + c(mf^*D - \sum r_j F_j) \\
&\quad + f^*(aD - K_X) - \sum(a_j - cr_j)F_j.
\end{aligned}
$$

The first three summands are nef and big if $b \geq cm + a$ and $c \geq 0$. Therefore we need to choose c in such a way that

$$\sum(a_j - cr_j)F_j = A - \Delta - F,$$

where A is effective and f-exceptional and F is irreducible. Moreover, because of restriction problems with fractional divisors (3.10), F should not be among the components of Δ.

Thus we need to choose c such that $\min\{a_j - cr_j\} = -1$. Then F is the union of those F_j such that $a_j - cr_j = -1$. Unfortunately, there may be several such components. In order to eliminate this possibility, we perturb everything by a very small linear combination $\sum p_j F_j$. This is a small but useful technical point.

3.10. So we need to worry about the restriction of $\mathbb{Q}$-divisors and their round-ups to smooth hypersurfaces F of a smooth Y. We only restrict divisors $D = L + \sum d_i D_i$ where either $F \neq D_i$ for any i, or $F = D_j$ for some j for which d_j is an integer. In the latter case, we absorb $d_j D_j$ into L before restricting. In either case, we only consider situations in which the sum of the remaining D_i meets F in a simple normal crossing divisor. Then round-up commutes with restriction.

3.2 Basepoint-free Theorem

The following proof is taken almost verbatim from [Rei83c].

Step 1. In this step, we establish that $|mD| \neq \emptyset$ for every $m \gg 0$. Using (2.61) we construct a log resolution $f : Y \to X$ such that

(1) $K_Y \equiv f^*(K_X + \Delta) + \sum a_j F_j$ with all $a_j > -1$,
(2) $f^*(aD - (K_X + \Delta)) - \sum p_j F_j$ is ample for some $a > 0$ and for suitable $0 < p_j \ll 1$. (The F_j need not be f-exceptional.)

On Y we write

$$
\begin{aligned}
f^*(aD - (K_X + \Delta)) &- \sum p_j F_j \\
&= af^*D + \sum(a_j - p_j)F_j - (f^*(K_X + \Delta) + \sum a_j F_j) \\
&\equiv af^*D + G - K_Y,
\end{aligned}
$$

where $G = \sum(a_j - p_j)F_j$. By assumption, $\lceil G \rceil$ is an effective f-exceptional divisor ($a_j > 0$ only when F_j is f-exceptional since Δ is effective), $af^*D + G - K_Y$ is ample, and $H^0(Y, mf^*D + \lceil G \rceil) = H^0(X, mD)$.

We can now apply Non-vanishing to get that

$$H^0(X, mD) > 0 \quad \text{for all } m \gg 0.$$

Step 2. For $s \in \mathbb{Z}_{>0}$ let $B(s)$ denote the reduced base locus of $|sD|$. Clearly $B(s^u) \subset B(s^v)$ for any positive integers $u > v$. Noetherian induction implies that the sequence $B(s^u)$ stabilizes, and we call the limit B_s. So either B_s is non-empty for some s or B_s and $B_{s'}$ are empty for two relatively prime integers s and s'. In the latter case, take u and v such that $B(s^u)$ and $B(s'^v)$ are empty, and use the fact that every sufficiently large integer is a linear combination of s^u and s'^v with non-negative coefficients to conclude that $|mD|$ is basepoint-free for all $m \gg 0$. So we must show that the assumption that some B_s is non-empty leads to a contradiction. We let $m = s^u$ such that $B_s = B(m)$ and assume that this set is non-empty.

Starting with the linear system obtained from the Non-vanishing Theorem, we can blow up further to obtain a new $f : Y \to X$ for which the conditions of Step 1 hold, and, for some $m > 0$,

$$f^*|mD| = |L| \text{ (moving part)} + \sum r_j F_j \text{ (fixed part)}$$

with $|L|$ basepoint-free. Therefore $\cup\{f(F_j) : r_j > 0\}$ is the base locus of $|mD|$. Note that $f^{-1} \operatorname{Bs} |mD| = \operatorname{Bs} |mf^*D|$. We obtain the desired contradiction by finding some F_j with $r_j > 0$ such that, for all $b \gg 0$, F_j is not contained in the base locus of $|bf^*D|$.

Step 3. For an integer $b > 0$ and a rational number $c > 0$ such that $b > cm + a$, we define divisors:

$$
\begin{aligned}
N(b,c) &:= bf^*D - K_Y + \sum(-cr_j + a_j - p_j)F_j \\
&\equiv (b - cm - a)f^*D \quad \text{(nef)} \\
&\quad + c(mf^*D - \sum r_j F_j) \quad \text{(basepoint-free)} \\
&\quad + f^*(aD - K_X - \Delta) - \sum p_j F_j \quad \text{(ample)}.
\end{aligned}
$$

Thus, $N(b,c)$ is ample for $b \geq cm + a$. If that is the case then, by the Vanishing Theorem, $H^1(Y, \lceil N(b,c) \rceil + K_Y) = 0$, and

$$\lceil N(b,c) \rceil = bf^*D + \sum \lceil -cr_j + a_j - p_j \rceil F_j - K_Y.$$

Step 4. c and the p_j can be chosen so that $\sum(-cr_j + a_j - p_j)F_j = A - F$ for some $F = F_{j'}$, where $\lceil A \rceil$ is effective and A does not have F as a component. In fact, we choose $c > 0$ so that

$$\min_j(-cr_j + a_j - p_j) = -1.$$

If this last condition does not single out a unique j, we wiggle the p_j slightly to achieve the desired uniqueness. This j satisfies $r_j > 0$ and $\lceil N(b,c) \rceil + K_Y = bf^*D + \lceil A \rceil - F$.

Now Step 3 implies that

$$H^0(Y, bf^*D + \lceil A \rceil) \to H^0(F, (bf^*D + \lceil A \rceil)|_F)$$

is a surjection for $b \geq cm + a$.

Note: if F_j appears in $\lceil A \rceil$, then $a_j > 0$, so F_j is f-exceptional. Thus $\lceil A \rceil$ is f-exceptional.

Step 5. Notice that

$$N(b,c)|_F = (bf^*D + A - F - K_Y)|_F = (bf^*D + A)|_F - K_F.$$

So we can apply the Non-vanishing Theorem on F to get

$$H^0(F, (bf^*D + \lceil A \rceil)|_F) \neq 0.$$

So $H^0(Y, bf^*D + \lceil A \rceil)$ has a section not vanishing on F. Since $\lceil A \rceil$ is f-exceptional and effective,

$$H^0(Y, bf^*D + \lceil A \rceil) = H^0(Y, bf^*D) = H^0(X, bD).$$

So, as in (3.9), $f(F)$ is not contained in the base locus of $|bD|$ for all $b \gg 0$. This completes the proof of the Basepoint-free Theorem. $\quad\square$

One of the most important applications of the Basepoint-free Theorem is to the finite generation of canonical rings:

Theorem 3.11. *Let* (X, Δ) *be a proper klt pair,* Δ *effective. Assume that* $K_X + \Delta$ *is nef and big. Then the canonical ring, defined as*

$$\oplus_{m=0}^{\infty} H^0(X, \mathcal{O}_X(mK_X + \lfloor m\Delta \rfloor)),$$

is finitely generated over $\mathbb{C}$.

Note. The canonical ring is indeed a ring since $\lfloor m_1 \Delta \rfloor + \lfloor m_2 \Delta \rfloor \leq \lfloor (m_1 + m_2)\Delta \rfloor$. The inequality is reversed for $\lceil m\Delta \rceil$, and so we would not get a ring using $\lceil m\Delta \rceil$.

Proof. By (3.3), there is an $r > 0$ such that $r\Delta$ is an integral divisor and $\mathcal{O}_X(rK_X + r\Delta)$ is generated by global sections. These sections define a morphism $f : X \to Z$ and there is an ample invertible sheaf L on Z such that $f^*L = \mathcal{O}_X(rK_X + r\Delta)$. Let $G_m = f_*\mathcal{O}_X(mK_X + \lfloor m\Delta \rfloor)$. Then

$$\oplus_{m=0}^{\infty} H^0(X, \mathcal{O}_X(mK_X + \lfloor m\Delta \rfloor)) = \oplus_{m=0}^{\infty} H^0(Z, G_m).$$

The G_m are coherent sheaves and $G_{m+r} \cong G_m \otimes L$ by the projection formula. Since L is ample, $R = \oplus_{m=0}^{\infty} H^0(Z, L^m)$ is a finitely generated ring over $\mathbb{C}$ and $\oplus_{m=0}^{\infty} H^0(Z, G_{j+rm})$ is a finitely generated R-module for every $0 \leq j < r$. Thus

$$\oplus_{m=0}^{\infty} H^0(Z, G_m) = \oplus_{j=0}^{r-1} \left(\oplus_{m=0}^{\infty} H^0(Z, G_{j+rm}) \right)$$

is a finitely generated ring over $\mathbb{C}$. $\quad\square$

The above result is a special case of the following:

Conjecture 3.12 (Abundance conjecture). *Let* (X, Δ) *be a proper log canonical pair,* Δ *effective. Then:*

(1) $\oplus_{m=0}^{\infty} H^0(X, \mathcal{O}_X(mK_X + \lfloor m\Delta \rfloor))$ *is a finitely generated ring.*
(2) *If* $K_X + \Delta$ *is nef then* $|m(K_X + \Delta)|$ *is basepoint free for some* $m > 0$.

3.13. As in (3.11) we see that if $K_X + \Delta$ is nef then (2) implies (1). In general, the minimal model program reduces (1) to (2). Frequently only (2) is called the abundance conjecture.

For surfaces this is a non-trivial result. The conjecture is also known to be true in dimension 3. The $\Delta = 0$ case is a culmination of a series of papers of Miyaoka and Kawamata. See [K+92] for a simplified proof and for references. The general log canonical case in dimension 3 is proved in [KMM94a].

Very little is known in higher dimensions.

3.3 The Cone Theorem

This proof of the Cone Theorem grew out of conversations among J. Kollár, T. Luo, K. Matsuki and S. Mori.

3.14 (Informal explanation). First we give an idea of the way the Rationality Theorem is used to get information about the cone of curves.

Let H be ample on X and choose $r = r(H)$ as in the Rationality Theorem. Since r is rational, $m(H + r(K_X + \Delta))$ is Cartier for some $m > 0$. Note that $m(H + r(K_X + \Delta))$ is nef but not ample. Thus $\overline{NE}(X) \cap (H + r(K_X + \Delta))^{\perp}$ is a face of $\overline{NE}(X)$. Starting with various ample divisors, we get various faces of $\overline{NE}(X)$. The proof of the Cone Theorem turns out to be a formal consequence of this observation. To be precise, the Cone Theorem follows immediately from the Rationality Theorem and the following abstract result.

Theorem 3.15. *Let* $N_{\mathbb{Z}}$ *be a free* $\mathbb{Z}$-*module of finite rank and* $N_{\mathbb{R}} = N_{\mathbb{Z}} \otimes_{\mathbb{Z}} \mathbb{R}$ *the base change to* $\mathbb{R}$. *Let* $\overline{NE} \subset N_{\mathbb{R}}$ *be a closed convex cone not containing a straight line. Let* K *be an element of the dual* $\mathbb{Q}$-*vector space* $N_{\mathbb{Q}}^*$ *such that* $(K \cdot C) < 0$ *for some* $C \in \overline{NE}$. *Assume that there exists* $a(K) \in \mathbb{Z}_{>0}$ *such that, for all* $H \in N_{\mathbb{Z}}^*$ *with* $H > 0$ *on* $\overline{NE} - \{0\}$,

$$r := \max\{t \in \mathbb{R} : H + tK \geq 0 \text{ on } \overline{NE}\}$$

is a rational number of the form $u/a(K)$ ($u \in \mathbb{Z}$). Then

$$\overline{NE} = \overline{NE}_{K \geq 0} + \sum \mathbb{R}_{\geq 0}[\xi_i],$$

for a collection of $\xi_i \in N_{\mathbb{Z}}$ with $(\xi_i \cdot K) < 0$ such that the $\mathbb{R}_{\geq 0}[\xi_i]$ do not accumulate in the halfspace $K < 0$.

Proof of (3.15) and the Cone Theorem: We may assume that $K := K_X + \Delta$ is not nef.

Step 1. For a nef divisor class L set $F_L = L^{\perp} \cap \overline{NE}$. If L is not ample, then by (1.18), $F_L \neq \{0\}$. Assume that $F_L \not\subset \overline{NE}_{K \geq 0}$ and let H be an arbitrary ample Cartier divisor. For $n \in \mathbb{Z}_{>0}$, let

$$r_L(n, H) := \max\{t \in \mathbb{R} : nL + H + \frac{t}{a(K)}K \text{ is nef}\}.$$

By the Rationality Theorem, $r_L(n, H)$ is a (non-negative) integer, and, since L is nef, $r_L(n, H)$ is a non-decreasing function of n. Now $r_L(n, H)$ stabilizes to a fixed $r_L(H)$ for $n \geq n_0$ since, if $\xi \in F_L \setminus \overline{NE}_{K \geq 0}$, then

$$r_L(n, H) \leq a(K)\frac{(H \cdot \xi)}{-(K \cdot \xi)}.$$

Also L and $D(nL, H) := na(K)L + a(K)H + r_L(H)K$ are both non-ample nef divisors for $n > n_0$, so $0 \neq F_{D(nL,H)} \subset F_L$ and $F_{D(nL,H)} \subset \overline{NE}_{K<0} \cup \{0\}$.

Step 2. We claim that, if $\dim F_L > 1$ and $F_L \not\subset \overline{NE}_{K \geq 0}$, then we can find an ample H with

$$\dim F_{D(nL,H)} < \dim F_L.$$

To see this, choose ample divisors H_i which give a basis for F_L^*. If $\dim F_L > 1$, the linear functions

$$\left(nL + H_i + \frac{r_L(H_i)}{a(K)}K\right)|_{F_L} = \left(H_i + \frac{r_L(H_i)}{a(K)}K\right)|_{F_L}$$

cannot all be identically zero on F_L. Thus $\dim F_{D(nL,H_i)} < \dim F_L$ for some i, proving our claim.

Repeating the argument over successively smaller faces, we obtain that for every nef L with $F_L \not\subset \overline{NE}_{K \geq 0}$ there is a nef L' such that $F_L \supset F_{L'}$, $\dim F_{L'} = 1$ and $F_L \subset \overline{NE}_{K<0} \cup \{0\}$.

Step 3. We claim that

$$\overline{NE} \quad \text{and} \quad \overline{NE}_{K\geq 0} + \sum_{\dim F_L = 1} F_L,$$

have the same closure.

To prove this, assume that the right-hand-side of the claimed equality is smaller. Then there is a divisor class $M \in N_{\mathbb{Z}}^*$ so that the hyperplane $M = 0$ misses the right-hand-side but not the left-hand-side. We can now argue exactly as in (1.24). A straightforward application of the Rationality Theorem, followed by Step 2, gives a contradiction.

Step 4. Next we show that the one-dimensional F_L do not accumulate in the halfspace $K < 0$. To see this, choose ample divisors $H_1, \ldots, H_d \in N_{\mathbb{Z}}^*$ such that they, together with K, form a basis of $N_{\mathbb{R}}^*$. For each one-dimensional F_L and i, take n_i such that $F_{D(n_i L, H_i)} = F_L$. Then, for ξ generating F_L and for all i,

$$\frac{(\xi \cdot H_i)}{(\xi \cdot K)} = \frac{\text{integer}}{a(K)}.$$

If the F_L accumulated somewhere in the halfspace $K < 0$, then the points of the projectivization $\mathbb{P}(N_{\mathbb{R}})$ to which they correspond would have to accumulate somewhere in the affine subset $U \subset \mathbb{P}(N_{\mathbb{R}})$ given by $K < 0$.

But the equation above rules out that possibility, because

$$\xi \in U \mapsto \frac{(\xi \cdot H_1)}{(\xi \cdot K)}, \ldots, \frac{(\xi \cdot H_d)}{(\xi \cdot K)}$$

is an affine coordinate system.

This shows that there are only finitely many F_L such that $F_L \cdot (K + \epsilon H) < 0$. Thus

$$\overline{NE}_{K+\epsilon H \geq 0} + \sum_{F_L \cdot (K+\epsilon H) < 0} F_L$$

is a closed cone. We have proved in Step 3 that its closure is $\overline{NE}$. Thus

$$\overline{NE} = \overline{NE}_{K+\epsilon H \geq 0} + \sum_{F_L \cdot (K+\epsilon H) < 0} F_L.$$

This proves (3.7.2), except that we do not yet know that each F_L is spanned by a curve in (3.7).

Remark 3.16. This is a convenient place to prove (3.6) as a side step. The proof of the Cone Theorem will continue with Step 5.

To avoid notational confusion, let M denote the nef and big divisor in (3.6). We can write $M \equiv m_0 K_X + \sum m_i H_i$ for some $m_i \in \mathbb{Q}$. Step 4 above shows that $-(\xi \cdot M)/(\xi \cdot (K_X + \Delta))$ are non-negative rational numbers with bounded denominators as ξ runs through all $(K_X + \Delta)$-negative extremal rays. The infimum of this set is thus a minimum, archived by an extremal ray ξ_M. Thus $r(M) = -(\xi_M \cdot M)/(\xi_M \cdot (K_X + \Delta))$ and $\xi_M \cdot (M + r(M) \cdot (K_X + \Delta)) = 0$. □

Step 5. It is not hard to prove that (3.7.1) is a formal consequence of (3.7.2) and of Step 4, see [Kol96, III.1.2]. We do not need this in the sequel.

Step 6. Here we prove that if $F \subset \overline{NE}(X)$ is a $(K_X + \Delta)$-negative extremal face then there is a nef Cartier divisor D such that $F_D = F$.

Let $\langle F \rangle$ be the linear span of F and $V \subset N_1(X)^*$ the set of linear functions which vanish on $\langle F \rangle$. F is spanned by extremal rays, and each of them is defined over $\mathbb{Q}$. Thus V is defined over $\mathbb{Q}$. Choose $\epsilon > 0$ such that $K_X + \Delta + \epsilon H$ is negative on F.

Since F is extremal, $\langle F \rangle \cap \overline{NE}(X) = F$. Thus

$$W_F := \overline{NE}(X)_{K_X + \Delta + \epsilon H \geq 0} + \sum_{\dim F_L = 1, F_L \not\subset F} F_L$$

is a closed cone which intersects $\langle F \rangle$ only at the origin and $\overline{NE}(X) = W_F + F$. Thus there is a hyperplane $(g = 0)$ which contains $\langle F \rangle$ but intersects W_F only at the origin.

The supporting functions of F are exactly those elements of V which are strictly positive on $W_F \setminus \{0\}$. This is an open condition which is non-empty since g is such. Thus F has a supporting function which is defined over $\mathbb{Q}$. A suitable multiple of it gives a supporting function which corresponds to a Cartier divisor; denote it by D.

Step 7. (3.7.3) is now an easy consequence of the Basepoint-free Theorem.

By assumption $-(K_X + \Delta)$ is strictly positive on F. This implies that $mD - (K_X + \Delta)$ is strictly positive on $\overline{NE}(X) \setminus \{0\}$ for $m \gg 1$. Hence by (1.18), $mD - (K_X + \Delta)$ is ample for $m \gg 1$. Thus $|mD|$ is basepoint free for $m \gg 1$ by (3.3). Let g_F be the Stein factorization of the morphism given by $|mD|$ for some $m \gg 1$. This is the contraction morphism associated to F.

Step 8. If $F \neq \{0\}$ then the corresponding contraction morphism is not an isomorphism by (1.18), thus at least one curve $C \subset X$ is contracted by g_F. This shows that each $(K_X + \Delta)$-negative extremal face contains the class of a curve.

If X is smooth then (1.13) can be used to show that there is a rational curve in F (we leave this as an exercise). The general case is proved in [Kaw91] using different methods.

Step 9. Let F be a $(K_X + \Delta)$-negative extremal face. Let D be any supporting Cartier divisor of F, $g_F : X \to Z$ the contraction of F and $W_F \subset \overline{NE}(X)$ as in Step 6. As we showed in Step 7, g_F is given by $|mD|$ for every $m \gg 1$. Thus both mD and $(m+1)D$ is a pull back of a Cartier divisor on Z. Their difference gives a Cartier divisor D_Z such that $g_F^* D_Z = D$.

Let L be any Cartier divisor on X such that $L|_F \equiv 0$. Since D is positive on $W_F \setminus \{0\}$, $L + mD$ is also positive on $W_F \setminus \{0\}$ for $m \gg 1$. Thus $L + mD$ is also a supporting Cartier divisor of F, hence as we just proved, there is a Cartier divisor M_Z such that $L + mD = g_F^* M_Z$. Set $L_Z := M_Z - mD_Z$ to obtain (3.7.4). $\qquad\square$

The following two corollaries examine the divisor theory of contractions.

Corollary 3.17. *Let (X, Δ) be a projective klt pair and $R \subset \overline{NE}(X)$ a $(K_X + \Delta)$-negative extremal ray with contraction morphism $g_R : X \to Z$. Let $C \subset X$ be a curve which generates R. We have an exact sequence*

$$0 \to \operatorname{Pic} Z \xrightarrow{L \mapsto g_R^* L} \operatorname{Pic} X \xrightarrow{M \mapsto (M \cdot C)} \mathbb{Z}.$$

In particular, $\rho(Z) = \rho(X) - 1$.

Proof. Let L be a line bundle on Z. Then $(g_R)_*(g_R^* L) = L \otimes (g_R)_* \mathcal{O}_X = L$, thus $L \mapsto g_R^* L$ is an injection. If M is a line bundle on X and $(N \cdot C) = 0$ then $M = g_R^* L$ for some L by (3.7.4). $\qquad\square$

Corollary 3.18. *Let (X, Δ) be a projective klt pair and $R \subset \overline{NE}(X)$ a $(K_X + \Delta)$-negative extremal ray with contraction morphism $g_R : X \to Z$. Assume that X is $\mathbb{Q}$-factorial and g_F is either a divisorial or a Fano contraction (2.14 Steps 3 and 5). Then Z is also $\mathbb{Q}$-factorial.*

Proof. Assume first that g_R is a divisorial contraction with exceptional divisor $E \subset X$. Then $(E \cdot R) \neq 0$. Let B be a Weil divisor on Z. There is a rational number s such that $((g_F)_*^{-1} B + sE \cdot R) = 0$. $m((g_F)_*^{-1} B + sE)$ is a Cartier divisor for some $m > 0$, and by (3.7.4) it is the pull back of a Cartier divisor M_Z on Z. Thus $mB \sim M_Z$ and so B is $\mathbb{Q}$-Cartier.

Assume next that g_R is a Fano contraction. Let B be a Weil divisor on

Z and $Z^0 \subset Z$ the smooth locus. Let $D \subset X$ be the closure of $g_F^{-1}(B|_{Z^0})$. D is disjoint from the general fiber of g_F, hence $(D \cdot R) = 0$. mD is a Cartier divisor for some $m > 0$ and by (3.7.4) it is the pull back of a Cartier divisor M_Z on Z. Thus $mB \sim M_Z$ and so B is $\mathbb{Q}$-Cartier. $\qquad \square$

3.4 The Rationality Theorem

The proof is taken from [Kaw84a] with simplifications and additions as in [Kol84]. See also [KMM87, 4.1].

The main idea is the following. Assume for simplicity that X is smooth, H is ample and $\Delta = 0$. We want to use the basepoint freeness method to show that certain linear systems $|pH + qK_X|$ are basepoint free. By looking at the proof carefully, we see that a crucial part is the vanishing $H^1(X, \mathcal{O}_X(pH + qK_X)) = 0$, which holds if $pH + (q-1)K_X$ is nef and big. If we can choose p, q such that

$$\frac{q-1}{p} < r < \frac{q}{p}, \qquad (3.1)$$

then the above vanishing holds, but $|pH + qK_X|$ is not base point free, since it is not even nef. Of course we need other vanishing results as well, and so we can prove only that if there are lots of pairs (p, q) satisfying (3.1), then we get a contradiction. We show in Step 3 how the existence of such pairs is connected with the arithmetic properties of r.

Step 1. In this step we reduce the general case of (3.5) to the special case when H is basepoint free. Assume that $a(K_X + \Delta)$ is Cartier. By (3.3), $H' := m(cH + da(K_X + \Delta))$ is basepoint free for $m \gg c \gg d > 0$. Solving the linear equation

$$H + r(H)(K_X + \Delta) = \lambda(H' + r(H')(K_X + \Delta))$$

gives that

$$r(H) = \frac{r(H') + mda}{mc}.$$

Thus $r(H)$ is rational iff $r(H')$ is. Assume furthermore that $r(H')$ has denominator v. Then $r(H)$ has denominator dividing mcv. Since m and c can be arbitrary sufficiently large integers, this implies that $r(H)$ has denominator dividing v.

Step 2. Suppose Y is a smooth projective variety, $\{D_i\}$ a finite collection of Cartier divisors and A a normal-crossing $\mathbb{Q}$-divisor with $\lceil A \rceil$ effective.

Consider the Hilbert polynomial

$$P(u_1, \ldots, u_k) = \chi\left(\sum u_i D_i + \lceil A \rceil\right).$$

Suppose that, for some values of the u_i, $\sum u_i D_i$ is nef and $\sum u_i D_i + A - K_Y$ is ample.

Then, for all integers $m \gg 0$, $\sum m u_i D_i + A - K_Y$ is still ample so that $H^i(\sum m u_i D_i + \lceil A \rceil) = 0$ for $i > 0$ by the Vanishing Theorem, and $\mathcal{O}_Y(\sum m u_i D_i + \lceil A \rceil)$ must have a section by the Non-vanishing Theorem. Therefore $\chi(\sum m u_i D_i + \lceil A \rceil) \neq 0$.

Thus $P(u_1, \ldots, u_k)$ is not identically zero and its degree is $\leq \dim Y$.

Step 3. We need a result about zeros of polynomials:

Lemma 3.19. *Let $P(x, y)$ be a non-trivial polynomial of degree $\leq n$, and assume that P vanishes for all sufficiently large integral solutions of $0 < ay - rx < \epsilon$ for some fixed positive integer a and positive ϵ for some $r \in \mathbb{R}$.*

Then r is rational, and in reduced form, r has denominator $\leq a(n+1)/\epsilon$.

Proof: First assume r irrational. Then an infinite number of integral points in the (x, y)-plane on each side of the line $ay - rx = 0$ are closer than $\epsilon/(n+2)$ to that line. So there is a large integral solution (x', y') with $0 < ay' - rx' < \epsilon/(n+2)$. But then $(2x', 2y'), \ldots, ((n+1)x', (n+1)y')$ are also solutions by hypothesis. So $(y'x - x'y)$ divides P, since P and $(y'x - x'y)$ have $(n+1)$ common zeroes. Choose a smaller ϵ and repeat the argument. Do this $n+1$ times to get a contradiction.

Now suppose $r = u/v$ (in lowest terms). For given j, let (x', y') be a solution of $ay - rx = aj/v$. (Note that an integral solution exists for any j.) Then $a(y' + ku) - r(x' + akv) = aj/v$ for all k. So, as above, if $aj/v < \epsilon$, $(ay - rx) - (aj/v)$ must divide P. So we can have at most n such values of j. Thus $a(n+1)/v \geq \epsilon$.

Step 4. Let ϵ be a positive number and H a nef and big Cartier divisor. Let $a \in \mathbb{Z}_{>0}$ be such that $a(K_X + \Delta)$ is also Cartier. Assume that $K_X + \Delta$ is not nef and let $r = \max\{t \in \mathbb{R} : H + t(K_X + \Delta) \text{ is nef}\}$.

For each (p, q), let $L(p, q)$ denote the base locus of the linear system $|pH + qa(K_X + \Delta)|$ on X (with reduced scheme structure). By definition, $L(p, q) = X$ iff $|pH + qa(K_X + \Delta)| = \emptyset$.

Claim 3.20. For (p, q) sufficiently large and $0 < aq - rp < \epsilon$, $L(p, q)$ is the same subset of X. We call this subset L_0.

Proof: Consider the following diagram of divisors on X:

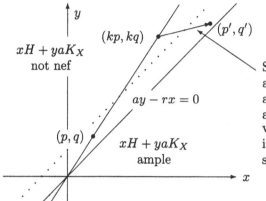

The above diagram shows that $L(p', q') \subset L(p, q)$, which proves the claim by the Noetherian condition on subvarieties.

For (p, q) as above, the linear system $|pH + qa(K_X + \Delta)|$ cannot be basepoint-free on X since $pH + qa(K_X + \Delta)$ is not nef. We let $I \subset Z \times Z$ be the set of (p, q) for which $0 < aq - rp < 1$ and $L(p, q) = L_0$. Let us emphasize that I contains all sufficiently large (p, q) with $0 < aq - rp < 1$.

Step 5. Suppose (X, Δ) is a klt pair. Let $g : Y \rightarrow X$ be a log resolution. Let $D_1 = g^* H$, $D_2 = g^*(a(K_X + \Delta))$ and define A by the formula $K_Y \equiv g^*(K_X + \Delta) + A$. $\lceil A \rceil$ is effective and g-exceptional.

Then we put $P(x, y) = \chi(x D_1 + y D_2 + \lceil A \rceil)$.

Since D_1 is nef and big, P is not identically zero by Riemann–Roch and by Step 2. Since $\lceil A \rceil$ is effective and g-exceptional,

$$H^0(Y, p D_1 + q D_2 + \lceil A \rceil) = H^0(X, pH + qa(K_X + \Delta)).$$

Step 6. Suppose now that the assertion of the Rationality Theorem that r is rational is false. If $0 < ay - rx < 1$, then $x D_1 + y D_2 + A - K_Y$ is numerically equivalent to the pull-back of the nef and big $\mathbb{Q}$-divisor $xH + (ay - 1)(K_X + \Delta)$. Thus, by the Vanishing Theorem,

$$H^i(Y, x D_1 + y D_2 + \lceil A \rceil) = 0 \quad \text{for } i > 0.$$

By Step 3, there must exist arbitrarily large (p, q) with $0 < aq - rp < 1$ for which $P(p, q) = h^0(Y, p D_1 + q D_2 + \lceil A \rceil) \neq 0$, since otherwise $P(x, y)$ would vanish 'too often' implying that r is rational for X and H. Thus $|pH + qa(K_X + \Delta)| \neq \emptyset$ for all $(p, q) \in I$ by Step 4. That is, $L_0 \neq X$.

Step 7. For $(p, q) \in I$, choose a log resolution $f : Y \to X$ such that:

(1) $K_Y \equiv f^*(K_X + \Delta) + \sum a_j F_j$, where $a_j > -1$.

(2) $f^*(pH + (qa - 1)(K_X + \Delta)) - \sum p_j F_j$ is ample for some sufficiently small, positive p_j. (This is possible by (2.61) since $pH + (qa-1)K_X$ is nef and big.)

(3) $f^*|pH + qa(K_X + \Delta)| = |L| + \sum r_j F_j$ for some non-negative integers r_j, where L is basepoint-free and $\sum r_j F_j$ is the fixed part.

Step 8. Let $(p, q) \in I$ be as chosen in Step 7. As before, we can choose rational $c > 0$ and $p_j > 0$ so that

$$\sum(-cr_j + a_j - p_j)F_j = A' - F,$$

with $\lceil A' \rceil$ effective, A' not involving F. By examining coefficients, we notice that F maps into some component B of the base locus $L(p, q) = f(\cup_{r_j > 0} F_j)$ of $|pH + qa(K_X + \Delta)|$. Define

$$
\begin{aligned}
N(p', q') &= f^*(p'H + q'a(K_X + \Delta)) + A' - F - K_Y \\
&\equiv f^*((p' - (1 + c)p)H + (q' - (1 + c)q)a(K_X + \Delta)) \\
&\quad + f^*((1 + c)pH + (1 + c)qa(K_X + \Delta)) \\
&\quad + \sum(-cr_j + a_j - p_j)F_j - K_Y \\
&\equiv cL \quad \text{(basepoint-free)} \\
&\quad + f^*((p' - (1 + c)p)H + (q' - (1 + c)q)a(K_X + \Delta)) \\
&\quad + f^*(pH + (qa - 1)(K_X + \Delta)) - \sum p_j F_j \quad \text{(ample)}.
\end{aligned}
$$

Notice that if p' and q' are big enough and $aq' - rp' < aq - rp$, then

$$
\begin{aligned}
(q' - (1 + c)q)a &< r(p' - (1 + c)p), &\text{so} \\
(p' - (1 + c)p)H + (q' - (1 + c)q)a(K_X + \Delta) &&\text{is nef.}
\end{aligned}
$$

Therefore $N(p', q')$ is ample. Thus, by the Vanishing Theorem, the following map is surjective

$$H^0(Y, f^*(p'H + q'a(K_X + \Delta)) + \lceil A' \rceil) \twoheadrightarrow$$
$$\twoheadrightarrow H^0(F, (f^*(p'H + q'a(K_X + \Delta)) + \lceil A' \rceil)|_F)$$

Step 9. By the adjunction formula, the restriction of the divisor $f^*(p'H + q'a(K_X + \Delta)) + A' - F - K_Y$ to F is the divisor

$$(f^*(p'H + q'a(K_X + \Delta)) + A')|_F - K_F.$$

As in Step **2**, the Hilbert polynomial

$$\chi(F, (f^*(p'H + q'a(K_X + \Delta)) + \lceil A' \rceil)|_F)$$

is not identically zero.

For $0 < aq' - rp' < aq - rp$,

$$(f^*(p'H + q'a(K_X + \Delta)) + A')|_F - K_F = N(p', q')|_F$$

is ample, so, in this strip,

$$\chi(F, (f^*(p'H + q'a(K_X + \Delta)) + \lceil A' \rceil)|_F) =$$
$$= h^0(F, (f^*(p'H + q'a(K_X + \Delta)) + \lceil A' \rceil)|_F).$$

So, by *Step 3* applied to the Hilbert polynomial on F with $\epsilon = aq - rp$, there must be arbitrarily large (p', q') such that $0 < aq' - rp' < aq - rp$ and

$$h^0(F, (f^*(p'H + q'a(K_X + \Delta)) + \lceil A' \rceil)|_F) \neq 0.$$

Step 10. We are now ready to derive a contradiction. By assumption $L(p, q) = L_0$. For (p', q') as in Step 9

$$H^0(Y, f^*(p'H + q'a(K_X + \Delta)) + \lceil A' \rceil) \rightarrow$$
$$\rightarrow H^0(F, (f^*(p'H + q'a(K_X + \Delta)) + \lceil A' \rceil)|_F) \neq 0$$

is surjective. Thus F is not a component of the base locus of $|f^*(p'H + q'a(K_X + \Delta)) + \lceil A' \rceil|$. Since $\lceil A' \rceil$ is f-exceptional and effective,

$$H^0(Y, f^*(p'H + q'a(K_X + \Delta)) + \lceil A' \rceil) = H^0(X, p'H + q'a(K_X + \Delta)),$$

and so, as in (3.9), this implies that $f(F)$ is not contained in $L(p', q')$. Thus $L(p', q')$ is a proper subset of $L(p, q) = L_0$, giving the desired contradiction.

Step 11. So now we know that r is rational. We next suppose that the assertion of the Rationality Theorem concerning the denominator of r is false. We proceed to a contradiction in much the same way.

Using Step **3** with $\epsilon = 1$, conclude as in Step 6 that there exist arbitrarily large (p, q) with $0 < aq - rp < 1$ such that

$$P(p, q) = h^0(Y, pD_1 + qD_2 + \lceil A \rceil) \neq 0,$$

since otherwise $P(x, y)$ would vanish 'too often'. Thus $|pH + qa(K_X + \Delta)| \neq \emptyset$ for all $(p, q) \in I$ by Step 4.

Choose $(p,q) \in I$ such that $aq - rp$ is the maximum, say it is equal to d/v. Choose a resolution f as in Step 7. In the strip $0 < aq' - rp' < d/v$, we have as before that

$$\chi(F, (f^*(p'H + q'a(K_X + \Delta)) + \lceil A' \rceil)|_F) =$$
$$= h^0(F, (f^*(p'H + q'a(K_X + \Delta)) + \lceil A' \rceil)|_F).$$

By Step 3, there exists (p', q') in the strip $0 < aq' - rp' < 1$ with $\epsilon = 1$ for which $h^0(F, (f^*(p'H + q'a(K_X + \Delta)) + \lceil A' \rceil)|_F) \neq 0$. But then $aq' - rp' < d/v = aq - rp$ automatically. The desired contradiction is then derived as in Steps 7–9. This completes the proof of the Rationality Theorem. $\qquad\square$

3.5 The Non-vanishing Theorem

The proof is taken from [Sho85].

Step 0. Let $f : X' \to X$ be a projective resolution of X. Write $K_{X'} - G' \equiv f^*(K_X - G)$. Then $(X', -G')$ is klt and $af^*D + G' - K_{X'} \equiv f^*(aD+G-K_X)$. As in (2.61) choose a small effective divisor F such that $af^*D+G'-K_{X'}-F$ is ample and $(X', -G'+F)$ is klt. Set $G'' := G'-F$. Then $f_*(G'') \leq G$ and so $h^0(X', mf^*D+\lceil G'' \rceil) \leq h^0(X, mD+\lceil G \rceil)$. This shows that it is sufficient to prove (3.4) under the following additional conditions, which we assume for the rest of the proof.

(*) X is smooth and projective, $(X, -G)$ is klt and $aD + G - K_X$ is ample for some $a > 0$.

Step 1. Since $(X, -G)$ is klt, the coefficient of every summand of G is greater than -1, that is, $\lceil G \rceil$ is effective.

If D is numerically trivial, then

$$h^0(X, mD + \lceil G \rceil) = \chi(X, mD + \lceil G \rceil) = \chi(X, \lceil G \rceil) = h^0(X, \lceil G \rceil) \neq 0,$$

where the middle equality holds by (2.57) and the two others by (2.47). So the assertion of the theorem is trivially satisfied. Thus for the rest of the section we can assume that D is not numerically trivial.

Step 2. Pick a point $x \in X$ which does not lie in the support of G. (We blow up this point first in the construction of f below.) We claim that we can pick a positive integer q_0 so that for every $q \geq q_0$ there is a $\mathbb{Q}$-divisor $M(q) \equiv (qD + G - K_X)$ with $\mathrm{mult}_x M(q) > 2 \dim X$.

To see that this is possible, let $d = \dim X$. Since D is nef, $(D^e \cdot A^{d-e}) \geq 0$ for any ample divisor A. Thus

$$\begin{aligned}
((qD + G - K_X)^d) &= (((q-a)D + aD + G - K_X)^d) \\
&\geq d(q-a)(D \cdot (aD + G - K_X)^{d-1}).
\end{aligned}$$

There is some curve C so that $(D \cdot C) > 0$ and some p such that $((p(aD + G - K_X))^{d-1})$ is represented by C plus an effective one-cycle. So $(D \cdot (aD + G - K_X)^{d-1}) > 0$. Thus the right-hand quantity goes to infinity with q. Then, by the Riemann–Roch formula and Serre vanishing

$$h^0(e(qD + G - K_X)) \geq \frac{e^d}{d!}(qD + G - K_X)^d + \text{ (lower powers of } e).$$

On the other hand, the number of conditions on $M(q, e) \in |e(qD + G - K_X)|$ that x be a point of multiplicity $> 2de$ on $M(q, e)$ is at most

$$\frac{e^d}{d!}(2d)^d + \text{ (lower powers of } e).$$

Since $(qD + G - K_X)^d \to \infty$ as $q \to \infty$, we have more sections than conditions. Thus we can find $M(q, e) \in |e(qD + G - K_X)|$ such that $\text{mult}_x M(q, e) > 2de$. Set $M(q) := M(q, e)/e$. This proves the claim.

Step 3. Let $f = f(q) : Y \to X$ be some log resolution of $(X, G + M(q))$ such that f dominates the blow up $B_x X$ of $x \in X$. Set

  (1) $K_Y + f^*G \equiv f^*K_X + \sum b_j F_j$, where $b_j > -1$ by assumption,
  (2) $(1/2)f^*(aD + G - K_X) - \sum p_j F_j$ is ample for suitable $0 < p_j \ll 1$,
  (3) $f^*M(q) = \sum r_j F_j$ with F_0 corresponding to the exceptional divisor of the blow-up of x.

Step 4. We define

$$N(b, c) := bf^*D + \sum(-cr_j + b_j - p_j)F_j - K_Y.$$

As before, we want to make $N(b, c)$ ample. We calculate:

$$\begin{aligned}
N(b, c) &= bf^*D + \sum(-cr_j + b_j - p_j)F_j - K_Y \\
&\equiv bf^*D - cf^*(qD + G - K_X) - \sum p_j F_j + f^*G - f^*K_X \\
&= (b - a - c(q - a))f^*D + (1 - c)f^*(aD + G - K_X) - \sum p_j F_j \\
&= (b - a - c(q - a))f^*D \quad \text{(nef if } b - a - c(q - a) \geq 0) \\
&\quad + (\tfrac{1}{2} - c)f^*(aD + G - K_X) \quad \text{(nef if } \tfrac{1}{2} \geq c) \\
&\quad + \tfrac{1}{2}f^*(aD + G - K_X) - \sum p_j F_j \quad \text{(ample)}.
\end{aligned}$$

Thus as long as $1/2 \geq c$ and $b \geq a + c(q - a)$, $N(b, c)$ is ample.

Step 5. Pick $c = \min\{(1 + b_j - p_j)/r_j\}$, where the minimum is taken over those j such that $r_j > 0$. Then $c > 0$. As before, we wiggle the p_j so that this minimum is achieved for only one value j' of j and set $F = F_{j'}$. Since $x \notin \operatorname{Supp} G$, $b_0 = d - 1$, $r_0 > 2d$, and therefore $c < (1 + (d-1) - p_0)/2d < 1/2$. Thus, $c < 1/2$, and so, for $b \geq a + c(q - a)$, $N(b, c)$ is ample.

Step 6. The rest of the story is as in the proofs of the Basepoint-free and Rationality Theorems. Write

$$N(b, c) = bf^*D + A - F - K_Y.$$

Let $f^*G = \sum g_j F_j$. If F_j is not f-exceptional, then $b_j = g_j$. The coefficient of F_j in A is $(-cr_j + b_j - p_j) < b_j$, and therefore

$$\lceil A \rceil \leq f^* \lceil G \rceil + (f\text{-exceptional divisors}).$$

Adding exceptional divisors to a pull back does not increase H^0, thus

$$H^0(Y, bf^*D + \lceil A \rceil) \subset H^0(Y, bf^*D + f^* \lceil G \rceil) = H^0(X, bD + \lceil G \rceil).$$

Since $N(b, c)$ is ample,

$$H^1(Y, bf^*D + \lceil A \rceil - F) = H^1(Y, bf^*D + \lceil A - F \rceil) = 0,$$

so $H^0(X, bD + \lceil G \rceil) \neq 0$ if we show that

$$H^0(F, (bf^*D + \lceil A \rceil)|_F) \neq 0.$$

This last inequality can be achieved by making an induction on $\dim X$. We can assume that we have already proved the Non-vanishing Theorem for varieties of dimension $< \dim X$. Applying the induction assumption to F, we complete the proof of the theorem. $\qquad \square$

3.6 Relative versions

In this section, we state the relative versions of the Basepoint-free and Cone Theorems. In many cases these are much easier to apply than the global versions.

3.21. The following is the most primitive form of the relative setting.

(0) $f : X \to Y$ is a morphism of projective varieties over an algebraically closed field of characteristic zero. In this case, the relative Basepoint-free and Cone Theorems follow from the absolute theorems via the relative Kleiman's criterion (1.44). We give the arguments at the end of this section.

The more general relative versions are known in two different cases.

(1) $f : X \to Y$ is a proper morphism of varieties over an algebraically closed field of characteristic zero. The proofs in this case follow along the same lines as the global versions, with only minor changes. See [KMM87] for a detailed treatment.

(2) $f : X \to Y$ is a projective morphism of complex analytic spaces. In this case we have to assume that Y is a suitably small neighbourhood of a compact set, to avoid cases like when X is the blow up of Y at an infinite set of points. This case is treated in [Nak87].

Theorem (3.24) and similar results should also hold when $f : X \to Y$ is a proper morphism of complex analytic spaces which is bimeromorphic to a projective (or even to a Kähler) morphism, but we do not know any good reference. The relevant vanishing results are proved in [Tak94].

Definition 3.22. Let $f : X \to Y$ be a proper morphism satisfying one of the conditions (3.21.0–2), X irreducible. A Cartier divisor D on X is called f-*big* if rank $f_* \mathcal{O}_X(kD) > c \cdot k^n$ for some $c > 0$ and $k \gg 1$, where n is the dimension of the general fiber of f (compare with (2.59)).

If $f_* \mathcal{O}_X(D) \neq 0$ then the natual homomorphism $f^* f_* \mathcal{O}_X(D) \to \mathcal{O}_X(D)$ induces a rational map $\phi_{|D|/Y} : X \dashrightarrow \mathbb{P}_Y(f_* \mathcal{O}_X(D))$ over Y. D is called f-*free* if $f^*(f_* \mathcal{O}_X(D)) \twoheadrightarrow \mathcal{O}_X(D)$. In this case, $\phi_{|D|/Y}$ is a morphism.

The following properties of f-bigness can be proved similarly to (2.60).

Lemma 3.23. *Let* $f : X \to Y$ *be a projective morphism and* D *a Cartier divisor on* X. *Then the following are equivalent:*

(1) D *is* f-*big*,
(2) $mD \sim A + E$ *where* A *is* f-*ample and* E *effective for some* $m > 0$,
(3) *the rational map* $\phi_{|mD|/Y}$ *is birational for some* $m > 0$, *and*
(4) $\phi_{|mD|/Y}$ *is generically finite for some* $m > 0$.

If Y *has a big divisor* M, *this is also equivalent to*

(5) $D + mf^*M$ *is big for* $m \gg 0$. $\qquad\square$

Theorem 3.24 (Relative Basepoint-free Theorem). *Let* (X, Δ) *be a klt pair,* Δ *effective, and* $f : X \to Y$ *a proper morphism satisfying one of the conditions (3.21.0–2). Let* D *be an* f-*nef Cartier divisor such that* $aD - K_X - \Delta$ *is* f-*nef and* f-*big for some* $a > 0$. *Then* bD *is* f-*free for all* $b \gg 0$.

Theorem 3.25 (Relative Cone theorem). *Let* (X, Δ) *be a klt pair,* Δ *effective, and* $f : X \to Y$ *a projective morphism satisfying one of the conditions (3.21.0–2). Then:*

(1) *There are (countably many) rational curves* $C_j \subset X$ *such that* $f(C_j) = point$, $0 < -(K_X + \Delta) \cdot C_j \leq 2 \dim X$, *and*

$$\overline{NE}(X/Y) = \overline{NE}(X/Y)_{(K_X + \Delta) \geq 0} + \sum \mathbb{R}_{\geq 0}[C_j].$$

(2) *For any* $\epsilon > 0$ *and* f-*ample divisor* H,

$$\overline{NE}(X/Y) = \overline{NE}(X/Y)_{(K_X + \Delta + \epsilon H) \geq 0} + \sum_{\text{finite}} \mathbb{R}_{\geq 0}[C_j].$$

(3) *Let* $F \subset \overline{NE}(X/Y)$ *be a* $(K_X + \Delta)$-*negative extremal face. Then there is a unique morphism* $\text{cont}_F : X/Y \to Z/Y$ *such that* $(\text{cont}_F)_* \mathcal{O}_X = \mathcal{O}_Z$ *and an irreducible curve* $C \subset X$ *is mapped to a point by* cont_F *iff* $[C] \in F$. cont_F *is called the contraction of* F. Z *is projective over* Y.

(4) *Let* F *and* $\text{cont}_F : X \to Z$ *be as in (3). Let* L *be a line bundle on* X *such that* $(L \cdot C) = 0$ *for every curve* C *with* $[C] \in F$. *Then there is a line bundle* L_Z *on* Z *such that* $L \cong \text{cont}_F^* L_Z$.

In the rest of this section we prove the relative versions when $f : X \to Y$ is a morphism of projective varieties.

3.26 (Proof of (3.24) in Case (3.21.0)). We are going to replace D with $D + f^*(\text{ample})$ and change Δ to make D nef and $aD - K_X - \Delta$ ample keeping (X, Δ) klt. (Note that $(f\text{-nef}) + \nu f^*(\text{ample})$ need not be nef for any $\nu > 0$ in general (1.46).) Once this is done, we can apply the absolute Basepoint-free Theorem (3.3) to D and (3.24) follows.

Since $aD - K_X - \Delta$ is f-big, $a(D + f^*H) - K_X - \Delta$ is big for some ample divisor H on Y (3.23). There exists an effective $\mathbb{Q}$-divisor E on X such that $a(D + f^*H) - K_X - \Delta - \epsilon E$ is ample if $0 < \epsilon \ll 1$ (2.61). Since $(X, \Delta + \epsilon E)$ is klt if $0 < \epsilon \ll 1$ (2.35.2), we may rename $\Delta + \epsilon E$ as Δ, and $D + f^*H$ as D. By doing this, $F := aD - K_X - \Delta$ becomes ample while all other conditions remain satisfied.

So there are only a finite number of extremal rays $\mathbb{R}_{\geq 0}[\ell_j]$ of $\overline{NE}(X)$ in $D_{<0}$ by $aD \equiv K_X + \Delta + F$ and (3.7.2). Note that ℓ_j is not in a fiber of f because $(\ell_j \cdot D) < 0$ and D is f-nef. Thus for $m \gg 0$, $(\ell_j \cdot D + mf^*H) > 0$ for the finitely many ℓ_j. Then $D + mf^*H$ is nef (3.7.2) and $a(D + mf^*H) - K_X - \Delta$ is ample, as required. $\square$

Proposition 3.27. *Let* $f : X \to Y$ *be a morphism of projective varieties. Then the following hold.*

(1) f *induces an* $\mathbb{R}$-*linear map* $f_* : N_1(X) \to N_1(Y)$.
(2) $f_* NE(X) \subset NE(Y)$ *and* $f_* \overline{NE}(X) \subset \overline{NE}(Y)$.
(3) $N_1(X/Y) \subseteq \ker f_*$, *and the equality does not hold in general.*
(4) $\overline{NE}(X/Y) = \overline{NE}(X) \cap N_1(X/Y)$.

Proof. (1) and the first parts of (2) and (3) are obvious. The second part of (2) follows from the first by the continuity of f_*. The second part of (3) is shown by the example (1.46.1), where $\rho(X/Y) = \rho(Y) = 1$ and $\rho(X) \geq 3$. Only (4) remains to be proved.

First, '$\subseteq$' is obvious. Assuming $\overline{NE}(X/Y) \subsetneq \overline{NE}(X) \cap N_1(X/Y)$, we derive a contradiction. There exists a linear function H on $N_1(X/Y)$ such that $H_{>0} \supset \overline{NE}(X/Y) \setminus \{0\}$ and $(H \cdot \xi) < 0$ for some $\xi \in \overline{NE}(X) \cap N_1(X/Y)$. Perturbing the coefficients of H a little, we can assume that H is a $\mathbb{Q}$-Cartier divisor on X. The relative Kleiman criterion (1.44) implies that H is f-ample. $H + f^* M$ is ample on X for some ample divisor M on Y by (1.45). Then $0 < (H + f^* M \cdot \xi) = (H \cdot \xi) < 0$, a contradiction. Thus (4) is proved. □

3.28 (Proof of (3.25) in Case (3.21.0)). First we prove (3.25.1). Let $\xi \in \overline{NE}(X/Y)$. As in (3.7.1), we can write $\xi = \eta + \sum r_j [C_j]$ where $\eta \in \overline{NE}(X)_{(K_X + \Delta)) \geq 0}$ and $r_j > 0$. Then $0 = f_* \xi = f_* \eta + \sum r_j f_* [C_j]$ with $f_* \eta, f_* [C_j] \in \overline{NE}(Y)$ (3.27.2). Since Y is projective, $\overline{NE}(Y)$ does not contain a straight line (1.18). Hence $f(C_j)$ is a point, that is, $[C_j] \in \overline{NE}(X/Y)$ for every j and $\eta = \xi - \sum r_j [C_j] \in N_1(X/Y)$. Thus $\eta \in \overline{NE}(X) \cap N_1(X/Y) = \overline{NE}(X/Y)$. The rest of (3.25) is easy. □

3.7 Running the MMP

In (2.14) we gave an outline of the minimal model program for smooth projective varieties. In this section we generalize this to pairs (X, Δ) and show how the theorems proved in this chapter apply to the minimal model program.

The MMP may work in many cases beyond projective varieties over $\mathbb{C}$, and we state the outlines of the program in a setting as general as possible.

3.29 (Choice of a class of pairs). Let X be either a normal scheme over a field or a normal complex analytic space. In choosing the class of pairs (X, Δ) that we work with, we have three objectives in mind.

(1) It should include the case when X is smooth and Δ an effective snc divisor with coefficients 1 or less.
(2) The steps of the MMP should not lead us out of the class.
(3) The steps of the MMP should exist.

The largest class where these may hold is probably the class of all lc pairs. (2) and (3) present technical difficulties, even in cases when they are known to be true. Thus we opt for a smaller class: all pairs (X, Δ) where X is a $\mathbb{Q}$-factorial projective variety and (X, Δ) is dlt.

This is the smallest class which satisfies conditions (1) and (2). (This follows relatively easily from (2.44).) (3) is known in dimensions 2 and 3.

The $\mathbb{Q}$-factoriality assumption is a very natural one if we start with smooth varieties, and it makes many proofs easier. On the other hand, it is a rather unstable condition in general. It is not local in the Euclidean (or étale) topology, and it is very hard to keep track of when we pass from a variety to a divisor in an inductive proof.

3.30 (The birational geometry of pairs). Once we established the class of pairs we work with, we face the natural question:

When are two pairs (X_i, Δ_i) birational?

Clearly, X_1 and X_2 should be birational. Let $\phi : X_1 \dashrightarrow X_2$ be a birational map. We still have to establish a relationship between Δ_1 and Δ_2. The simplest choice would be to require $\Delta_2 = \phi_* \Delta_1$. This seems reasonable if ϕ is a morphism, or, more generally, when ϕ^{-1} has no exceptional divisors. However, if $E \subset X_2$ is an exceptional divisor of ϕ^{-1} then there appears no particular reason to assign the coefficient 0 to E. In fact, from many points of view it is more natural to assign the coefficient 1 to all exceptional divisors of ϕ^{-1}.

We do not know of any definition which establishes a good notion of birationally equivalent pairs. Even if $\phi : X_1 \to X_2$ is a morphism and $\Delta_2 = \phi_* \Delta_1$, the two pairs should not be considered 'birational'. This can be seen from the following example:

Let $X_2 = \mathbb{P}^2$ and $\Delta_2 = (x_0 x_1 x_2 = 0)$. Let $\phi : X_1 \to X_2$ be the blow up of $(1 : 0 : 0)$ and set $\Delta_1 = \phi_*^{-1} \Delta_2$. Then $\Delta_2 = \phi_* \Delta_1$.

Computing global sections of the multiples of $K + \Delta$ we see that $h^0(X_2, \mathcal{O}_{X_2}(m(K_{X_2} + \Delta_2))) = 1$, but $h^0(X_1, \mathcal{O}_{X_1}(m(K_{X_1} + \Delta_1))) = 0$ for every $m > 0$.

Thus at the moment we do not have a good working definition of a minimal model of a pair. It is better to establish first the framework of the minimal model program for pairs, and then define the concept of minimal models to fit our needs in the next section.

In most applications we need to run the MMP over a fixed variety. This consists of the following steps:

3.31 (Relative MMP for Q-factorial dlt pairs).
We start with a pair $(X, \Delta) = (X_0, \Delta_0)$ where X is either a normal scheme over a field or a normal complex analytic space. Let $f_0 : X \to S$ be a projective morphism.

The aim is to set up a recursive procedure which creates intermediate pairs (X_i, Δ_i) and projective morphisms $f_i : X_i \to S$. After some steps it should stop with a final pair (X^*, Δ^*) and $f^* : X^* \to S$.

Step 0 (Initial datum). Assume that we already constructed (X_i, Δ_i) and $f_i : X_i \to S$ with the following properties:

(1) X_i is Q-factorial,
(2) (X_i, Δ_i) is dlt,
(3) f_i is projective.

Step 1 (Preparation). If $K_{X_i} + \Delta_i$ is f_i-nef, then we go directly to Step 3.2. If $K_{X_i} + \Delta_i$ is not f_i-nef then we establish two results:

(1) (Cone Theorem) $\overline{NE}(X_i/S) = \overline{NE}(X_i/S)_{K_{X_i} + \Delta_i \geq 0} + \sum \mathbb{R}^+[C_i]$.
(2) (Contraction Theorem) Any $(K_{X_i} + \Delta_i)$-negative extremal ray $R_i \subset \overline{NE}(X_i/S)$ can be contracted. Let $\mathrm{cont}_{R_i} : X_i \to Y_i$ denote the corresponding contraction (2.5, 3.25). It sits in a commutative diagram

$$
\begin{array}{ccc}
X_i & \xrightarrow{\mathrm{cont}_{R_i}} & Y_i \\
f_i \searrow & & \swarrow g_i \\
& S &
\end{array}
$$

Step 2 (Birational transformations). If $\mathrm{cont}_{R_i} : X_i \to Y_i$ is birational, then we produce a new pair (X_{i+1}, Δ_{i+1}) as follows.

(1) (Divisorial contraction) If cont_{R_i} is a divisorial contraction as in (2.5.2), then set $X_{i+1} := Y_i$, $f_{i+1} := g_i$ and $\Delta_{i+1} := (\mathrm{cont}_{R_i})_*(\Delta_i)$.
(2) (Flipping contraction) If cont_{R_i} is a flipping contraction as in (3.33), then set $(X_{i+1}, \Delta_{i+1}) := (X_i^+, \Delta_i^+)$ (the flip of cont_{R_i}) and $f_{i+1} := g_i \circ f_i^+$.

In both cases we prove that X_{i+1} is Q-factorial, f_{i+1} is projective and (X_{i+1}, Δ_{i+1}) is dlt. Then we go back to Step 0 with (X_{i+1}, Δ_{i+1}) and start anew.

Step 3 (Final outcome). We expect that eventually the procedure stops, and we get one of the following two possibilities:

(1) (Fano contraction) If cont_{R_i} is a Fano contraction as in (2.5.1) then set $(X^*, \Delta^*) := (X_i, \Delta_i)$ and $f^* := f_i$.
(2) (Minimal model) If $(K_{X_i} + \Delta_i)$ is f_i-nef then again set $(X^*, \Delta^*) := (X_i, \Delta_i)$ and $f^* := f_i$.

We hope that the special properties of (X^*, Δ^*) can be used in further attempts to understand (X^*, Δ^*) and hence also (X, Δ).

3.32. The rest of the section is devoted to explaining which parts of the program follow from the already established results, and which parts remain to be done. We assume that $f_0 : X \to S$ satisfies one of the conditions (3.21.1–2), and consider the following points in detail:

(1) We used a concept of flip which is more general than (2.8). The general definition is given in (3.33).
(2) The Cone Theorem is generalized to our setting in (3.35).
(3) In (3.36) and (3.37) we prove that the $\mathbb{Q}$-factoriality condition is preserved in the course of the MMP.
(4) In (3.44) we prove that the dlt condition is preserved in the course of the MMP. More generally, we also consider the problem for terminal, canonical, klt and lc pairs (3.42) and (3.43).
(5) Studying the outcome of the minimal model program helps us to establish a good definition of relative minimal models of pairs. This turns out to be quite delicate and is done in the next section.
(6) In (3.45) we consider the question of whether the MMP stops. This is not known in general. Some special cases are treated in later chapters.
(7) The biggest open question is the existence of flips. The rest of the book is essentially devoted to a few cases of this problem.

Definition 3.33. Let X be a normal scheme (or complex analytic space) and D a $\mathbb{Q}$-divisor on X such that $K_X + D$ is $\mathbb{Q}$-Cartier. We do not assume for the moment that D is effective.

A $(K + D)$-*flipping contraction* is a proper birational morphism $f : X \to Y$ to a normal scheme (or complex analytic space) Y such that $\mathrm{Ex}(f)$ has codimension at least two in X and $-(K_X + D)$ is f-ample.

A normal scheme (or complex analytic space) X^+ together with a proper birational morphism $f^+ : X^+ \to Y$ is called a $(K + D)$-*flip* of f if

(1) $K_{X^+} + D^+$ is $\mathbb{Q}$-Cartier, where D^+ is the birational transform of D on X^+,

(2) $K_{X^+} + D^+$ is f^+-ample, and

(3) $\mathrm{Ex}(f^+)$ has codimension at least two in X^+.

By a slight abuse of terminology, the induced rational map $\phi : X \dashrightarrow X^+$ is also called a $(K + D)$-flip. We see in (6.4) that a $(K + D)$-flip is unique and the main open question is its existence. A $(K+D)$-flip gives a commutative diagram:

$$X \xrightarrow{\ \phi\ } X^+$$
$$-(K_X + D) \text{ is } f\text{-ample} \searrow \quad \swarrow (K_{X^+} + D^+) \text{ is } f^+\text{-ample}$$
$$Y$$

The terminology in the literature is not uniform. The above operation is sometimes called a D-flip.

Especially when $\dim X = 3$, a curve $C \subset \mathrm{Ex}(f)$ is called a *flipping curve* and a curve $C^+ \subset \mathrm{Ex}(f^+)$ is called a *flipped curve*.

In the course of a MMP, the flips we encounter have other useful properties. The following condition is frequently very convenient.

Definition 3.34. A proper birational morphism $f : X \to Y$ is called *extremal* if

(1) X is $\mathbb{Q}$-factorial, and

(2) if B, B' are Cartier divisors then there are $a, a' \in \mathbb{Z}$ (not both zero) such that $aB \sim_f a'B'$.

Assume that (X, Δ) is klt and X is $\mathbb{Q}$-factorial. Let $f : X \to Y$ be the contraction of a $(K_X + \Delta)$-negative extremal ray R. Pick $[C] \in R$. We can choose $a, a' \in \mathbb{Z}$ (not both zero) such that $a(B \cdot C) = a'(B' \cdot C))$. Then $aB \sim_f a'B'$ by (3.25.4), thus f is extremal.

Theorem 3.35. *The Relative Cone Theorem (3.25) holds for (X, Δ) and $f : X \to S$ where X is $\mathbb{Q}$-factorial, f is projective and (X, Δ) is dlt.*

Proof. First we prove that (3.25.2) holds even if (X, Δ) is dlt. Choose $0 < \delta \ll 1$ such that $H' := \delta\Delta + \epsilon H$ is f-ample. $(X, (1 - \delta)\Delta)$ is klt by (2.43), thus by (3.25.2)

$$\overline{NE}(X/S) = \overline{NE}(X/S)_{(K_X + (1-\delta)\Delta + H') \geq 0} + \sum_{\text{finite}} \mathbb{R}_{\geq 0}[C_j],$$

which is equivalent to

$$\overline{NE}(X/S) = \overline{NE}(X/S)_{(K_X + \Delta + \epsilon H) \geq 0} + \sum_{\text{finite}} \mathbb{R}_{\geq 0}[C_j].$$

As we remarked in (3.15. Step 5), this implies (3.25.1) in the dlt case.

This argument also shows that any $(K_X + \Delta)$-negative relative extremal ray is also a $(K_X + (1 - \delta)\Delta)$-negative relative extremal ray for $0 < \delta \ll 1$, and $(X, (1 - \delta)\Delta)$ is klt. Thus the klt case of (3.25.3–4) implies the same assertions in the $\mathbb{Q}$-factorial dlt case. $\quad\square$

By the same reasoning, using (3.17) and (3.18), we obtain that $\mathbb{Q}$-factoriality is preserved in the dlt cases:

Proposition 3.36. *Let (X, Δ) be a projective, $\mathbb{Q}$-factorial dlt pair and $g_R : X \to Y$ the contraction of a $(K_X + \Delta)$-negative extremal ray. Assume that g_R is either a divisorial or a Fano contraction. Then*

(1) *Y is $\mathbb{Q}$-factorial, and*
(2) *$\rho(Y) = \rho(X) - 1$.* $\quad\square$

$\mathbb{Q}$-factoriality is also preserved under flips:

Proposition 3.37. *Let (X, Δ) be a projective, $\mathbb{Q}$-factorial dlt pair and $g_R : X \to Y$ the flipping contraction of a $(K_X + \Delta)$-negative extremal ray with flip $g_R^+ : X^+ \to Y$. Then*

(1) *X^+ is $\mathbb{Q}$-factorial, and*
(2) *$\rho(X^+) = \rho(X)$.*

Proof. Since $\phi : X \dashrightarrow X^+$ is an isomorphism in codimension 1, it induces a natural isomorphism between the group of Weil divisors on X and the group of Weil divisors on X^+. Let D^+ be a Weil divisor on X^+. There is a rational number r such that $(R \cdot (D + r(K_X + \Delta))) = 0$. Choose $m \in \mathbb{Z}_{>0}$ such that $m(D + r(K_X + \Delta))$ is Cartier. By (3.17), there is a Cartier divisor D_Y such that $m(D + r(K_X + \Delta)) \sim g_R^* D_Y$. Thus

$$mD^+ = m\phi_* D \sim (g_R^+)^* D_Y - (mr)(K_{X^+} + \Delta^+)$$

is $\mathbb{Q}$-Cartier. $\quad\square$

The following general result is used to compare discrepancies in several different settings.

Lemma 3.38. *Consider a commutative diagram*

$$
\begin{array}{ccc}
X & \overset{\phi}{\dashrightarrow} & X' \\
f \searrow & & \swarrow f' \\
& Y &
\end{array}
$$

where X, X', Y are normal varieties and f, f' are proper and birational. Let Δ (resp. Δ') be a $\mathbb{Q}$-divisor on X (resp. X'). Assume that:

(1) $f_*\Delta = f'_*\Delta'$,
(2) $-(K_X + \Delta)$ *is $\mathbb{Q}$-Cartier and f-nef, and*
(3) $K_{X'} + \Delta'$ *is $\mathbb{Q}$-Cartier and f'-nef.*

Then for an arbitrary exceptional divisor E over Y, we have

$$
a(E, X, \Delta) \le a(E, X', \Delta').
$$

Strict inequality holds if either

(4) $-(K_X + \Delta)$ *is f-ample and f is not an isomorphism above the generic point of $\mathrm{center}_Y\, E$, or*
(5) $(K_{X'} + \Delta')$ *is f'-ample and f' is not an isomorphism above the generic point of $\mathrm{center}_Y\, E$.*

Proof. Let Z be a normal variety with birational morphisms $g : Z \to X$ and $g' : Z \to X'$ such that $\mathrm{center}_Z\, E$ is a divisor. Set $h := f \circ g = f' \circ g'$. Let $m > 0$ be a sufficiently divisible integer. We have linear equivalences

$$
-m\left(K_Z - \sum a(E_i, X, \Delta) E_i\right) \;\sim\; -m g^*(K_X + \Delta),
$$

$$
m\left(K_Z - \sum a(E_i, X', \Delta') E_i\right) \;\sim\; m(g')^*(K_{X'} + \Delta').
$$

Adding the two we obtain that

$$
H := \sum (a(E_i, X, \Delta) - a(E_i, X', \Delta')) E_i
$$

is h-nef and a sum of exceptional divisors by assumption (1). By (3.39) we obtain that all coefficients are non-positive. Moreover, if H is not numerically h-trivial over the generic point of $\mathrm{center}_Y\, E$ then the coefficient of E in H is negative. $\qquad\square$

The following is very useful in many situations:

Lemma 3.39. *Let $h : Z \to Y$ be a proper birational morphism between normal varieties. Let $-B$ be an h-nef $\mathbb{Q}$-Cartier $\mathbb{Q}$-divisor on Z. Then*

(1) B *is effective iff h_*B is.*

(2) *Assume that B is effective. Then for every $y \in Y$, either $h^{-1}(y) \subset \operatorname{Supp} B$ or $h^{-1}(y) \cap \operatorname{Supp} B = \emptyset$.*

Proof. If B is effective then so is $h_* B$. The main question is the converse.

By Chow's Lemma, there is a proper birational morphism $p : Z' \to Z$ such that $Z' \to Y$ is projective. Then B is effective iff $p^* B$ is. Thus we may assume to start with that h is projective. We may also assume that Y is affine. Write $B = \sum B^k$ where B^k is the sum of those irreducible components B_i of B such that $h(B_i)$ has codimension k in Y.

First we deal with the case when $\dim Y = 2$. $B = B^2 + B^1$ and B^1 is also h-nef. Hence $-B^2$ is an h-nef linear combination of exceptional curves. As before, the question of the effectivness of B^2 can be reduced to any resolution $U \to Z$. B^2 is effective by the following two results. The first one is a special case of the Hodge Index Theorem.

Lemma 3.40. *Let $f : Y \to X$ be a resolution of a normal surface with exceptional curves E_i. Then the intersection matrix $(E_i \cdot E_j)$ is negative definite.*

Proof. First we prove the case when X is projective. Let $D = \sum e_i E_i$ be a non-zero linear combination of exceptional curves, and assume that $(D^2) \geq 0$. Consider first the case when D is effective. Let H be an ample divisor on Y such that $H - K_Y$ is ample. $H^2(Y, \mathcal{O}_Y(nD + H)) = 0$ by Serre duality and

$$(nD + H \cdot nD + H - K_Y) \geq (nD + H \cdot nD) \geq n(D \cdot H).$$

As in (1.20), we conclude from the Riemann–Roch formula for surfaces that $h^0(Y, \mathcal{O}_Y(nD + H))$ goes to infinity with n. On the other hand,

$$H^0(Y, \mathcal{O}_Y(nD + H)) \subset H^0(X, \mathcal{O}_X(f_*(nD + H))) = H^0(X, \mathcal{O}_X(f_* H)),$$

a contradiction. If D is not effective, write $D = D_+ - D_-$ as a difference of two effective divisors without common irreducible components. Then $(D^2) \leq (D_+^2) + (D_-^2) < 0$.

The case when X is quasi-projective can be reduced to the one discussed by compactifying X.

The result also holds when X is an analytic surface, see [Gra62, p.367] for a proof. □

Lemma 3.41. *Let U be a smooth surface and $C = \cup C_i$ a set of proper curves on U. Assume that the intersection matrix $(C_i \cdot C_j)$ is negative*

*definite. Let $A = \sum a_i C_i$ be an $\mathbb{R}$-linear combination of the curves C_i.
Assume that $(A \cdot C_j) \geq 0$ for every j. Then*

    (1) *$a_i \leq 0$ for every i.*
    (2) *If C is connected, then either $a_i = 0$ for every i or $a_i < 0$ for
    every i.*

Proof. Write $A = A^+ - A^-$ where A^+, A^- are non-negative linear
combinations of the curves C_i and no curve appears in both with positive
coefficient.

Assume that $A^+ \neq 0$. The matrix $((C_i \cdot C_j))$ is negative definite, thus
$(A^+ \cdot A^+) < 0$. Hence there is a $C_i \subset \operatorname{Supp} A^+$ such that $(C_i \cdot A^+) < 0$.
C_i is not in $\operatorname{Supp} A^-$, so $(C_i \cdot A^-) \geq 0$. Thus $(C_i \cdot A) < 0$, a contradiction.
This shows (1).

Assume next that C is connected, and $\emptyset \neq \operatorname{Supp} A^- \neq \operatorname{Supp} C$. Then
there is a curve C_i such that $C_i \not\subset \operatorname{Supp} A^-$ but C_i intersects $\operatorname{Supp} A^-$.
Then $(C_i \cdot A) = -(C_i \cdot A^-) < 0$, a contradiction. $\qquad\qquad\square$

Going back to the proof of (3.39), let $S \subset Y$ be the complete intersection of $n - 2$ general hypersurfaces and $T := h^{-1}(S)$. Then $h : T \to S$
is a birational map between normal surfaces. $B|_T = B^2|_T + B^1|_T$, thus
B^2 is effective.

Let $H \subset X$ be a general very ample divisor. Set $A = B|_H$. Then $-A$
is h-nef, $A^i = B^{i+1}|_H$ if $i \geq 2$ and $A^1 = B^1|_H + B^2|_H$. B^1 is effective by
assumption and we have proved that B^2 is effective. Thus A^1 is effective.
By induction on the dimension, A is effective, hence so is B.

For $y \in Y$, $h^{-1}(y)$ is connected. Thus if $h^{-1}(y)$ intersects $\operatorname{Supp} B$ but
is not contained in it, then there is an irreducible curve $C \subset h^{-1}(y)$ such
that $(C \cdot B) > 0$. This is impossible since $-B$ is h-nef. $\qquad\qquad\square$

The first consequences of (3.38) show that the minimal model program
does not create worse singularities:

Corollary 3.42. *Let (X, Δ) be a pair and $\phi : X \dashrightarrow X^+$ a $(K_X + \Delta)$-
flip. If (X, Δ) is terminal (resp. canonical, klt or lc) then (X^+, Δ^+) is
also terminal (resp. canonical, klt or lc).*

Corollary 3.43. *Let (X, Δ) be a pair and $f : X \to Y$ a divisorial contraction of a $(K_X + \Delta)$-negative extremal ray with exceptional divisor E.*

    (1) *If (X, Δ) is klt (resp. lc) then so is $(Y, f_* \Delta)$.*
    (2) *If (X, Δ) is terminal (resp. canonical) and $E \not\subset \operatorname{Supp} \Delta$, then
    $(Y, f_* \Delta)$ is also terminal (resp. canonical).*

(3) *If X is terminal (resp. canonical) and $\Delta = 0$ then Y is also terminal (resp. canonical).*

Proof. We obtain (3.42) by applying (3.38) with $X' = X^+$. For (3.43), we use $X' = Y$.

The notions 'terminal' and 'canonical' are defined using the discrepancies of exceptional divisors only. E is exceptional over Y but not over X, so it has to be taken care of by hand. This accounts for the formulation of (2). □

Corollary 3.44. *Let (X, Δ) be a dlt pair. Let $g : X \dashrightarrow X'$ be either a divisorial contraction of a $(K_X+\Delta)$-negative extremal ray or a $(K_X+\Delta)$-flip. Let $\Delta' := g_*\Delta$.*
Then (X', Δ') is also dlt.

Proof. Let $Z \subset X$ be as in (2.37) and set $Z' := g(Z) \cup \mathrm{Ex}(g^{-1})$. $X' \setminus Z'$ is isomorphic to an open subset of $X \setminus Z$, thus $X' \setminus Z'$ is smooth and $\Delta'|_{X' \setminus Z'}$ is a snc divisor.

Let E be an exceptional divisor over X' such that $\mathrm{center}_{X'} E \subset Z'$. Then $\mathrm{center}_X E \subset Z \cup \mathrm{Ex}(g)$. Thus

$$a(E, X', \Delta') \geq a(E, X, \Delta) \geq -1.$$

If $\mathrm{center}_X E \subset Z$ then the second inequality is strict by the definition of dlt. If $\mathrm{center}_X E \subset \mathrm{Ex}(g)$ then the first inequality is strict by (3.38). □

3.45 (Termination of the MMP). Let (X_0, Δ_0) be a $\mathbb{Q}$-factorial dlt pair, and assume that we run the MMP to get pairs (X_i, Δ_i). If X_{i+1} is obtained by a divisorial contraction then $\rho(X_{i+1}) = \rho(X_i) - 1$. If it is obtained by a flip then $\rho(X_{i+1}) = \rho(X_i)$. Thus we have inequalities

$$\rho(X_0) \geq \rho(X_1) \geq \cdots \geq \rho(X_i) \geq \cdots$$

and an inequality is strict if we have a divisorial contraction. Thus the number of divisorial contractions is bounded by $\rho(X_0)-1$. This argument does not say anything about the number of flips.

(3.38) shows that at least some of the discrepancies increase under a flip, but it does not seem easy to get a bound on the number of flips in general from this observation. In many cases this is, however, possible. Some of these are treated in Chapter 6. See also [K+92].

3.46. The main remaining problem is the existence of $(K + \Delta)$-flips. This has been proved in the following cases:

(1) $\dim X = n$ and X admits a faithful $(\mathbb{C}^*)^n$-action [Rei83a] or certain other algebraic group actions [BK94];

(2) $\dim X = 3$, $(X, 0)$ is terminal [Mor88];

(3) $\dim X = 3$, (X, Δ) is klt [Sho92]. A new proof and a generalization to (X, Δ) lc is in [K$^+$92].

Chapter 7 is devoted to establishing the existence of the so-called semistable flips in dimension 3. From the classification of 3-dimensional flips [KM92] we know that almost all 3-dimensional flips are semi-stable, but this is rather hard to prove. The importance of semi-stable flips stems from their application to families of surfaces. This is explained in section 7.5.

In dimension 2 there are no flips, thus in this case the MMP is complete:

Theorem 3.47. *Let (X, Δ) be a 2-dimensional, $\mathbb{Q}$-factorial, projective and dlt pair. There is a sequence of at most $\rho(X) - 1$ contractions*

$$(X, \Delta) = (X_0, \Delta_0) \to (X_1, \Delta_1) \to \cdots \to (X_i, \Delta_i) = (X^*, \Delta^*)$$

such that one of the following holds:

(1) *(Log minimal model) $K_{X^*} + \Delta^*$ is nef.*

(2) *(Log ruled surface) There is a morphism onto a curve $g : X^* \to C$ such that $-(K_{X^*} + \Delta^*)$ is g-ample.*

(3) *(Log Del Pezzo) $-(K_{X^*} + \Delta^*)$ is ample.* □

3.8 Minimal and Canonical Models

The results of this section hold when X is a normal scheme over a field or a normal complex analytic space and $f : X \to S$ a proper morphism (which, in the analytic case, is bimeromorphic to a projective morphism).

The definition of a minimal model of a pair is designed to codify the outcome (3.31.3.2) of the MMP. One aspect of (2.13) is easy to generalize:

Definition 3.48. Let (X, Δ) be a pair and $f : X \to S$ a proper morphism. We say that (X, Δ) is f-*minimal* or *relatively minimal* if (X, Δ) is dlt and $K_X + \Delta$ is f-nef.

The definition (2.13) has another part, namely it tells when X is a minimal model of X'. As we saw in (3.30), the notion of birational equivalence of pairs is problematic, thus we still need to define when one pair is a minimal model of another. The following example shows that one has to proceed with some caution.

Example 3.49. Let S be a projective surface with a single singular point $P \in S$ and $f : S' \to S$ the minimal resolution. Assume that K_S is ample, $\text{Ex}(f)$ consists of a single smooth rational curve D and $(D^2) = -n$. Then $(S, 0)$ is klt and $K_{S'} \equiv f^* K_S - \frac{n-2}{n} D$. Run the MMP for (S', cD). If $c > \frac{n-2}{n}$ then in the first step we contract D and then we stop. If $c \leq \frac{n-2}{n}$ then $(D \cdot (K_{S'} + cD)) \geq 0$, thus we do not contract D and frequently S' is its own minimal model.

Thus $(S, 0)$ is minimal, but it is not always a minimal model of (S', cD).

Keeping this example in mind, we see that we have to compare the discrepancies of the minimal model with the discrepancies of the original pair.

Definition 3.50. Let (X, Δ) be a log canonical pair and $f : X \to S$ a proper morphism. A pair (X^w, Δ^w) sitting in a diagram

$$
\begin{array}{ccc}
X & \overset{\phi}{\dashrightarrow} & X^w \\
 f \searrow & & \swarrow f^w \\
 & S &
\end{array}
$$

is called a *weak canonical model* of (X, Δ) over S if

  (1) f^w is proper,
  (2) ϕ^{-1} has no exceptional divisors,
  (3) $\Delta^w = \phi_* \Delta$,
  (4) $K_{X^w} + \Delta^w$ is f^w-nef, and
  (5) $a(E, X, \Delta) \leq a(E, X^w, \Delta^w)$ for every ϕ-exceptional divisor $E \subset X$.

If (X, Δ) is dlt then a weak canonical model $(X^m, \Delta^m) = (X^w, \Delta^w)$ is called a *minimal model* of (X, Δ) over S if in addition to (1–5) we have

  (5^m) $a(E, X, \Delta) < a(E, X^m, \Delta^m)$ for every ϕ-exceptional divisor $E \subset X$.

A weak canonical model $(X^c, \Delta^c) = (X^w, \Delta^w)$ is called a *canonical model* of (X, Δ) over S if in addition to (1–5) we have

  (4^c) $K_{X^c} + \Delta^c$ is f^c-ample.

Assume that we start with (X, Δ), $f : X \to S$ and we run the MMP over S. If the program works, we obtain a pair (X^*, Δ^*) and a projective morphism $f^* : X^* \to S$. Let $\phi : X \dashrightarrow X^*$ be the resulting birational map. Then (1–3) hold. (4) holds if we ended up with the minimal model case (3.31.3.2). Finally, ϕ is the composite of divisorial contractions and flips, hence using (3.38) step–by–step we get (5).

In (2.13) we defined the notion of minimal model in case X is smooth and $\Delta = 0$. The above definition is a generalization of (2.13). Indeed, there is no problem with (1–4). If E is a ϕ-exceptional divisor, then $a(E, X, \Delta) = 0$ since $\Delta = 0$ and $a(E, X^m, \Delta^m) > 0$ since X^m is terminal. Thus (5^m) also holds.

In (3.50) we have not assumed anything about the singularities of the pair (X^w, Δ^w). The next result shows that they are at least as good as the singularities of the pair (X, Δ):

Proposition 3.51. *Let (X^w, Δ^w) be a weak canonical model of (X, Δ). Then $a(E, X, \Delta) \leq a(E, X^w, \Delta^w)$ for every divisor E over X.*

Proof. Consider any diagram

$$
\begin{array}{ccc}
& Y & \\
{\scriptstyle g}\swarrow & & \searrow{\scriptstyle h} \\
X & \overset{\phi}{\dashrightarrow} & X^w \\
{\scriptstyle f}\searrow & & \swarrow{\scriptstyle f^w} \\
& S &
\end{array}
$$

where (X^w, Δ^w) is a weak canonical model and center$_Y$ E is a divisor. Write

$$
\begin{aligned}
K_Y &\equiv g^*(K_X + \Delta) + E_1, \quad \text{and} \\
K_Y &\equiv h^*(K_{X^w} + \Delta^w) + E_2.
\end{aligned}
$$

Notice that $a(E, X^w, \Delta^w) - a(E, X, \Delta)$ is the coefficient of E in $E_2 - E_1$. Set

$$
B := g^*(K_X + \Delta) - h^*(K_{X^w} + \Delta^w) \equiv E_2 - E_1.
$$

Then $-B$ is g-nef and $g_* B = g_*(E_2 - E_1)$ is effective by (3.50.5). Thus $E_2 - E_1$ is effective by (3.39), and we are done. □

The existence of minimal and canonical models is essentially equivalent to the MMP. Next we consider how unique these models are:

Theorem 3.52. *Let (X, Δ) be a lc pair and $f : X \to S$ a proper morphism.*

(1) *A canonical model* (X^c, Δ^c) *is unique and*

$$X^c = \operatorname{Proj}_S \oplus_{m \geq 0} f_* \mathcal{O}_X(mK_X + \lfloor m\Delta \rfloor).$$

(2) *Any two minimal models of* (X, Δ) *are isomorphic in codimension one.*

Proof. Consider the diagram used in the proof of (3.51) and set $\Delta_Y := g_*^{-1}\Delta + \operatorname{red}(\operatorname{Ex}(g))$; that is, all exceptional divisors of g are added with coefficient one. Since (X, Δ) and (X^w, Δ^w) are both lc, we obtain that

$$
\begin{aligned}
K_Y + \Delta_Y &\equiv g^*(K_X + \Delta) + \text{(effective g-exceptional divisor)}, \quad \text{and} \\
K_Y + \Delta_Y &\equiv h^*(K_{X^w} + \Delta^w) + \text{(effective h-exceptional divisor)}.
\end{aligned}
$$

Consider first the case when $(X^w, \Delta^w) = (X^c, \Delta^c)$. Then

$$
\begin{aligned}
(X^w, \Delta^w) &= \operatorname{Proj}_S \oplus_{m \geq 0} f_*^w \mathcal{O}_{X^w}(mK_{X^w} + \lfloor m\Delta^w \rfloor) \\
&= \operatorname{Proj}_S \oplus_{m \geq 0} (f \circ g)_* \mathcal{O}_Y(mK_Y + \lfloor m\Delta_Y \rfloor) \\
&= \operatorname{Proj}_S \oplus_{m \geq 0} f_* \mathcal{O}_X(mK_X + \lfloor m\Delta \rfloor).
\end{aligned}
$$

This implies that the canonical model is unique.

Assume next that $\phi_i : X \dashrightarrow X_i^m$ are two minimal models. We need to show that ϕ_1 and ϕ_2 have the same exceptional divisors. We can choose $g : Y \to X$ such that $h_i : Y \to X_i^m$ are both morphisms. We obtain that

$$K_Y + \Delta_Y \equiv h_i^*(K_{X_i^m} + \Delta_i^m) + Z_i,$$

where Z_i is effective and $\operatorname{Supp} Z_i$ contains all exceptional divisors of ϕ_i by (3.50.5). Combining the two formulas we get

$$h_1^*(K_{X_1^m} + \Delta_1^m) - h_2^*(K_{X_2^m} + \Delta_2^m) \equiv Z_2 - Z_1.$$

Applying (3.39) to $h_1 : Y \to X_1^m$ (resp. $h_2 : Y \to X_2^m$) we obtain that $Z_2 - Z_1$ (resp. $Z_1 - Z_2$) is effective. Thus $Z_1 = Z_2$ and so ϕ_1 and ϕ_2 have the same exceptional divisors. □

Corollary 3.53. *Let* (X, Δ) *and* (X', Δ') *be lc pairs,* $p : X' \to X$ *and* $f : X \to S$ *proper morphisms. Assume that* $K_{X'} + \Delta' - p^*(K_X + \Delta)$ *is effective and p-exceptional.*

Then (X, Δ) *and* (X', Δ') *have the same canonical models over S.*

Proof. By assumption $p_* \mathcal{O}_{X'}(mK_{X'} + \lfloor m\Delta' \rfloor) = \mathcal{O}_X(mK_X + \lfloor m\Delta \rfloor)$. Thus

$$(f \circ p)_* \mathcal{O}_{X'}(mK_{X'} + \lfloor m\Delta' \rfloor) = f_* \mathcal{O}_X(mK_X + \lfloor m\Delta \rfloor).$$

Now apply (3.52). □

Corollary 3.54. *Let $f_i : X_i \to S$, $i = 1, 2$ be proper morphisms such that X_i is terminal and K_{X_i} is f_i-nef. Then any birational map $\phi : X_1 \dashrightarrow X_2$ (over S) is an isomorphism in codimension 1.*

Proof. Let $g_i : X' \to X_i$ be a common resolution of singularities. Then $K_{X'} - g_i^* K_{X_i}$ is effective, hence by (3.53) X_i is a minimal model of X' for $i = 1, 2$. Thus ϕ is an isomorphism in codimension 1 by (3.52). $\square$

4

Surface Singularities of the Minimal Model Program

In this chapter we discuss various classes of surface singularities that are important for the minimal model program.

The first section contains an essentially complete description of all log canonical surface pairs. These results provide important signposts toward the higher dimensional theory. Many of these results are used later.

The aim of section 2 is to classify Du Val singularities by equations. These are precisely the canonical surface singularities. From this point of view they were classified by [DV34]. Du Val singularities appear naturally in many different contexts, see [Dur79] for a survey.

Simultaneous resolution of Du Val singularities is considered in section 3. This result was established by [Bri66, Bri71]. Our presentation is modelled on [Tyu70].

The natural setting for sections 2–3 is local analytic geometry. In all cases it is possible to work with algebraic varieties, but only at the price of some rather artificial constructions. In applying these results we therefore have to rely on some basic comparison theorems between analytic and algebraic geometry.

Section 4 considers elliptic surface singularities. Their theory was developed by [Rei76, Lau77]. These results are crucial for the treatment of 3-dimensional canonical singularities given in section 5.3.

In the last section we construct miniversal deformation spaces for isolated hypersurface singularities.

While all these results are very useful, and for the moment indispensable, for the 3-dimensional minimal model program, they are a hindrance from the point of view of the general theory. Many of the methods reducing 3-dimensional questions to surface problems, and the 2-dimensional results used in the process, do not generalize to higher dimensions. Thus, if the minimal model program works in all dimensions, its proof must, by necessity, proceed along different lines.

4.1 Log Canonical Surface Singularities

The aim of this section is to describe in detail the local structure of 2–dimensional pairs (X, Δ) which are log canonical. In this case a complete classification is known. We cover most of it, except for some finer points which we do not need later. The arguments work in the algebraic and analytic settings, in most cases without any change.

Notation 4.1. Let X be either a normal algebraic surface or a normal complex analytic surface and Δ a $\mathbb{Q}$-divisor on X. In these cases one can define the notion of discrepancy, without assuming that $K_X + \Delta$ is $\mathbb{Q}$-Cartier, as follows.

Let $f : Y \to X$ be a resolution of X with exceptional divisor $E = \sum E_i$. The intersection matrix $(E_i \cdot E_j)$ is negative definite by (3.40), thus the system of linear equations

$$E_j \cdot \left(\sum_i a_i E_i\right) = E_j \cdot (K_Y + f_*^{-1}\Delta) \qquad \forall j$$

has a unique solution. We write this as

$$K_Y + f_*^{-1}\Delta \equiv \sum a_i E_i.$$

The number $a(E_i, X, \Delta) := a_i$ is called the *discrepancy* of E_i with respect to (X, Δ). If $K_X + \Delta$ is $\mathbb{Q}$-Cartier then this coincides with the earlier definition (2.22).

We say that (X, Δ) is *numerically log canonical* if Δ is effective and $a(E_i, X, \Delta) \geq -1$ for every exceptional curve E_i and f. We say that (X, Δ) is *numerically dlt* if Δ is effective and there is a finite set $Z \subset X$ such that $X \setminus Z$ is smooth, $\Delta|_{X \setminus Z}$ is snc and $a(E, X, \Delta) > -1$ for every exceptional curve E which maps to Z.

It turns out that for surfaces (X, Δ) numerically lc (resp. dlt) implies that $K_X + \Delta$ is $\mathbb{Q}$-Cartier, thus (X, Δ) is also lc (resp. dlt) (cf. 4.11). Thus the notion numerically lc is used only for temporary convenience.

The following two immediate consequences of (3.41) will be used repeatedly:

Corollary 4.2. *Let U be a smooth surface and $C = \cup C_i$ a connected proper curve on U. Assume that the intersection matrix $(C_i \cdot C_j)$ is negative definite. Let $A = \sum a_i C_i$ and $B = \sum b_i C_i$ be $\mathbb{Q}$-linear combinations* *of the curves C_i. Assume that $(B \cdot C_j) \leq (A \cdot C_j)$ for every j. Then either $a_i = b_i$ for every i, or $a_i < b_i$ for every i.* □

Corollary 4.3. *Notation as in (4.1). Assume that $K_Y + f_*^{-1}\Delta$ is f-nef. Then either $a_i = 0$ for every i, or $a_i < 0$ for every i.* □

The classification of canonical surface singularities leads to the notion of Du Val singularities, first studied in the series of papers [DV34]. We study them in more detail in section 2. For now, the following definition is the most convenient.

Definition 4.4. A normal surface singularity $(0 \in X)$ with minimal resolution $f : Y \to X$ is called a *Du Val* singularity iff $K_Y \cdot E_i = 0$ for every exceptional curve $E_i \subset Y$. (We prove in (4.20) that K_X is Cartier.)

Theorem 4.5. *Notation as in (4.1). Then*

   (1) (X, Δ) *is terminal iff X is smooth and $\mathrm{mult}_0 \Delta < 1$.*
   (2) (X, Δ) *is canonical iff either*
      (smooth) X is smooth and $\mathrm{mult}_0 \Delta \leq 1$, or
      (Du Val) $(0 \in X)$ is a Du Val singularity and $\Delta = 0$.

Proof. Let $f : Y \to X$ be the minimal resolution of X and $E = \cup E_i$ the exceptional set. If (X, Δ) is canonical then $a_i \geq 0$. By (4.3), $a_i \leq 0$, thus $a_i = 0$ for every i. If (X, Δ) is terminal then this is possible only if there are no exceptional curves, hence X is smooth. Blowing up $0 \in X$ shows that $\mathrm{mult}_0 \Delta \leq 1$, and < 1 in the terminal case. Induction on the number of blow ups shows that this condition is also sufficient.

Assume next that $E \neq \emptyset$. Then $K_Y + f_*^{-1}\Delta$ is numerically trivial. If $f_*^{-1}\Delta \neq 0$ then $(E_i \cdot f_*^{-1}\Delta) > 0$ for some i, hence $(E_i \cdot K_Y) < 0$. This is impossible. Thus $f_*^{-1}\Delta = 0$ and $(K_Y \cdot E_i) = 0$ for every i.

The fact that K_X is Cartier will be established later (4.20). □

Definition 4.6. Let $C = \cup C_i$ be a collection of proper curves on a smooth surface U. The *dual graph* Γ of C is defined as follows:

   (1) The vertices of Γ are the curves C_i.
   (2) The vertex C_i is labelled by $b_i = -(C_i \cdot C_i)$. The labelling is omitted if the value of $(C_i \cdot C_i)$ is not known or not important. (In many situations one may also want to keep track of the genus or the singularities of C_i, or some other data. For us these are mostly irrelevant.)

(3) The vertices C_i and C_j for $i \neq j$ are connected with $(C_i \cdot C_j)$ edges.

If D_i are additional (not necessarily proper) curves on U, we define the *extended dual graph* Γ^* by adding each D_i as a vertex (with no label) and connecting D_i and C_j with $(D_i \cdot C_j)$ edges.

A *cycle* is a graph whose vertices and edges can be ordered $C_1, \ldots, C_n$ and $e_1, \ldots, e_n$ $(n \geq 2)$ such that e_i connects C_i and C_{i+1} $(n+1$ meaning 1). For $n = 2$ we get two vertices and two edges connecting them, for $n = 1$ we get a loop.

A *tree* is a connected graph which does not contain a subgraph which is a cycle.

A vertex C_i of a tree is called a *fork* if there are at least three edges from C_i. The connected components of a tree minus a fork are called the *branches* of the fork.

A tree without forks is called a *chain*.

Theorem 4.7. *Let $(0 \in X, \Delta)$ be a numerically log canonical pair, $f : Y \to X$ the minimal resolution and Γ the dual graph of the exceptional curves $E = \cup E_i$ of f. Then exactly one of the following holds:*

1. *There is only one exceptional curve $E = E_1$ which is a smooth elliptic curve or a nodal cubic and $\Delta = 0$. (X, Δ) is not numerically dlt.*

For the remaining cases every E_i is a smooth rational curve and $\sum E_i$ is a snc divisor.

2. *Γ is a cycle of smooth rational curves and $\Delta = 0$. (X, Δ) is not numerically dlt.*
3. *$\Delta = 0$ and Γ is one of the following. (There are $m \geq 5$ exceptional curves.)*

(X, Δ) is not numerically dlt.

4. *Γ is a tree with exactly one fork, which has three branches.*
5. *(Cyclic quotient) Γ is a (possibly empty) chain: $\circ - \cdots - \circ$.*

Proof. Set $A = \sum a_i E_i$ as in (4.1) and let

$$d_i := (A \cdot E_i) = ((K_Y + f_*^{-1}\Delta) \cdot E_i) = b_i - 2 + 2p_a(E_i) + (E_i \cdot f_*^{-1}\Delta).$$

f is the minimal resolution, hence $d_i \geq 0$ for every i. In what follows we write down linear combinations $B = \sum b_i E_i$ such that $(B \cdot E_i) \leq d_i$ for every i. By (4.2) this implies that $a_j \leq b_j$ and if equality holds for some j then $B = A$.

(1) Assume that $p_a(E_1) \geq 1$ and set $B = -E_1$. Then $(B \cdot E_i) \leq 0 \leq d_i$ if $i \neq 1$ and $(B \cdot E_1) = b_1 \leq b_1 - 2 + 2p_a(E_1) \leq d_1$. Thus if (X, Δ) is lc, then equality holds everywhere. So E_1 is the only curve, $p_a(E_1) = 1$ and $f_*^{-1}\Delta = 0$. Two more blow ups show that the case when E_1 is a cuspidal cubic is not numerically lc.

From now on we may assume that all the E_i are smooth rational curves.

(2) Assume that $(E_1 \cdot E_2) \geq 2$. Set $B = -E_1 - E_2$. Computing as above we find that $E = E_1 \cup E_2$, $f_*^{-1}\Delta = 0$ and $(E_1 \cdot E_2) = 2$. Two more blow ups show that the case when E_1 and E_2 are tangent at a point is not numerically lc. This gives the $n = 2$ case of (2).

Now we know that in Γ two vertices are connected by at most one edge. Assume that Γ is not a tree. Then it contains a cycle, that is, a collection of curves $E_1, \ldots, E_n$ as in (2). Let $B = -E_1 - \cdots - E_n$. Computing as above gives (2). One blow up shows that the case when $n = 3$ and the E_i meet at a point is not numerically lc.

(3) Thus Γ is a tree. Assume that it contains at least two forks. Then it contains a subgraph as follows:

Let $E_1, \ldots, E_4$ be the vertices marked $\diamond$, and $E_5, \ldots, E_{4+n}$ ($n \geq 1$) the vertices marked $\circ$. Let $B = -\frac{1}{2}(E_1 + \cdots + E_4) - (E_5 + \cdots + E_{4+n})$. Then

$$(B \cdot E_j) \begin{cases} \leq 0 & \text{if } j \geq n + 5 \\ = b_j - 2 & \text{if } n + 4 \geq i \geq 5, \\ = b_j/2 - 1 & \text{if } j = 1, \ldots, 4. \end{cases}$$

Thus we see that (X, Δ) is numerically lc iff there are no other curves, $f_*^{-1}\Delta = 0$ and $b_j = 2$ for $j = 1, \ldots, 4$. This gives (3).

The remaining possibilities are listed under (4) and (5).   □

Note 4.8 (Traditional terminology). The singularities in (1) with $E = E_1$ smooth are called simple elliptic. (2) and the nodal case of (1) together are called cusps. Cusp singularities occur at the boundary of certain modular varieties [Hir71]; the boundary points traditionally have been called cusps.

One can prove that the singularities in (3) are quotients of cusps by the 2 element group. Quotients of a smooth point by a cyclic group give all singularities listed under (5). Quotients by non-cyclic groups give all the log terminal singularities in (4). The log canonical ones with one fork arise as quotients of simple elliptic points. See [Bri68, KSB88, Lam86] for proofs.

Remark 4.9. The converse of (4.7) also holds. More precisely, the following are true:

(1) Let Y be a smooth analytic surface and $E \subset Y$ an irreducible proper curve. Assume that E is either elliptic or a nodal cubic. If $(E^2) < 0$ then E is contractible $f : (E \subset Y) \to (0 \in X)$ and X has a log canonical singularity at 0. Moreover, the analytic isomorphism type of the germ $(0 \in X)$ is determined by E and (E^2).

(2) Let Y be a smooth analytic or algebraic surface and $E = \sum E_i \subset Y$ a connected proper curve. Assume that every E_i is a smooth rational curve and $(E_i^2) \le -2$. Assume furthermore that the dual graph of E is one of the graphs obtained in (4.7.2–5) and the intersection matrix $(E_i \cdot E_j)$ is negative definite. Then E is contractible $f : (E \subset Y) \to (0 \in X)$ and X has a log canonical singularity at 0. Moreover, the analytic isomorphism type of the germ $(0 \in X)$ is determined by the dual graph of E, with the exception of the dual graph

where the isomorphism type is determined by the dual graph and the cross ratio of the four intersection points on the fork.

We do not use these results in the sequel. The contractibility of the klt cases when Y is projective is proved next. For proofs of the other results see [Art62, Art66, Lau73].

Proposition 4.10. *Let Y be a smooth projective surface and $E_i \subset Y$ irreducible curves forming an snc divisor $\cup E_i$. Assume that the intersection matrix $(E_i \cdot E_j)$ is negative definite and there are rational numbers $0 \geq a_i > -1$ such that $(K_Y - \sum_i a_i E_i \cdot E_j) = 0$ for every j. Then there is a birational morphism $f : Y \to X$ to a normal projective surface X contracting E_i's and no other curves. Furthermore, X is $\mathbb{Q}$-factorial.*

Proof. Let $\Delta_Y := -\sum_i a_i E_i$, then (Y, Δ_Y) is klt. Let H be ample on Y such that $H - K_Y - \Delta_Y$ is also ample. There are positive $b_i \in \mathbb{Q}$ such that $(H + \sum_i b_i E_i \cdot E_j) = 0$ for every j. By replacing H with a suitable multiple, we may assume $b_i \in \mathbb{Z}$. Let $L := H + \sum_i b_i E_i$. Then $L - K_Y - \Delta_Y$ is big. We claim that L and $L - K_Y - \Delta_Y$ are nef. Indeed $(L \cdot E_j) = (L - K_Y - \Delta_Y \cdot E_j) = 0$ and, for other curves $C \subset Y$, $(C \cdot L) \geq (C \cdot H) > 0$ and

$$(C \cdot L - K_Y - \Delta_Y) \geq (C \cdot H - K_Y - \Delta_Y) > 0$$

since $\sum_i b_i E_i$ is effective. By (3.3) $|mL|$ is free for $m \gg 1$. Taking the Stein factorization of the induced morphism, we obtain a birational $f : Y \to X$ contracting exactly the curves E_j. Furthermore, X is normal and $f_* H = f_* L$ is $\mathbb{Q}$-Cartier. By Zariski's Main Theorem [Har77, V.5.2], f does not depend on the choice of L. Every curve C can be written as the difference of two ample divisors $C = H_1 - H_2$, thus $f_* C$ is $\mathbb{Q}$-Cartier for any curve C on Y. $\square$

The next result implies that numerically dlt is equivalent to dlt.

Proposition 4.11. *Let (X, Δ) be a numerically dlt pair. Then every Weil divisor on X is $\mathbb{Q}$-Cartier.*

Proof. First we prove the case when X is a projective surface.

$(X, \emptyset)$ is also numerically dlt by (4.2). Let $f : Y \to X$ be the minimal resolution with exceptional curves $E = \sum E_i$. By (4.7) E is an snc divisor. Let $K_Y \equiv_f \sum a_i E_i$. Then $0 \geq a_i > -1$, the first inequality by (4.3) and the second by the numerically dlt assumption. Since f is the contraction of E, X is $\mathbb{Q}$-factorial by (4.10).

In the general algebraic case, when X is not projective, we can either compactify X or use (3.24) in the argument of (4.10). If X is an analytic surface the result follows from (4.12) and (4.13).   □

The following is a very special case of (5.22).

Theorem 4.12. *Let* $(0 \in X, \Delta)$ *be a numerically dlt surface pair and* $f : Y \to X$ *the minimal resolution. Then* $R^1 f_* \mathcal{O}_Y = 0$.

Proof. $(X, \emptyset)$ is also numerically dlt by (4.2). Let $E = \sum E_i$ be the exceptional curves. By (4.7) E is an snc divisor. Write $K_Y \equiv \sum a_i E_i$. Then $0 \geq a_i > -1$, the first inequality by (4.3) and the second by the dlt assumption. $\mathcal{O}_Y \equiv K_Y + \sum (-a_i) E_i$. If X is algebraic, we can use (2.68) to conclude that $R^1 f_* \mathcal{O}_Y = 0$.

In the analytic case one can either prove the analogue of (2.68), or proceed as in [Art66].   □

Later we use the following result for non-reduced complex spaces as well.

Lemma 4.13. *Let* U *be a complex analytic space such that* $H^1(U, \mathcal{O}_U) = 0$. *Let* $\mathbb{Z}_U$ *denote the constant sheaf with values in* $\mathbb{Z}$. *Then*

(1) *There is an injection* $\mathrm{Pic}(U) \hookrightarrow H^2(U, \mathbb{Z}_U)$.
(2) *Let* $C = \cup C_i \subset U$ *be a compact curve such that* $\mathrm{red}(U)$ *retracts to* C. *Then a line bundle* L *on* U *is determined up to isomorphism by the values* $(L \cdot C_i)$.

Proof. Notice that the exponential sequence

$$0 \to \mathbb{Z}_U \xrightarrow{2\pi\sqrt{-1}} \mathcal{O}_U \xrightarrow{\exp} \mathcal{O}_U^* \to 1$$

is exact even if $\mathcal{O}_U$ has nilpotent elements. Taking cohomologies gives

$$H^1(U, \mathcal{O}_U) \to H^1(U, \mathcal{O}_U^*) \cong \mathrm{Pic}(U) \to H^2(U, \mathbb{Z}_U).$$

If $\mathrm{red}(U)$ retracts to C then $H^2(U, \mathbb{Z}_U) = H^2(C, \mathbb{Z}_C) = \oplus \mathbb{Z}[C_i]$. This shows (2).   □

Corollary 4.14. *Let* $f : Y \to (P \in X)$ *be a resolution of the germ of a surface singularity such that* $R^1 f_* \mathcal{O}_Y = 0$ *and let* $I \subset \mathcal{O}_{P,X}$ *be the maximal ideal. Then* $I\mathcal{O}_Y = \mathcal{O}_Y(-E)$ *for some effective Cartier divisor* $E \subset Y$ *and* $I^\nu = f_* \mathcal{O}_Y(-\nu E)$ *for all* $\nu > 0$.

Proof. Let $\mathcal{O}_Y(-E)$ be the divisorial part of $I\mathcal{O}_Y$. Then $\mathcal{O}_Y(-E) \supset I\mathcal{O}_Y$ and $\Sigma := \mathrm{Supp}\, \mathcal{O}_Y(-E)/I\mathcal{O}_Y$ is a finite set. Thus $f^* I \to \mathcal{O}_Y(-E)$ is surjective on $Y \setminus \Sigma$ and $\mathcal{O}_Y(-E)$ is f-nef. By (4.13), $\mathcal{O}_Y(-E)$ is f-free.

We have $I \subseteq f_* \mathcal{O}_Y(-E) \subsetneq \mathcal{O}_X$ by $E \neq \emptyset$ and $I \mathcal{O}_Y \subset \mathcal{O}_Y(-E)$. Hence $f_* \mathcal{O}_Y(-E) = I$ and $I \mathcal{O}_Y = \mathcal{O}_Y(-E)$ by the f-freeness of $\mathcal{O}_Y(-E)$.

We claim $f_* \mathcal{O}_Y(-\nu E) = I^\nu$ and $R^1 f_* \mathcal{O}_Y(-\nu E) = 0$ for all $\nu > 0$. We prove it by induction on ν. Let $\alpha, \beta \in I$ be general elements pulling back to global sections $f^*\alpha, f^*\beta$ of $\mathcal{O}_Y(-E)$ such that $(f^*\alpha = 0) \cap (f^*\beta = 0) = \emptyset$. Then $\phi := (f^*\alpha, f^*\beta) : \mathcal{O}_Y^{\oplus 2} \to \mathcal{O}_Y(-E)$ is a surjection, and we have an exact sequence

$$0 \to \mathcal{O}_Y(E) \to \mathcal{O}_Y^{\oplus 2} \xrightarrow{\phi} \mathcal{O}_Y(-E) \to 0$$

by comparing the determinants. Hence there are exact sequences

$$F_\nu : 0 \to \mathcal{O}_Y(-(\nu-1)E) \to \mathcal{O}_Y(-\nu E)^{\oplus 2} \to \mathcal{O}_Y(-(\nu+1)E) \to 0.$$

From F_0, we have a surjection $R^1 f_* \mathcal{O}_Y^{\oplus 2} \twoheadrightarrow R^1 f_* \mathcal{O}_Y(-E)$ hence the claim for $\nu = 1$. Let $k \geq 1$. If we assume the claim for $\nu \leq k$, then by F_k we have $f_* \mathcal{O}_Y(-(k+1)E) = (\alpha, \beta) I^k$ and $R^1 f_* \mathcal{O}_Y(-(k+1)E) = 0$. Since $f_* \mathcal{O}_Y(-(k+1)E) \supseteq I^{k+1}$, the claim for $\nu = k+1$ follows. □

Theorem 4.15. *Let $(0 \in X, \Delta)$ be a log canonical pair. Assume that $\lfloor \Delta \rfloor \neq 0$. Let $f : Y \to X$ be the minimal resolution with exceptional curves $E = \cup E_i$ and let Δ_j be the irreducible analytic branches of $\lfloor f_*^{-1} \Delta \rfloor$. Let Γ^* be the extended dual graph of the E_i and Δ_j. Then exactly one of the following holds, where the Δ_j are denoted by • and m denotes the number of exceptional curves.*

(1) $\lfloor \Delta \rfloor$ *has two analytic branches,* $\Delta = \lfloor \Delta \rfloor$, $m \geq 0$ *and* Γ^* *is*

$$\bullet - \circ - \cdots - \circ - \bullet.$$

(X, Δ) *is dlt iff X is smooth and Δ has two branches crossing transversally.*

(2) $\lfloor \Delta \rfloor$ *has one analytic branch,* $\Delta = \lfloor \Delta \rfloor$, $m \geq 3$ *and* Γ^* *is*

$$\begin{array}{c} \circ\, 2 \\ | \\ \bullet - \circ - \circ \cdots \circ - \circ - \quad \circ \\ | \\ \circ\, 2 \end{array}$$

(X, Δ) *is not dlt.*

(3) $\lfloor \Delta \rfloor$ *has one analytic branch,* $m \geq 0$ *and* Γ^* *is*

$$\bullet - \circ - \cdots - \circ.$$

In this case $(X, \lfloor \Delta \rfloor)$ is plt.

Proof. We are already in cases 4 or 5 of (4.7), thus Γ is a tree with at most one fork.

Set $A = \sum a_i E_i$ as in (4.1). As before, we write down linear combinations $B = \sum b_i E_i$ such that $(B \cdot E_i) \leq (A \cdot E_i)$ for every i.

(1) Assume that $\lfloor \Delta \rfloor$ has two analytic branches. Then we get a sub-
   • graph

$$\bullet - \circ - \cdots - \circ - \bullet,$$

where $E_1, \ldots, E_n$ $(n \geq 1)$ are the curves denoted by $\circ$. Set $B = -E_1 - \cdots - E_n$. This gives (1).

(2) Assume that $\lfloor \Delta \rfloor$ has one analytic branch and Γ^* has a fork. Then we have a subgraph

B contains the curves marked $\circ$ with coefficient -1, and the curves marked $\diamond$ with coefficient $-1/2$. Computing as in (3) of (4.7) we get (2).

(3) This leaves only case (3). Let $E = E_1 \cup \cdots \cup E_n$ and set $B = -(E_1 + \cdots + E_n)$. Then $(B \cdot E_i) \geq ((K_Y + \lfloor f_*^{-1}\Delta \rfloor) \cdot E_i)$ for every i and the inequality is strict for the curve on the right end of the chain. Thus by (4.2), $a_i > -1$ for every i, hence $(X, \lfloor \Delta \rfloor)$ is plt. □

Theorem 4.16. *Let $(0 \in X, \Delta)$ be a log canonical pair, $f : Y \to X$ the minimal resolution and Γ the dual graph of the exceptional curves $E = \cup E_i$. Assume that Γ is a tree with one fork and three branches with lengths $n_1 - 1, n_2 - 1, n_3 - 1$. Then*

$$\frac{1}{n_1} + \frac{1}{n_2} + \frac{1}{n_3} \geq 1,$$

and strict inequality holds if $(0 \in X, \Delta)$ is dlt.

Unfortunately, the converse of this statement does not hold, and the self-intersection numbers of the curves have to satisfy certain properties depending on the length of the branches. These are described in detail in many places; see, for instance, [K+92] Chapter 3.

Proof. A branch of Γ is described by a diagram

$$\diamond - \circ - \cdots - \circ$$

where $\diamond$ denotes the fork. Let m be the length of the branch. Construct B by assigning the coefficients as follows:

$$\underset{\diamond}{\overset{-1}{}} - \underset{\circ}{\overset{-\frac{m}{m+1}}{}} - \underset{\circ}{\overset{-\frac{m-1}{m+1}}{}} - \cdots - \underset{\circ}{\overset{-\frac{1}{m+1}}{}} .$$

Then

$$(B \cdot E_j) \begin{cases} = b_j - 3 + \sum 1/n_i & \text{if } E_j \text{ is the fork,} \\ \leq b_j - 2 & \text{otherwise.} \end{cases}$$

Thus if $\sum 1/n_i \leq 1$ then $(B \cdot E_j) \leq d_j$ with at least one strict inequality if $\sum 1/n_i < 1$. $\qquad\square$

Theorem 4.17. *Let* $(0 \in X)$ *be a normal surface singularity,* $f : Y \to X$ *the minimal resolution and* Γ *the dual graph of the exceptional curves* $E = \cup E_i$ *of* f. *Assume that*

(1) Γ *is a chain, or*

(2) Γ *is a tree with one fork and three branches with lengths* $n_i - 1$ *such that* $\sum 1/n_i > 1$ *and all curves on the branches have self-intersection* -2.

Then $(0 \in X)$ *is klt.*

Proof. In the first case set $B = -(E_1 + \cdots + E_n)$. Then $(B \cdot E_i) \geq d_i$ for every i and the inequality is strict for the curves on the ends of the chain. Thus by (4.2), $a_i > -1$ for every i, hence $(0 \in X)$ is klt.

In the second case choose B as in (4.16) and argue as before. $\qquad\square$

For completeness sake we state another characterization of log terminal surface singularities, though we do not use it later. It can easily be derived from (4.21) and (5.20).

Proposition 4.18. *For a normal surface germ* $(0 \in X)$, *the following are equivalent:*

(1) $(0 \in X)$ *is log terminal,*

(2) $(0 \in X)$ *is a quotient of* $(0 \in \mathbb{C}^2)$ *by a finite group which acts freely in codimension 1,*

(3) $(0 \in X)$ *is a quotient of* $(0 \in \mathbb{C}^2)$ *by a finite group.* $\qquad\square$

4.19. The proofs in this section work in any characteristic, except for (4.9), (4.12) and (4.18). (4.12) holds in every characteristic; this can be proved using [Art66]. (4.9) and (4.18) fail in positive characteristic even for Du Val singularities, see [Lip69].

4.2 Du Val Singularities

The aim of this section is to describe Du Val singularities by explicit equations.

Theorem 4.20. *Let $(0 \in X)$ be the germ of a normal surface singularity. The following are equivalent:*

(1) $(0 \in X)$ *is canonical.*
(2) $(0 \in X)$ *is Du Val.*
(3) $(0 \in X)$ *is analytically isomorphic to a singularity defined by one of the equations*

> *A.* $x^2 + y^2 + z^{n+1} = 0$;
> *D.* $x^2 + y^2 z + z^{n-1} = 0$;
> *E.* $x^2 + y^3 + z^4 = 0$, $x^2 + y^3 + yz^3 = 0$, *or* $x^2 + y^3 + z^5 = 0$.

Remark 4.21. Du Val singularities come up naturally in many different contexts, and they can be characterized in numerous ways [Dur79]. We just mention two of these, without proof:

(1) Du Val singularities are the quotients of $\mathbb{C}^2$ by finite subgroups of $SL(2, \mathbb{C})$,
(2) Du Val singularities are exactly the simple singularities (that is, only finitely many other singularities can be obtained by small perturbations of a Du Val singularity).

Furthermore, being Du Val is an open condition. That is if $(f = 0)$ is Du Val and g is arbitrary then $(f + \epsilon g)$ is Du Val for $|\epsilon| \ll 1$. This can be seen by analysing (4.25). This also follows from (5.24) and (5.42). We do not need this result in the sequel, though it is important conceptually.

Proof. We already saw that canonical implies Du Val (4.5). If $(0 \in X)$ is one of the singularities listed in (3), then explicit computation of the resolution shows that it is Du Val. This can be established using the following two steps:

(1) If $(0 \in X)$ is a surface double point and $p : B_0 X \to X$ is the blow up of the origin, then $K_{B_0 X} = p^* K_X$.

(2) If $(0 \in X)$ is one of the singularities listed in (3) then every singular point of $B_0 X$ is also on the list (3).

We strongly encourage the reader to perform these computations and to verify that one obtains the exceptional curve configurations described in (4.22).

Next we establish that if $(0 \in X)$ is Du Val then it is canonical. This is implied by (3.3), but the following direct proof may be simpler. Let $f : Y \to X$ be the minimal resolution with exceptional curves $E = \cup E_i$. By (4.12) $H^1(Y, \mathcal{O}_Y) = 0$. $\mathcal{O}_Y$ and $\mathcal{O}_Y(K_Y)$ are two line bundles on Y which have zero intersection number with any E_i. Thus by (4.13), $\mathcal{O}_Y \cong \mathcal{O}_Y(K_Y)$, in particular $\mathcal{O}_Y(K_Y)$ is generated by a single global section $\sigma \in H^0(Y, \mathcal{O}_Y(K_Y))$. Thus $f_* \mathcal{O}_Y(K_Y)$ is generated by $f_* \sigma$, hence it is locally free. We have a natural map

$$f_* \mathcal{O}_Y(K_Y) \hookrightarrow \mathcal{O}_X(K_X),$$

which is an isomorphism outside 0. Since $f_* \mathcal{O}_Y(K_Y)$ is locally free, this implies that $f_* \mathcal{O}_Y(K_Y) = \mathcal{O}_X(K_X)$. Thus K_X is Cartier and $f^* \mathcal{O}_X(K_X) = \mathcal{O}_Y(K_Y)$, which proves that $(0 \in X)$ is canonical.

The remaining implication is a special case of the next more precise result:

Theorem 4.22. *Every Du Val singularity has embedding dimension 3. Up to a local analytic change of coordinates, the following is a complete list of Du Val singularities:*

A. *The singularity* A_n $(n \geq 1)$ *has equation* $x^2 + y^2 + z^{n+1} = 0$ *and dual graph with* n *vertices:*

$$\circ - \cdots - \circ$$

D. *The singularity* D_n $(n \geq 4)$ *has equation* $x^2 + y^2 z + z^{n-1} = 0$ *and dual graph with* n *vertices:*

E. *The singularity* E_6 *(resp.* E_7*, resp.* E_8*) has equation* $x^2 + y^3 + z^4 = 0$*, (resp.* $x^2 + y^3 + yz^3 = 0$*, resp.* $x^2 + y^3 + z^5 = 0$*) and dual graph with 6 (resp. 7, resp. 8) vertices:*

If Y is a smooth surface and $E \subset Y$ a collection of proper curves whose dual graph is listed above, then $E \subset Y$ is the minimal resolution of a surface $0 \in X$ which has the corresponding Du Val singularity at 0.

Proof. Let $(0 \in X)$ be a Du Val singularity with minimal resolution $f : Y \to X$. Let $E = \cup E_i$ be the exceptional curves and Γ their dual graph. First we establish that Γ is one of the graphs listed above.

$(0 \in X)$ is lt, thus by (4.7), Γ is either a chain (type A) or it has one fork with three branches of lengths $n_i - 1$ such that

$$\frac{1}{n_1} + \frac{1}{n_2} + \frac{1}{n_3} > 1.$$

The only sets of integers satisfying this condition are $(2, 2, m)$ (type D) and $(2, 3, 3)$, $(2, 3, 4)$, $(2, 3, 5)$ (type E).

Explicit computation of the minimal resolutions shows that the equations correspond to the dual graphs as indicated above.

In order to get the equations, we proceed in two steps. First we identify a hypersurface section of $(0 \in X)$. This gives us a certain class of equations, which satisfy further properties if $(0 \in X)$ is canonical. After that we still have to make appropriate coordinate changes to achieve the required normal forms.

Lemma 4.23. *Let $(0 \in X)$ be a Du Val singularity. Then $(0 \in X)$ is a double point of embedding dimension 3.*

Proof. Let $f : Y \to X$ be the minimal resolution. We write down explicitly a divisor $Z + \bar{C}$ which turns out to be the pull back of a hyperplane section of $0 \in X$.

In the following diagrams $\bullet$ denotes the irreducible components of $\bar{C}$. The numbers next to a $\circ$ indicate the multiplicity of that curve in Z. The choice of $\bar{C}$ is not unique; let $\bar{C}_1$ and $\bar{C}_2$ be two choices of $\bar{C}$ which are disjoint.

$$A_n : \bullet - \overset{1}{\circ} - \cdots - \overset{1}{\circ} - \bullet \qquad D_n : \begin{array}{c} \overset{1}{\circ} - \overset{2}{\circ} - \cdots - \overset{2}{\circ} - \overset{1}{\circ} \\ | \qquad \qquad | \\ 1 \circ \qquad \qquad \bullet \end{array}$$

$$E_6 : \begin{array}{c} \overset{1}{\circ} - \overset{2}{\circ} - \overset{3}{\circ} - \overset{2}{\circ} - \overset{1}{\circ} \\ | \\ 2 \circ - \bullet \end{array} \qquad E_7 : \begin{array}{c} \bullet - \overset{2}{\circ} - \overset{3}{\circ} - \overset{4}{\circ} - \overset{3}{\circ} - \overset{2}{\circ} - \overset{1}{\circ} \\ | \\ 2 \circ \end{array}$$

$$E_8: \qquad \overset{2}{\circ} - \overset{4}{\circ} - \overset{6}{\circ} - \overset{5}{\circ} - \overset{4}{\circ} - \overset{3}{\circ} - \overset{2}{\circ} - \bullet$$
$$\overset{|}{\underset{3 \ \circ}{}}$$

One can check case by case that $(Z + \bar{C}_i) \cdot E_j = 0$ for every exceptional curve E_j and $(Z \cdot \bar{C}_i) = 2$. Thus by (4.13), $\mathcal{O}_Y(-Z - \bar{C}_i) \cong \mathcal{O}_Y$. Hence the section

$$1 \in H^0(Y, \mathcal{O}_Y) \cong H^0(Y, \mathcal{O}_Y(-Z - \bar{C}_i)) \subset H^0(Y, \mathcal{O}_Y) = H^0(X, \mathcal{O}_X)$$

gives a function g_i on X such that $C_i = (g_i = 0)$ is a hypersurface section of $0 \in X$ and $f^* C_i = Z + \bar{C}_i$.

The multiplicity of X at 0 divides the local intersection number of any two hypersurface sections. It can be computed by the projection formula:

$$(C_1 \cdot C_2)_X = (Z + \bar{C}_1 \cdot \bar{C}_2)_Y = (Z \cdot \bar{C}_2)_Y = 2.$$

C_i is thus a reduced curve singularity of multiplicity 2 (or 1), thus it is planar, hence X has embedding dimension 3 (or 2).

Note. This result is the 2-dimensional version of (5.30). Another proof, using general principles, is as follows. Let $(0 \in X)$ be a canonical surface singularity and $0 \in C \subset X$ a general hypersurface section with normalization $\pi : \bar{C} \to C$. By (5.30) we know that $\pi_* \omega_{\bar{C}} \supset m_{0,C} \omega_C$. This is equivalent to $m_{0,C} \mathcal{O}_{\bar{C}} \subset \mathcal{O}_C$ (this may need some local duality theory). The latter implies that C is an ordinary node or cusp. $\qquad\square$

We thus know that $(0 \in X)$ is defined in $\mathbb{C}^3$ by an equation of the form $F = x^2 + zf(x, y, z)$. The classification of equations up to analytic coordinate change is studied in [AGZV85], and this can be used to complete the proof of (4.22). Below we present a somewhat simpler version of this approach.

4.24. We repeatedly use four methods:

(1) The Weierstrass preparation theorem.
(2) The elimination of the y^{n-1}-term from the polynomial $a_n y^n + a_{n-1} y^{n-1} + \ldots$ by a coordinate change $y \mapsto y - a_{n-1}/na_n$ when a_n is invertible.
(3) Hensel's lemma in the following form: let $f(y, z)$ be a power series with leading term $f_d(y, z)$. Assume that $f_d = gh$ where g and h do not have common factors. Then $f = GH$ where g (resp. h) is the leading term of G (resp. H).

(4) Let M_1, M_2, M_3 be multiplicatively independent monomials in the variables x, y, z. Then any power series of the form $M_1 \cdot (\text{unit}) + M_2 \cdot (\text{unit}) + M_3 \cdot (\text{unit})$ is equivalent to $M_1 + M_2 + M_3$ by a suitable coordinate change $x \mapsto x \cdot (\text{unit}), y \mapsto y \cdot (\text{unit}), z \mapsto z \cdot (\text{unit})$.

Let f_d denote the degree d homogeneous part of a power series f.

4.25. The proof is in several steps. We successively reduce the equation to simpler and simpler forms.

Step 1. If $\text{mult}_0 F = 1$ then we have a smooth point. Thus assume in the sequel that $\text{mult}_0 F = 2$.

Step 2. Applying (4.24.1) and then (4.24.2) to x^2 we reduce the equations to the form

$$F = (\text{unit}) \cdot (x^2 + f(y, z)).$$

Step 3. If $\text{mult}_0 f \leq 2$ then apply (4.24.1) and (4.24.2) to f to get the form

$$F = (\text{unit}) \cdot (x^2 + (\text{unit}) \cdot (y^2 + z^m \cdot (\text{unit})) \quad \text{for some } m \geq 2.$$

(4.24.4) gives $F = x^2 + y^2 + z^m$; these are the A cases.

Assume next that $\text{mult}_0 f = 3$, equivalently, $f_3 \neq 0$.

Step 4. Assume that f_3 is not a cube. Then $f_3 = lq$ where l is linear and does not divide q. By (4.24.3) $f = LQ$ and we can choose L as our coordinate z. Thus $f = z(ay^2 + \ldots)$ and $a \neq 0$ since l does not divide q. Applying (4.24.1) and then (4.24.2) to y^2 we obtain the form

$$(\text{unit}) \cdot (x^2 + (\text{unit}) \cdot z(y^2 + z^m \cdot (\text{unit}))) \quad \text{for some } m \geq 2.$$

By (4.24.4) the equation becomes

$$x^2 + z(y^2 + z^m) \quad \text{for some } m \geq 2.$$

This gives the D cases.

Step 5. We are left with the cases when f_3 is a cube. (4.24.1) and (4.24.2) give

$$f = y^3 \cdot u + yz^a \cdot u_a + z^b \cdot u_b,$$

where $a \geq 3, b \geq 4, u$ is a unit and u_a, u_b are either units or zero.

Step 6. We claim that the singularity

$$X := (x^2 + y^3 \cdot u + yz^a \cdot u_a + z^b \cdot u_b = 0), \quad (\text{where } u(0) \neq 0)$$

is Du Val iff either $a \leq 3$ and $u_a(0) \neq 0$ or $b \leq 5$ and $u_b(0) \neq 0$.

Proof. Assume that $a \geq 4$ and $b \geq 6$. Let Y be defined as

$$(p^2 + q^3 \cdot u(pr^3, qr^2, r) + qr^{a-4} \cdot u_a(pr^3, qr^2, r) + r^{b-6} \cdot u_b(pr^3, qr^2, r) = 0).$$

Then

$$\pi : (p, q, r) \mapsto (x = p \cdot r^3, y = q \cdot r^2, z = r)$$

gives a birational morphism $\pi : Y \to X$ with irreducible exceptional divisor $(r = 0) \subset Y$. Y is smooth at general points of $(r = 0)$. A local generator of ω_X is given by $(1/x)(dy \wedge dz)$.

$$\pi^* \frac{dy \wedge dz}{x} = \frac{1}{r} \frac{dq \wedge dr}{p},$$

thus it has a pole along $r = 0$ and X is not canonical.

The converse is seen by examining three cases. Two of them can be treated together:

Step 7. $a \geq b - 1$, $b = 4, 5$ and

$$f = y^3 \cdot (\text{unit}) + yz^a \cdot v_1 + z^b \cdot (\text{unit}).$$

If $a \geq b$ then $yz^a \cdot v_1$ can be absorbed into the last term and we are done by (4.24.4). Thus assume $a = b - 1$. Applying (4.24.2) to z^b and moving multiples of $y^i z^{b-i}$ ($i \geq 3$) into $y^3 \cdot (\text{unit})$, we get

$$f = y^3 \cdot (\text{unit}) + y^2 z^{b-2} \cdot v_2 + z^b \cdot (\text{unit}).$$

Using (4.24.2) similarly for y^3 we get

$$f = y^3 \cdot (\text{unit}) + z^b \cdot (\text{unit}),$$

because $2(b - 2) \geq b$. (4.24.4) gives the equations for E_6 or E_8.

Step 8. $b \geq 5$ and

$$f = y^3 \cdot (\text{unit}) + yz^3 \cdot (\text{unit}) + z^b \cdot v_3.$$

Blow up the origin via the substitutions $y = y_1 z_1$ and $z = z_1$. We get

$$\bar{f} = y_1^3 \cdot (\text{unit}) + y_1 z_1 \cdot (\text{unit}) + z_1^{b-3} \cdot v_3.$$

$\text{mult}_0 \bar{f} = 2$ and $\bar{f}_2$ is not a square. By (4.24.3) $\bar{f}$ is reducible and so is f. Since $f_3 = y^3 \cdot (\text{constant})$, one of the factors of f is of the form $y + $ higher terms. Choosing this as our new coordinate y, we transform f to the form

$$f = y(y^2 \cdot (\text{unit}) + yz^2 \cdot v_4 + z^3 \cdot (\text{unit})).$$

Applying (4.24.2) to z^3 we get

$$f = y(y^2 \cdot (\text{unit}) + z^3 \cdot (\text{unit})).$$

Finally (4.24.4) gives $f = y^3 + yz^3$, which is E_7.

Step 9. $\text{mult}_0 f \geq 4$.

We show that this is not a Du Val singularity. As in Step 6, set $Y :=$ $(p^2 + r^{-4} \cdot f(qr, r))$. Then

$$\pi : (p, q, r) \mapsto (x = p \cdot r^2, y = q \cdot r, z = r)$$

maps Y to X. As in Step 6 we obtain that X is not Du Val.

Finally, the last part of (4.22) follows from (4.10) and the already proved results.   □

4.3 Simultaneous Resolution for Du Val Singularities

The aim of this section is to study simultaneous resolution of flat families of Du Val singularities.

Definition 4.26. Let $f : X \to S$ be a morphism of schemes or analytic spaces. A *simultaneous resolution* of f is a commutative diagram

$$\begin{array}{ccc} \bar{X} & \xrightarrow{p} & X \\ \bar{f} \downarrow & & \downarrow f \\ S & = & S \end{array}$$

where p is proper, $\bar{f}$ is smooth and for every $s \in S$ the induced morphism $p_s : \bar{X}_s \to X_s$ is birational.

If S is a point, then a simultaneous resolution of $f : X \to S$ is the same as a resolution of X. In most cases a simultaneous resolution does not exist:

Example 4.27. Let $f : X \to C$ be a flat morphism to a smooth curve such that f is smooth over $C \setminus \{0\}$ for some $0 \in C$. f does not have a simultaneous resolution in any of the following cases:

(1) X_0 is singular and X is smooth.

(2) X_0 is a reduced singular curve.

(3) $\dim X_0 \geq 3$ and X_0 has only isolated hypersurface singularities. (X is factorial by [Gro68].)

The main result of the section is the following.

Theorem 4.28. *[Bri71, Tyu70] Let $f : (x \in X) \to (0 \in S)$ be a flat morphism of pointed analytic space germs such that X_0 is a surface with a Du Val singularity at x. Then there is a finite and surjective morphism $g : S' \to S$ such that $f' : X' := X \times_S S' \to S'$ has a simultaneous resolution*

$$\begin{array}{ccc} \bar{X}' & \overset{p}{\to} & X' \\ \bar{f}' \downarrow & & \downarrow f' \\ S' & = & S'. \end{array}$$

Moreover, p is projective and $\bar{X}'_{s'}$ is the minimal resolution of $X'_{s'}$ for every $s' \in S'$.

There are several proofs of this result. The original approach of [Bri66] gives a construction of simultaneous resolutions using an explicit description of all deformations of Du Val singularities. Du Val singularities can be related to the corresponding semisimple complex Lie groups, and the study of the unipotent elements provides another demonstration of (4.28) [Bri71]. The method of [Art74] proceeds via a general study of the 'stack' of simultaneous resolutions. This approach is the most general, but it is technically rather demanding. Here we present a proof following [Tyu70], which uses an explicit simultaneous resolution. We start with the resolutions and work our way down to the singularities. This method was also used by [Pin80] in certain cases.

Our construction uses the miniversal deformation spaces of Du Val singularities to be discussed in detail in section 4.5.

Definition 4.29. Let $0 \in X_0 = (f(x_1, \ldots, x_m) = 0) \subset \mathbb{C}^m$ be an isolated hypersurface singularity at the origin. By the local Nullstellensatz,

$$\mathbb{C}[[x_1, \ldots, x_m]]/(f, \partial f/\partial x_1, \ldots, \partial f/\partial x_m).$$

is a finite dimensional $\mathbb{C}$-vector space. Its dimension is called the *Tyurina number* of X_0 and it is denoted by $\tau(X_0)$.

Example 4.30. Explicit computation shows that $\tau(A_n) = n$, $\tau(D_n) = n$ and $\tau(E_n) = n$.

The next result shows that in order to prove (4.28), it is sufficient to construct a simultaneous resolution for one sufficiently large deformation.

Proposition 4.31. *Let $0 \in X_0 \subset \mathbb{C}^3$ be a Du Val singularity. Assume that there is a flat morphism of pointed analytic germs $\bar{f} : (x \in \bar{X}) \to (0 \in \bar{S})$ with the following properties:*

(1) $\bar{X}_0 \cong X_0$;
(2) $\dim \bar{S} = \tau(X_0)$;
(3) $\bar{X}_0$ is the only fiber with a singularity isomorphic to X_0;
(4) $\bar{f}$ has a simultaneous resolution.

Then (4.28) holds for X_0.

Proof. Let $u : (x \in \mathbf{X}) \to (0 \in \mathbf{U})$ be the miniversal deformation space of X_0 (4.59). By definition, there is a morphism $u(\bar{f}) : \bar{S} \to \mathbf{U}$. (4.31.3) implies that $u(\bar{f})^{-1}(0) = 0$, thus $u(\bar{f})$ is finite (see, for instance, [GR84, p.63]). $\bar{S}$ and $\mathbf{U}$ have the same dimension and $\mathbf{U}$ is irreducible as shown by the explicit construction of $\mathbf{U}$ in (4.61). Thus $u(\bar{f})$ is surjective.

Let $f : X \to S$ be any flat deformation of X_0. We get a morphism $u(f) : S \to \mathbf{U}$. Let $S' := \bar{S} \times_\mathbf{U} S$ and $f' : X' \to S'$ the induced deformation. $S' \to S$ is surjective and a simultaneous resolution of $f' : X' \to S'$ is obtained by pulling back the simultaneous resolution of $\bar{f} : (x \in \bar{X}) \to (0 \in \bar{S})$.

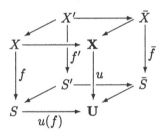

In order to prove (4.28), we thus need to construct an example as in (4.31) for every Du Val singularity. Our constructions give global examples, where all the singularities sit on rational surfaces.

Construction 4.32. Let Z be a smooth surface and $C \subset Z$ a smooth curve. Let $P_1, \ldots, P_n \in C$ be a collection of closed points (repetitions allowed). We define a sequence of surfaces as follows:

(1) Set $Z_0 = Z$ and $r_0 = id$.
(2) Assume that we have already defined birational maps $r_i : Z_i \to Z_0$. $(r_i)_*^{-1}C \cong C$, thus we can identify P_{i+1} with a point $\bar{P}_{i+1}$ of Z_i. Let $Z_{i+1} \to Z_i$ be the blow up of $\bar{P}_{i+1}$ and $r_{i+1} : Z_{i+1} \to Z_i \to Z_0$ the composite.

Remark. If Z is a surface and $P_1, \ldots, P_n \in Z$ a collection of closed points, then in general one cannot define the blow up of Z at these points. For two coincident points, there is no unique choice for the second blow up. In our constructions we could work with the space of all surfaces obtained from Z by successive blow ups. Fixing a curve C as above simplifies most of the technical details.

The following properties of this construction are obvious.

Lemma 4.33. *Notation as above. Then*

(1) $r_n : Z_n \to Z_0$ *does not depend on the order of the points P_i.*

(2) *Every exceptional curve of r_n is either a (-1)-curve or a (-2)-curve.*

If we let the points P_i vary, the resulting surfaces fit together into a family:

Construction 4.34. Let Z be a smooth surface and $C \subset Z$ a smooth curve. For every $n \geq 0$, we define a sequence of varieties and morphisms as follows:

(1) Set $S = C^n$, $Y_0 = Z \times S$ and let $p_0 : Y_0 \to S$ be the projection. Let $D_0 \subset Y_0$ be the divisor $D_0 = C \times S \subset Z \times S$.

(2) Let $s_i : C^n \to Y_0$ be the section $s_i(P_1, \ldots, P_n) = P_i$ and set $B_i = \operatorname{im}(s_i)$. ($B_i$ is the ith diagonal of $C^n \times C \subset C^n \times Z$.) Then $B_i \subset D_0$, the B_i are smooth and $\sum B_i \subset D_0$ is a normal crossing divisor in D_0.

(3) Assume that we have already defined birational maps $r_i : Y_i \to Y_0$. Set $D_i := (r_i)_*^{-1} D_0$ and assume that $r_i : D_i \to D_0$ is an isomorphism. Then B_j can be identified with a subscheme $\bar{B}_j \subset D_i$. Let $Y_{i+1} \to Y_i$ be the blow up of B_{i+1}, $r_{i+1} : Y_{i+1} \to Y_i \to Y_0$ the composite and $p_{i+1} = p_0 \circ r_{i+1}$.

Y_n depends on the choice of the order of the blow ups, but the fibers of p_n do not:

Lemma 4.35. *Notation as above. Then the fiber of $Y_n \to C^n$ over $(P_1, \ldots, P_n)$ is the blow up of Z at the points $P_1, \ldots, P_n \in C$ as in (4.32).* □

For suitable choices of $C \subset Z$, we obtain simultaneous resolutions for various Du Val singularities.

Example 4.36 (A_n-case). Let $0 \in \mathbb{P}^2$ be a point and $Z = B_0\mathbb{P}^2$ with projection $q : B_0\mathbb{P}^2 \to \mathbb{P}^2$. Let $C \subset Z$ be the birational transform of a line through 0 and $P_0 \in C$ the intersection point of C with the exceptional curve of q. Construct $p_n : Y_n \to S$ as in (4.34). Set $V := \mathbb{P}^2$, $v := q \circ r_n : Y_n \to Y_0 \to V$ and $L_r := v^*\mathcal{O}_V(r)(D_n)$.

Example 4.37 (D_n-case). Let $Z = \mathbb{P}^1 \times \mathbb{P}^1$ with second projection $q : Z \to \mathbb{P}^1$. Let $C \subset Z$ be a smooth curve of type $(2,1)$ and $P_0 \in C$ a ramification point of q. Construct $p_n : Y_n \to S$ as in (4.34). Set $V := \mathbb{P}^1$, $v := q \circ r_n : Y_n \to Y_0 \to V$ and $L_r := v^*\mathcal{O}_V(r)(D_n)$.

Example 4.38 (E_n-case). Let $Z = \mathbb{P}^2$ and $C \subset Z$ a smooth cubic and $P_0 \in C$ a flex. Construct $p_n : Y_n \to S$ as in (4.34). Set $V :=$ point, $v := q \circ r_n : Y_n \to Y_0 \to V$ and $L_r := v^*\mathcal{O}_V(r)(D_n)$ (it does not depend on r).

Theorem 4.39. *The above constructions have the following properties:*

(1) *$p_n : Y_n \to S$ is smooth and projective.*

(2) *$R^i(p_n)_*L_r^s = 0$ for $i = 1, 2$, $s \geq 1$ and $r \gg 1$. (In fact, $r \geq 1$ works except in the E_8 case.)*

(3) *L_r^s is p_n-relatively basepoint free for $s \gg 1$ and $r \gg 1$.*

(4) *This gives a morphism $F_r : Y_n \to \bar{Y}_n$ and a commutative diagram*

$$
\begin{array}{ccccc}
Y_n & \overset{F_r}{\to} & \bar{Y}_n & \overset{\bar{v}}{\to} & V \times C^n \\
p_n \downarrow & & \bar{p}_n \downarrow & & \downarrow \\
S & = & S & = & C^n
\end{array}
$$

(5) *$\bar{p}_n : \bar{Y}_n \to S$ is flat, projective, and every fiber is a surface with only Du Val singularities.*

(6) *$p_n : Y_n \to S$ is a simultaneous resolution of $\bar{p}_n : \bar{Y}_n \to C^n$.*

(7) *A_n-case: The only fiber of $\bar{p}_n$ with an A_n-singularity is the one over $(P_0, \ldots, P_0) \in S$.*

(8) *D_n-case ($n \geq 4$): The only fibers of $\bar{p}_n$ with a D_n-singularity are the ones over $(P_0, \ldots, P_0) \in S$, where P_0 is a ramification point of $q : C \to \mathbb{P}^1$.*

(9) *E_n-case ($n = 6, 7, 8$): There is a neighbourhood $(P_0, \ldots, P_0) \in S^0 \subset S$ such that the only fiber of $\bar{p}_n$ over S^0 with an E_n-singularity is the one over $(P_0, \ldots, P_0)$.*

(10) *Set $Z_n(0) := p_n^{-1}(P_0, \ldots, P_0)$ and $\bar{Z}_n(0) := \bar{p}_n^{-1}(P_0, \ldots, P_0)$. Let $\Gamma \subset Z_n(0)$ be the exceptional curve of $F_r : Z_n(0) \to \bar{Z}_n(0)$. Then $\mathrm{Pic}\, Y_n \to \mathrm{Pic}\, \Gamma$ is surjective.*

Remark 4.40. (1) The configuration of the curves on $Z_n(0)$ is shown by the following diagrams.

$$A_n : \diamond - \circ - \cdots - \circ - \bullet - C'$$

$$D_n : \quad \begin{array}{c} \circ - \circ - \cdots - \circ - \circ - \bullet - C' \\ | \\ \diamond \end{array}$$

$$E_n : \quad \begin{array}{c} \circ - \circ - \circ - \cdots - \circ - \circ - \bullet - C' \\ | \\ \diamond \end{array}$$

In the diagrams, $\bullet$ is the (-1)-curve obtained by the last blow up, $C' = (r_n)_*^{-1}C$ and $\diamond$ is the (-2)-curve $(r_n)_*^{-1}G$, where $G \subset Z$ is the exceptional curve in the A_n-case, $\mathbb{P}^1 \times q(P_0)$ in the D_n-case or the line touching C at P_0 in the E_n-case.

(2) It is true that in the E_n-cases the only fibers of $\bar{p}_n$ over S with an E_n-singularity are the fibers over $(P, \ldots, P)$ where P is a flex. This can be proved by a careful case analysis. It is quite interesting to see which are the possible configurations of Du Val singularities that occur on the fibers. For E_6, the answer corresponds to classical results about singular cubic surfaces, cf. [Hen11, Fur86].

Proof. (1) follows from (4.35). By cohomology and base change, in order to see (2), (3) and (4) it is sufficient to show that they hold fiberwise. Let $C \subset Z$ be one of the examples. For any collection $P_1, \ldots, P_n \subset C$ we get $r_n : Z_n \to Z$ and $v : Z_n \to Z \to V$. The restriction of D_n to Z_n is the curve $C_n = (r_n)_*^{-1}(C)$. Write $M_r = v^* \mathcal{O}_V(r)(C_n)$. We check that the conditions of the vanishing and basepoint–free theorems are satisfied.

M_r is effective for $r \geq 0$ and it is nef and big if $M_r \cdot C_n > 0$. In the A, D-cases this holds for $r \gg 1$. For the E-cases $(C_n \cdot C_n) = 9 - n \geq 1$.

In all cases C_n is linearly equivalent to $-K_{Z_n} \otimes v^* \mathcal{O}_V(t)$ for some t, so $-K_{Z_n}$ is v-nef.

$$M_r^s = \mathcal{O}(K_{Z_n}) \otimes M_r^s(-K_{Z_n}),$$

and $M_r^s(-K_{Z_n})$ is nef and big for $s > 0$ and $r \gg 1$. Thus $H^i(Z_n, M_r^s) = 0$ for $i, s > 0$ and $r \gg 1$ by the Vanishing Theorem (2.70) and M_r^s is generated by global sections for $r, s \gg 1$ by the Basepoint-free Theorem (3.3).

The fibers of $\bar{p}_n$ are the images of the surfaces Z_n under the linear system $|M_r^s|$. For $s \gg 1$ we may assume that

$$H^0(\bar{Z}_n, (\bar{M}_r^s)^t) = H^0(Z_n, (M_r^{st})) \quad \text{for all } t \geq 1.$$

Thus the Hilbert polynomials of the fibers of $\bar{p}_n$ agree with the Hilbert polynomial $\chi(Z_n, M_r^{st})$. Hence $\bar{p}_n$ is flat (cf. [Har77, III.9.9]). K_{Y_n} is numerically trivial on the fibers of F_r, hence the fibers of $\bar{p}_n$ have Du Val singularities. This gives (5), which in turn implies (6).

In order to get an A_n-configuration in the exceptional set of $Z_n \to \mathbb{P}^2 = V$, the exceptional set must contain a connected subset on $(n+1)$ curves (there has to be at least one (-1)-curve intersecting). This can happen only if all blown up points lie over $0 \in \mathbb{P}^2$. This shows (7).

Similarly, in the D-case we obtain that all points must lie in the same fiber F of v. If $C \cap F$ consists of two points, we get an A_{n-1}-configuration, which shows (8).

The analysis of the E-case is harder. Let $P_1(t), \ldots, P_n(t) \in C$ be points depending holomorphically on $t \in \mathbb{C}$ such that $P_i(0) = P_0$. Let

$$\mathbb{P}^2 = Z \xleftarrow{r_n(t)} Z_n(t) \to \bar{Z}_n(t)$$

be the corresponding surfaces. Assume that $\bar{Z}_n(t)$ has an E_n-point $Q(t) \in \bar{Z}_n(t)$ for every t. We need to show that $P_i(t) = P_0$ for every i, t.

The Picard number of $Z_n(t)$ is $n+1$ and the Picard number of $\bar{Z}_n$ is at least 1. Thus $Z_n(t) \to \bar{Z}_n(t)$ has at most n exceptional curves (3.40). This means that $Q(t)$ is the unique singular point of $\bar{Z}_n(t)$, thus $T \mapsto Q(t)$ is holomorphic. This shows that the family $\bar{Z}_n(t) : t \in \mathbb{C}$ can be resolved by repeatedly blowing up sections. In particular, the exceptional curves $J_1(t), \ldots, J_n(t)$ of $Z_n(t) \to \bar{Z}_n(t)$ can be so numbered such that each $J_i(t)$ depends holomorphically on t.

Explicit computation shows that if all the points $P_1, \ldots, P_n$ coincide then we get an E_n configuration only if P_1 is a flex. The birational transform of the flex tangent and $n-1$ of the exceptional curves of $r_n(0)$ form the E_n configuration. We can assume that $J_1(0)$ is the birational transform of the flex tangent and $J_2(0), \ldots, J_n(0)$ are exceptional. Thus $r_n(t)(J_1(t))$ is a line in Z and $J_2(t), \ldots, J_n(t)$ are $r_n(t)$-exceptional. So $r_n(t) : Z_n(t) \to Z(t)$ has only one exceptional (-1)-curve and all the points $P_1(t), \ldots, P_n(t)$ coincide. By the above computations each $P_i(t)$ is also a flex, thus $P_i(t) = P_0$ identically.

To see (10) note that $\Gamma \subset Z_n(0)$ is a configuration of type A_n (resp. D_n, E_n) in the cases A_n (resp. D_n, E_n). The Picard group of $Z_n(0)$ is generated by the Picard group of Z and the exceptional curves. Thus

Pic $Y_n \to$ Pic $Z_n(0)$ is surjective. Hence we need to show that Pic $Z_n(0) \to$ Pic Γ is surjective.

Let S be any smooth surface and $\Gamma \subset S$ a configuration of curves of type A_n (resp. D_n, E_n). Assume that there are curves C (resp. C, C') in S such that C intersects Γ in a single point on one of the end curves in case A (resp. C, C' intersect Γ in a single point on different end curves in cases D, E). It is easy to see that Pic $S \to$ Pic Γ is surjective. (In some cases C' is not needed, and for E_8 we do not even need C.)

For us the (-1)-curve of $Z_n(0) \to Z$ serves as C. For C' we can choose the fiber of q through P_0 in case D and a line in $\mathbb{P}^2$ in case E. This proves (10). $\qquad\square$

4.41. Proof of (4.28). It is sufficient to check the conditions of (4.31). We have constructed $\bar{p}_n : \bar{Y}_n \to S$. By (4.39,7-8-9) there is a fiber with a singular point of type A_n, D_n or E_n. Let $x \in \bar{Y}_n$ be the corresponding point. We claim that the germ around x satisfies the conditions of (4.31). (4.31.1) holds by construction, (4.31.2) is implied by (4.30), (4.31.3) follows from (4.39.7-8-9) and (4.31.4) from (4.39.6). $\qquad\square$

Remark 4.42. The above explicit description of the deformations of Du Val singularities gives an easy way to describe the possible singularities in nearby fibers. For instance, it gives straightforward constructions of the adjacencies $D_n \to A_{n-1}$, $E_n \to D_{n-1}$ and $E_n \to A_{n-1}$, which are not obvious from the equations.

The construction of (4.28) can be globalized, though this is not clear from what we have done so far. For completeness sake we mention the result, although we do not need it in the sequel.

Theorem 4.43. *[Bri71, Art74] Let $f : X \to S$ be a flat family of surfaces such that X_s has only Du Val singularities for every $s \in S$. Then there is a finite and surjective morphism $g : S' \to S$ and a simultaneous resolution of $g^*f : X \times_S S' \to S'$:*

$$\begin{array}{ccccc} \bar{X}' & \xrightarrow{p} & X \times_S S' & \xrightarrow{f^*g} & X \\ \bar{f}' \downarrow & & g^*f \downarrow & & \downarrow f \\ S' & = & S' & \xrightarrow{g} & S. \end{array} \quad\square$$

4.4 Elliptic Surface Singularities

The aim of this section is to discuss the structure theory of elliptic surface singularities. These results are used in (5.35) to study canonical threefold singularities. The main theorem (4.57) is due to [Lau77, Rei76].

Notation 4.44. In this section $(0 \in X)$ denotes a germ of a normal surface with a single singular point 0. There is no difference between the algebraic and the complex analytic cases. We usually think of X as a given representative, which is always assumed affine in the algebraic case and Stein in the analytic case.

Let $f : Y \to X$ be the minimal resolution with exceptional curves $E = \sum E_i$. As in (4.1), we can write

$$K_Y \equiv \sum a_i E_i, \quad \text{and } a_i \le 0.$$

Set $Z := \sum \lfloor -a_i \rfloor E_i$ and $\Delta_Y := \sum \{-a_i\} E_i$. Then $K_Y + \Delta_Y \equiv -Z$. X is log terminal iff $Z = 0$. Even if X is not log terminal, it may happen that $\operatorname{Supp} Z \ne E$. If K_X is Cartier and X is not Du Val, then $\operatorname{Supp} Z = E$ by (4.3).

Our first aim is to show that the study of cohomological properties of Y can frequently be reduced to the investigation of Z.

Proposition 4.45. *Notation as above. Let L be an f-nef line bundle on Y. Then:*

(1) $H^0(Y, L) \twoheadrightarrow H^0(Z, L|_Z)$ *is surjective.*
(2) $H^1(Y, L) \cong H^1(Z, L|_Z)$.
(3) $L \cong \mathcal{O}_Y$ *iff $L \equiv_f 0$ and $L|_Z \cong \mathcal{O}_Z$.*
(4) $f_* \omega_Y(Z) = \omega_X$.
(5) $\omega_X / f_* \omega_Y \cong H^0(Z, \omega_Z)$.
(6) $H^1(Y, \mathcal{O}_Y)$ *is dual to $\omega_X / f_* \omega_Y$.*

Proof. Consider the exact sequence

$$0 \to L(-Z) \to L \to L|_Z \to 0,$$

and apply f_*. $L(-Z) \equiv K_Y + \Delta_Y + L$, thus $R^1 f_* L(-Z) = 0$ by (2.68). This gives (1) and (2).

If $Z = 0$ then X is lt and (3) follows from (4.13). Next assume that $L \equiv_f 0$, $Z \ne 0$ and $L|_Z \cong \mathcal{O}_Z$. By (1) the constant 1 section of $L|_Z \cong \mathcal{O}_Z$ lifts to a section σ of L. Since $L \equiv_f 0$, σ generates L near E, thus $L \cong \mathcal{O}_Y$.

Any section of ω_X gives a rational section of $\omega_Y(Z)$, with possible poles along E. Thus (4) follows from (4.46). This in turn implies (5) using the sequence

$$0 \to f_*\omega_Y \to f_*\omega_Y(Z) \to H^0(Z, \omega_Z) \to R^1 f_*\omega_Y = 0.$$

Finally (6) is a consequence of (2) and (5). □

Lemma 4.46. *Notation as above. Let L be a line bundle on Y such that $L \equiv -M - \Delta$ where M is f-nef, $\lfloor \Delta \rfloor = 0$ and Δ is supported on E. Then any section σ of $L|_{Y \setminus E}$ extends to a section of L.*

Proof. σ extends to a rational section, and correspondingly $L \cong \mathcal{O}_Y(B + C)$ where $\operatorname{Supp} B \subset E$ and every irreducible component of C is finite over X. We need to prove that B is effective. $-B - \Delta \equiv M + C$ and $M + C$ is f-nef. Thus by (3.41), $-\Delta \leq B$, hence B is effective. □

Using the cycle Z one can give a numerical characterization of the cases when K_X is Cartier:

Proposition 4.47. *Notation as in (4.44). The following are equivalent:*

(1) K_X *is Cartier.*
(2) $\Delta_Y = 0$ *and if $Z' \subsetneq Z$ is a divisor then $h^1(\mathcal{O}_{Z'}) < h^1(\mathcal{O}_Z)$.*

Proof. If $\Delta_Y = 0$ and $Z = 0$ then X is Du Val, thus we may assume that $Z \neq 0$. If K_X is Cartier then $\Delta_Y = 0$. We have an exact sequence

$$0 \to \omega_Y \to \omega_Y(Z') \to \omega_{Z'} \to 0,$$

which shows that

$$H^0(Y, \omega_Y(Z'))/H^0(Y, \omega_Y) \cong H^0(Z', \omega_{Z'}). \tag{4.1}$$

Thus

$$
\begin{array}{ccccc}
h^1(\mathcal{O}_{Z'}) & < & h^1(\mathcal{O}_Z) & \text{iff} & \text{(by duality)} \\
h^0(\omega_{Z'}) & < & h^0(\omega_Z) & \text{iff} & \text{(by (4.1))} \\
H^0(\omega_Y(Z')) & \subsetneq & H^0(\omega_Y(Z)). & &
\end{array}
$$

The latter holds for every $Z' \subsetneq Z$ iff $\omega_Y(Z)$ is generated by global sections at all generic points of Z. Since $\omega_Y(Z)$ is numerically f-trivial, this holds iff $\omega_Y(Z) \cong \mathcal{O}_Y$.

If K_X is Cartier then $\omega_Y(Z) \cong f^*\mathcal{O}_X(K_X) \cong \mathcal{O}_Y$. Conversely, assume that $\omega_Y(Z) \cong \mathcal{O}_Y$. Then $\omega_X = f_*\omega_Y(Z) \cong f_*\mathcal{O}_Y = \mathcal{O}_X$, thus K_X is Cartier. □

Definition 4.48. Notation as in (4.44). $(0 \in X)$ is called *elliptic* if K_X is Cartier and $R^1 f_* \mathcal{O}_Y \cong \mathbb{C}$. By (4.45) the latter is equivalent to $f_* \omega_Y = m_{0,X} \omega_X$.

Lemma 4.49. *Assume that $(0 \in X)$ is elliptic and let Z be as in (4.44). Then either Z is an irreducible and reduced curve of arithmetic genus 1, or every irreducible component $E_i \subset E$ is a smooth rational curve with $(E_i \cdot (-Z + E_i)) = -2$.*

Proof: $H^1(Y, \mathcal{O}_Y) \cong H^1(Z, \mathcal{O}_Z)$ by (4.45.2), so we are done if Z is irreducible and reduced. Otherwise $E_i \subsetneq Z$ hence by (4.47), $h^1(\mathcal{O}_{E_i}) < h^1(\mathcal{O}_Z) = 1$. Thus $h^1(\mathcal{O}_{E_i}) = 0$ and $E_i \cong \mathbb{P}^1$. The last statement follows from adjunction since $-Z \equiv K_Y$.   □

So far we have transformed several questions about Y to problems about Z. Next we study sections of line bundles on Y by reducing the problems first to Z and then to a zero dimensional subscheme of Z. The ultimate aim is to understand rings of the form $\oplus_{m \geq 0} H^0(Y, L^m)$.

Lemma 4.50. *Let V be a proper (possibly non-reduced) curve such that $H^1(\mathcal{O}_V) = 0$ and L a nef line bundle on V. Then*

(1) *L is generated by global sections, and*
(2) *$H^1(V, L) = 0$.*

Proof: Let V_i be the irreducible components of $\mathrm{red}(V)$ and set $m_i = \deg(L|_{V_i})$. Pick general points $P_i \in V_i$ and Cartier divisors $D_i \subset V$ such that $D_i \cap V_i = P_i$. Then $L' := \mathcal{O}_V(\sum m_i D_i)$ is a line bundle on V such that L and L' have the same degree on every V_i. Thus $L \cong L'$ by (4.13). This shows that L is generated by global sections, except possibly at the points P_i. We obtain (1) by varying the points P_i.

(2) follows by taking H^1 of the surjection $H^0(V, L) \otimes \mathcal{O}_V \twoheadrightarrow L$.   □

Proposition 4.51. *Assume that $(0 \in X)$ is elliptic and let Z be as in (4.44). Let L be an f-nef line bundle on Y such that $(L \cdot Z) > 0$. Then:*

(1) *$H^1(Y, L) = H^1(Z, L|_Z) = 0$.*
(2) *There exists $s \in H^0(Z, L|_Z)$ such that $(s = 0)$ is a 0-dimensional subscheme, disjoint from the singular locus of $\mathrm{red}(Z)$.*
(3) *Let $C \subsetneq Z$ be an irreducible component of Z such that $(L \cdot C) > 0$ and set $Z' = Z - C$. Then $H^0(Z, L|_Z) \twoheadrightarrow H^0(Z', L|_{Z'})$ is surjective.*

Proof: If Z is irreducible and reduced, then this follows from (4.45.1). Otherwise $H^1(\mathcal{O}_{Z'}) = 0$ by (4.47), hence by (4.50) $L|_{Z'}$ is generated by global sections and $H^1(Z', L|_{Z'}) = 0$. Consider the exact sequence

$$0 \to L(-Z')|_C \to L|_Z \to L|_{Z'} \to 0.$$

By (4.47), $C \cong \mathbb{P}^1$ and $\deg L(-Z')|_C = (L \cdot C) - 2 \geq -1$. Therefore $H^1(C, L(-Z')|_C) = 0$, and from the corresponding cohomology sequence we obtain that $H^1(Z, L|_Z) = 0$ and $H^0(Z, L|_Z) \twoheadrightarrow H^0(Z', L|_{Z'})$ is onto, which proves (1) and (3). A general section of $H^0(Z', L|_{Z'})$ lifts back to $s \in H^0(L|_Z)$ such that $(s = 0) \cap Z'$ is 0-dimensional and disjoint from the singular locus of $\mathrm{red}(Z')$. This proves (2).

If $\mathrm{Supp}\, Z' = \mathrm{Supp}\, Z$ then $L|_Z$ is generated by global sections, but not in general. $\qquad\square$

Notation 4.52. Let $(0 \in X)$ be an elliptic surface singularity and L an f-nef line bundle on Y with a section $s \in H^0(Z, L|_Z)$ such that $\{s = 0\}$ is a 0-dimensional subscheme, disjoint from the singular locus of $\mathrm{red}(Z)$ and $s|_{\mathrm{red}(Z)}$ has no multiple zeros.

Set $V := (s = 0)$. Then $A := \mathcal{O}_V$ is a semilocal ring with radical m and $A = \oplus_{i=1}^r A_i, m = \oplus_{i=1}^r m_i$ where (A_i, m_i) are local Artin $\mathbb{C}$-algebras. Let $\mathrm{socle}(m) := \{a \in m | ma = 0\}$ denote the socle of m. We note that A_i is of the form $\mathbb{C}[t]/(t^a)$ and $\mathrm{socle}((t)) = (t^{a-1})$. Set

$$W_L := \mathrm{im}[H^0(Z, L|_Z) \to A \otimes L] \subset A \otimes L.$$

W_L is a linear subspace of $A \otimes L$, which is not an A-submodule in general.

The study of the images W_L provides a key to understanding global sections of line bundles on Y.

Lemma 4.53. *Notation and assumptions as in (4.52). Then:*

(1) *If $A \neq 0$ then $\mathrm{codim}(W_L, A \otimes L) = 1$.*
(2) *If $r \geq 2$ then the projections $W_L \to A \otimes L \to (A/A_j) \otimes L$ are surjective.*
(3) *If $m \neq 0$ then the projection $W_L \to A \otimes L \to (A/\mathrm{socle}(m)) \otimes L$ is surjective.*
(4) *If $\dim_{\mathbb{C}} A \geq 2$ then W_L generates $A \otimes L$ as an A-module.*

Proof. The section s gives a sequence

$$0 \to \mathcal{O}_Z \to L|_Z \to A \otimes L \to 0,$$

which in turn gives $H^0(Z, L|_Z) \to A \otimes L \to H^1(\mathcal{O}_Z) \to H^1(L|_Z)$ with

$H^1(\mathcal{O}_Z) \cong \mathbb{C}$ and $H^1(L|_Z) = 0$ (4.51.1). This shows (1). For every j we have a sequence

$$0 \to \mathcal{O}_Z(A_j) \to L|_Z \to (A/A_j) \otimes L \to 0.$$

Here $H^1(Z, \mathcal{O}_Z(A_j)) = 0$ by (4.51.1), and this shows (2).

Finally assume that $m \neq 0$, thus socle$(m) \neq 0$. Let $C \subset Z$ be an irreducible curve such that $(s = 0)$ has a non-reduced point on C. Then $(L \cdot C) > 0$ and Z is not reduced along C by (4.51.2). Set $Z' = Z - C$. Let s' be the restriction of s to $L|_{Z'}$ and $D' := (s' = 0) \subset Z'$. We get an exact sequence

$$0 \to \mathcal{O}_{Z'} \xrightarrow{s'} L|_{Z'} \to L|_{D'} \to 0,$$

which gives a surjection $H^0(Z', L|_{Z'}) \twoheadrightarrow H^0(D', L|_{D'})$ by (4.47). Composed with the surjection of (4.51.3), $H^0(Z, L) \twoheadrightarrow H^0(D', L|_{D'})$ is a surjection.

$(A/\text{socle}(m)) \otimes L$ is a quotient of $H^0(D', L|_{D'})$, and this proves (3).

If (4) fails then all elements of W_L vanish at a point of Spec A, but (2) and (3) show that this cannot happen. $\square$

The following is the main technical result of this section.

Proposition 4.54. *[Lau77, Rei76] Assume that $(0 \in X)$ is elliptic and let Z be as in (4.44). Let L be a nef line bundle on Z and set $k = \deg_Z L$. Then:*

(1) *If $k \geq 2$, then L is generated by its global sections.*
(2) *If $k \geq 3$, then $\oplus_{n\geq0}H^0(Z, L^{\otimes n})$ is generated by its elements of degree 1. More precisely,*

$$\oplus_{n\geq0}H^0(Z, L^{\otimes n}) \cong \mathbb{C}[x_1, \cdots, x_k]/I,$$

where $\deg x_i = 1$ and I is generated by elements of degree 2 and 3.
(3) *If $k = 2$, then $\oplus_{n\geq0}H^0(Z, L^{\otimes n})$ is generated by its elements of degree 1 and 2. More precisely,*

$$\oplus_{n\geq0}H^0(Z, L^{\otimes n}) \cong \mathbb{C}[x, y, z]/(z^2 + q_4(x, y)),$$

where $\deg(x, y, z) = (1, 1, 2)$ and q_4 is homogeneous of degree 4.
(4) *If $k = 1$, then $\oplus_{n\geq0}H^0(Z, L^{\otimes n})$ is generated by its elements of degree 1,2 and 3. More precisely,*

$$\oplus_{n\geq0}H^0(Z, L^{\otimes n}) \cong \mathbb{C}[x, y, z]/(z^2 + y^3 + ayx^4 + bx^6),$$

where $\deg(x, y, z) = (1, 2, 3)$ and $a, b \in \mathbb{C}$.

Proof. Let $R_Z(n) := H^0(Z, L^{\otimes n})$ and $R_Z := \oplus_{n \geq 0} R_Z(n)$ be the corresponding graded $\mathbb{C}$-algebra. Let $s \in H^0(Z, L)$ be a general section as in (4.51). Let $V = (s = 0)$, $A = \mathcal{O}_V$ and $(A, m) = \oplus_{i=1}^r (A_i, m_i)$ be as in (4.52). We have $\dim_{\mathbb{C}} A = k$ by the definition of A. Set $R_V(n) = H^0(V, L^{\otimes n})$ and $R_V = \oplus_{n \geq 0} R_V(n)$. Note that $R_V(n) = AT^n$ and $R_V = A[T]$ where T is any section of $L|_V$ generating $A \otimes L$.

We have exact sequences

$$0 \to R_Z(n-1) \xrightarrow{s} R_Z(n) \to R_V(n) \to H^1(Z, L^{n-1}) \to H^1(Z, L^n).$$

Hence by (4.51.1), R_Z/sR_Z is a graded $\mathbb{C}$-subalgebra of $A[T]$ such that

$$R_Z/sR_Z(n) = \begin{cases} AT^n & \text{for} & n \geq 2 \\ W_L & \text{for} & n = 1 \\ \mathbb{C} & \text{for} & n = 0. \end{cases}$$

Hence if $k = 1$ then $A = \mathbb{C}$ and $R_Z/sR_Z \cong \mathbb{C}[T^2, T^3] \cong \mathbb{C}[y, z]/(y^3 - z^2)$ with $\deg y = 2$, $\deg z = 3$. Thus (4) follows from (4.55) modulo an obvious coordinate change.

Assume that $k \geq 2$. Then V has either at least two points or a non-reduced point. In either case W_L generates $A \otimes L$ by (4.53.4), which proves (1). Thus we can assume our T comes from a global section t of L. Then $t \in W_L$ and $t(R_Z/sR_Z(n)) = (R_Z/sR_Z)(n+1)$ for $n \geq 2$.

Thus if $k = 2$ it is easy to see that $R_Z/(s, t) = \mathbb{C} \oplus \mathbb{C}u$ with $\deg u = 2$. Thus $R_Z/(s, t) \cong \mathbb{C}[z]/(z^2)$ and (3) follows from (4.55).

Assume $k \geq 3$. We derive a contradiction assuming that R_Z is not generated by $R_Z(1)$.

By (4.55), R/sR_Z is not generated by $R_Z/sR_Z(1)$. Furthermore by $t(R_Z/sR_Z(n)) = (R_Z/sR_Z)(n+1)$ for $n \geq 2$, this means that

$$t(R_Z/sR_Z(1)) \subseteq (R_Z/sR_Z(1))^2 \subsetneq R_Z/sR_Z(2).$$

Set $W_L = Bt$. Then $1 \in B$ and $B \subset A$ is a linear subspace of codimension 1. The above condition becomes $Bt^2 \subseteq (Bt)^2 \subsetneq At^2$, and it implies $Bt^2 = B \cdot Bt^2$. Thus we have $B^2 = B$. Hence B is a $\mathbb{C}$-subalgebra of A, and (4.53.2-3) imply the following.

(1) If $r \geq 2$ then the projections $B \to A \to A/A_j$ are surjective.
(2) If $m \neq 0$ then the projection $B \to A \to A/\operatorname{socle}(m)$ is surjective.

We will derive a contradiction out of these. First we claim that B is local. Indeed if otherwise, B has non-zero idempotents e_1, e_2 such that $e_1 e_2 = 0$ and $e_1 + e_2 = 1$. Then $B = Be_1 \oplus Be_2$ and $A = Ae_1 \oplus Ae_2$. By $\operatorname{codim}_A B = 1$, we get $Be_1 = Ae_1$ and $Be_2 \subsetneq Ae_2$ after possibly switching e_1, e_2. This contradicts the property (1) above. Thus B is local as claimed.

Next we note that B and hence A are not reduced because B is local and $\dim_{\mathbb{C}} B = k - 1 \geq 2$. By the property (2) above, we see that $\operatorname{Spec} A \to \operatorname{Spec} B$ is bijective and hence $r = 1$. We have $A \simeq \mathbb{C}[x]/(x^k)$. By the property (2) again and $k - 1 \geq 2$, we see $\xi = x + c x^{k-1} \in B$ for some $c \in \mathbb{C}$. Then $x = \xi - c\xi^{k-1} \in B$ and $A = B$, a contradiction.

Hence R_Z is generated by $R_Z(1)$. We also have $R_Z/(s,t)(n) = 0$ for $n \geq 3$. Thus $R_Z/(s,t)$ is generated by $R_Z/(s,t)(1)$ and its relation ideal contains all the forms of degree ≥ 3. This means that the relation ideal I of R_Z is generated by $I(2)$ and $I(3)$. Thus (2) follows from (4.55). $\square$

Lemma 4.55. *Let $R = \oplus_{i \geq 0} R(i)$ be a finitely generated graded $\mathbb{C}$-algebra with $R(0) = \mathbb{C}$ and $s \in R(1)$ a non-zero divisor. Let $x_1, \cdots, x_n \in R$ be homogeneous elements inducing a minimal set of generators of the graded $\mathbb{C}$-algebra R/sR. Let $\phi : \mathbb{C}[X_1, \cdots, X_n] \to R/sR$ be the $\mathbb{C}$-algebra map $X_i \mapsto x_i \mod sR$. Then*

(1) *$x_1, \cdots, x_n$, and s form a minimal set of generators of R.*
(2) *If $\psi : \mathbb{C}[X_1, \cdots, X_n, Y] \to R$ is the $\mathbb{C}$-algebra map $X_i \mapsto x_i$ and $Y \mapsto s$, then $(\ker \psi)/Y(\ker \psi) \simeq \ker \phi$. In particular a minimal set of homogeneous generators of $\ker \phi$ lift to the one for $\ker \psi$.*

Proof. (1) follows by Nakayama's Lemma in each degree, and (2) follows by the Snake Lemma applied to the commutative diagram.

$$
\begin{array}{ccccccccc}
0 & \to & \mathbb{C}[X,Y] & \overset{Y}{\to} & \mathbb{C}[X,Y] & \to & \mathbb{C}[X] & \to & 0 \\
 & & \downarrow \psi & & \downarrow \psi & & \downarrow \phi & & \\
0 & \to & R & \overset{s}{\to} & R & \to & R/sR & \to & 0
\end{array}
$$

$\square$

Definition 4.56. Fix $w_1, \ldots, w_n \in \mathbb{Z}_{>0}$. Let $R := K[x_1, \ldots, x_n]$. For a monomial in R set $w(\prod x_i^{m_i}) = \sum m_i w_i$. More generally, for $f \in R$, we set $w(f) = \min_{a_M \neq 0} w(M)$ where we write $f = \sum_M a_M M$ as the sum of monomials M. (Note that $w(0) = \infty$.) We obtain ideals $m^w(n) = \{f \in R | w(f) \geq n\}$. The *weighted blow up* of $\mathbb{A}^n$ with weights w_i is defined as $B_0^w \mathbb{A}^n := \operatorname{Proj}_R \oplus_{n \geq 0} m^w(n)$. For any $X \subset \mathbb{A}^n$ this defines $B_0^w X$ as the birational transform of X in $B_0^w \mathbb{A}^n$.

We are now in a position to prove the main result of the section.

Theorem 4.57. *[Lau77, Rei76] Let $(0 \in X)$ be an elliptic surface singularity, $f : Y \to X$ its minimal resolution. $\omega_Y = f^*\omega_X(-Z)$ for some effective cycle Z supported on the exceptional curve. Set $k = -(Z \cdot Z)$.*

(1) *Assume $k \geq 3$. Then $(0 \in X)$ has multiplicity k and embedding dimension k. Choose any embedding $(0 \in X) \hookrightarrow (0 \in \mathbb{A}^k)$. Let x_i be the coordinates on $\mathbb{A}^k$ and w the weight $w(x_1, \cdots, x_k) = (1, \cdots, 1)$.*

(2) *Assume $k = 2$. Then $(0 \in X)$ has multiplicity 2 and embedding dimension 3. After an analytic coordinate change it can be given by an equation*
$$z^2 + q(x, y) = 0 \quad \text{where} \quad \text{mult}_0 \, q = 4.$$
Let w be the weight $w(x, y, z) = (1, 1, 2)$.

(3) *Assume $k = 1$. Then $(0 \in X)$ has multiplicity 2 and embedding dimension 3. After an analytic coordinate change it can be given by an equation*
$$z^2 + y^3 + yq_4(x) + q_6(x) = 0 \quad \text{where} \quad \text{mult}_0 \, q_i \geq i.$$
Let w be the weight $w(x, y, z) = (1, 2, 3)$.

Let $g : \bar{Y} := B_0^w X \to X$ denote the weighted blow up with the weight w. Then $\bar{Y}$ has only Du Val singularities, it is dominated by Y via $h : Y \to \bar{Y}$ and $K_{\bar{Y}} \sim \mathcal{O}_{\bar{Y}}(1) \sim -h_* Z$. If $k = 2$, $B_0^w X$ is also the normalization of the standard blow up $B_0 X$.

Remark 4.58. Although the grading defined by the weights introduced above depends on the choice of the coordinates, the ideals $m^w(n)$ are independent of the choices made, cf. (5.37).

Proof. The proof is in several steps.

Step 1. f^* gives a natural isomorphism $H^0(X, \mathcal{O}_X) \cong H^0(Y, \mathcal{O}_Y)$. Set $L = \mathcal{O}_Y(-Z) \cong \omega_Y$.
$$H^0(Y, L^n) = H^0(Y, \mathcal{O}_Y(-nZ)) =: I_n \subset H^0(Y, \mathcal{O}_Y)$$

gives an ideal for every n. From the sequence

$$
\begin{array}{cccc}
0 \to & H^0(Y, \mathcal{O}_Y(-(n+1)Z)) & \to & H^0(Y, \mathcal{O}_Y(-nZ)) & \to \\
& H^0(Z, L^n|_Z) & \to & H^1(Y, \mathcal{O}_Y(-(n+1)Z)) & = 0,
\end{array}
$$

we conclude the following isomorphism for the corresponding graded ring:

$$\oplus_{n \geq 0} I_n/I_{n+1} \cong \oplus_{n \geq 0} H^0(Z, L^n|_Z).$$

The structure of the latter ring is described in (4.54). We use this to get information about $H^0(X, \mathcal{O}_X)$.

Since $H^0(Z, \mathcal{O}_Z) = \mathbb{C}$, we see that $f^* m_{0,X} = H^0(Y, \mathcal{O}_Y(-Z))$. This implies that $f^*(m_{0,X}^n) \subset I_n$, but we do not have equality in general.

Step 2. Assume that $k \geq 3$. Then by (4.54.2) $\oplus_{n \geq 0} I_n / I_{n+1}$ is generated by I_1 / I_2. Thus $I_1^n = I_n$ and so $f^*(m_{0,X}^n) = I_n$ for every $n \geq 0$.

This shows that the number of generators of $m_{0,X}$ is $h^0(Z, L|_Z) = k$. The multiplicity of X is also k from the equalities

$$\dim_{\mathbb{C}}(m_{0,X}^n / m_{0,X}^{n+1}) = h^0(Z, L^n|_Z) = kn.$$

Step 3. If $k \leq 2$ then we have three generators x, y, z for $\oplus_{n \geq 0} I_n / I_{n+1}$. Let the same letter denote an arbitrary lift of these to elements of $H^0(X, \mathcal{O}_X)$. Then x, y, z generate $m_{0,X} \cong I_1$, thus X has embedding dimension at most 3. The equations also lift back to equations modulo higher order terms. The indicated normal forms can now be achieved using the methods of (4.24).

Step 4. Set $B^*X := \mathrm{Proj}_X \oplus_{n \geq 0} f_* \mathcal{O}_Y(nK_Y)$. $\mathcal{O}_Y(nK_Y)$ is generated by global sections for $n \geq 3$, thus we have a morphism $p : Y \to B^*X$ which is given by the global sections of $\mathcal{O}_Y(nK_Y)$ for $n \gg 1$. $K_{B^*X} = p_* K_Y$ and $K_Y \equiv p^* K_{B^*X}$, which shows that B^*X has Du Val singularities. Also, $K_{B^*X} = \mathcal{O}_{B^*X}(1)$ from the Proj construction.

If $k \geq 3$ then $f_* \mathcal{O}_Y(nK_Y) = m_{0,X}^n$, thus $B^*X = B_0 X$. This completes the proof of (1).

Step 5. In the $k = 1, 2$ cases we proved that $m^w(n) = f_* \mathcal{O}_Y(nK_Y)$. Thus $B^*X = \mathrm{Proj}_X \oplus_{n \geq 0} m^w(n)$.

In the $k = 2$ case it is easy to see that B^*X is also the normalization of the ordinary blow up. $\qquad\square$

4.5 Deformations of Hypersurface Singularities

The aim of this section is to construct miniversal deformation spaces for isolated hypersurface singularities. While this result is used only for Du Val singularities in this book, the proofs remain unchanged in the more general setting. For a more detailed discussion of this and related subjects, see [Art76, Loo84, AGZV85].

Definition 4.59. Let $0 \in X_0$ be a germ of a complex analytic space. A *deformation* of $0 \in X_0$ is a flat morphism of pointed analytic space germs $f : (0 \in X) \to (0 \in S)$ such that $f^{-1}(0) \cong X_0$.

A deformation $u : (0 \in \mathbf{X}) \to (0 \in \mathbf{U})$ of $0 \in X_0$ is called a *versal* deformation of $(0 \in X_0)$ if the following holds.

Let $f : (0 \in X) \to (0 \in S)$ be any deformation of $0 \in X_0$. Then there is a morphism $u(f) : (0 \in S) \to (0 \in \mathbf{U})$ such that $f : (0 \in X) \to (0 \in S)$ is isomorphic to the pull back of u by $u(f)$

$$
\begin{array}{ccccc}
X & \cong & S \times_{\mathbf{U}} \mathbf{X} & \to & \mathbf{X} \\
f \downarrow & & \downarrow u(f)^* u & & \downarrow u \\
S & = & S & \overset{u(f)}{\to} & \mathbf{U}
\end{array}
$$

in such a way that the isomorphism $X \cong S \times_{\mathbf{U}} \mathbf{X}$ is compatible with the identification of X_0.

We say that $u : (0 \in \mathbf{X}) \to (0 \in \mathbf{U})$ is *miniversal* if in addition the tangent map $du(f) : T_0 S \to T_0 \mathbf{U}$ is uniquely determined by f. This implies that there is no analytic curve $0 \in D \subset \mathbf{U}$ such that $\mathbf{X}|_D$ is isomorphic to the product $X_0 \times D$.

Theorem 4.60. *[Gra72] Miniversal deformation spaces exist for any isolated singularity, and they are unique up to local analytic isomorphism.* $\qquad\square$

In general it is very hard to determine miniversal deformation spaces. Fortunately, miniversal deformation spaces are easy to write down explicitly for any isolated hypersurface singularity:

Theorem 4.61. *Let* $0 \in X_0 = (f(x_1, \ldots, x_m) = 0) \subset \mathbb{C}^m$ *be an isolated hypersurface singularity at the origin. Choose convergent power series* $g_1, \ldots, g_n$ $(n = \tau(X_0))$ *such that they give a basis of*

$$
\mathcal{O}_{0,\mathbb{C}^m} / (f, \partial f / \partial x_1, \ldots, \partial f / \partial x_m).
$$

Then

$$
\begin{array}{ccc}
\mathbf{X} := (f(x_1, \ldots, x_m) + \sum t_i g_i(x_1, \ldots, x_m) = 0) & \subset & \mathbb{C}^m \times \mathbb{C}^n \\
\downarrow & & \downarrow \\
\mathbf{U} := \mathbb{C}^n & = & \mathbb{C}^n
\end{array}
$$

is a miniversal deformation of X_0.

As a first step of the proof, we reformulate the statement in terms of equations. Thus (4.61) is equivalent to the following.

Theorem 4.62. *Notation as above. Let* $(0 \in T)$ *be a germ of a complex space and* $F(x, t) \in \mathcal{O}_{0,\mathbb{C}^m \times T}$ *a convergent power series such that*

$F(x, 0) = f(x)$. *Then there exist* $a_j(t) \in (t)\mathcal{O}_{0,T}$ *(j = 1, \cdots, n) and* $b_i(x, t), d(x, t) \in (t)\mathcal{O}_{0,\mathbb{C}^m \times T}$ *(i = 1, \cdots, m) such that*

$$(1 + d(x, t)) \cdot F(x, t) = f(x - b) + \sum_j a_j(t) g_j(x - b),$$

where $x = {}^t(x_1, \cdots, x_m)$ *and* $b = {}^t(b_1, \cdots, b_m)$. *Moreover, the* $a_j(t)$ *are unique modulo* $(t)^2 \mathcal{O}_{0,T}$.

The main part of the proof consists of two lemmas. First we establish (4.62) modulo higher order terms. Then we prove that sufficiently high order terms can always be eliminated.

Lemma 4.63. *Notation as above. Let* $J \subset \mathcal{O}_{0,\mathbb{C}^m \times T}$ *be an ideal such that* $\mathcal{O}_{0,\mathbb{C}^m \times T}/J$ *is finite over* $\mathcal{O}_{0,T}$. *Then there are convergent power series* $a_j(t) \in \mathcal{O}_{0,T}$ *and* $b_i(x, t), d(x, t) \in (t)\mathcal{O}_{0,\mathbb{C}^m \times T}$ *such that*

$$(1 + d(x, t)) \cdot F(x, t) \equiv f(x - b) + \sum_j a_j(t) g_j(x - b) \quad \mathrm{mod}\ J\mathcal{O}_{0,\mathbb{C}^m \times T}.$$

Proof. Let $v_k \in \mathcal{O}_{0,\mathbb{C}^m \times T}$ $(k = 1, \cdots, M)$ be representatives of generators of the $\mathcal{O}_{0,T}$-module $\mathcal{O}_{0,\mathbb{C}^m \times T}/J$ and $v = {}^t(v_1, \cdots, v_M)$. We introduce the vector $\eta = (\eta_1, \cdots, \eta_M)$ and the $m \times M$ matrix $\xi = (\xi_{i,k})$ where the entries $\eta_k, \xi_{i,k}$ are independent variables.

For simplicity of notation, set $f^+ = f(x - \xi \cdot v)$, $f_i^+ = \partial f / \partial x_i (x - \xi \cdot v)$, $g_j^+ = g_j(x - \xi \cdot v)$, $\mathcal{O} = \mathcal{O}_{0,\mathbb{C}^{m+mM+n} \times T}$ and $\mathcal{O}^+ = \mathcal{O}_{0,\mathbb{C}^{mM+n} \times T}$.

First we claim that the natural map

$$f^+(\mathcal{O}/J\mathcal{O}) \oplus \bigoplus_i f_i^+(\mathcal{O}/J\mathcal{O}) \oplus \bigoplus_j g_j^+ \mathcal{O}^+ \twoheadrightarrow \mathcal{O}/J\mathcal{O},$$

is a surjection. Indeed, $\mathcal{O}/J\mathcal{O}$ is finite over $\mathcal{O}^+$ by the assumption on J, hence by the Nakayama Lemma, surjectivity can be checked over the point $0 \in T$, where it is obvious.

Second, note that

$$(1 + \eta \cdot v) \cdot F(x, t) - (1 + \eta \cdot v) f^+ - (f_1^+, \cdots, f_m^+) \cdot \xi \cdot v \in (t)\mathcal{O} + (t, \eta, \xi)^2 \mathcal{O},$$

where (t, η, ξ) (resp. (t)) is the maximal ideal of $\mathcal{O}^+$ (resp. $\mathcal{O}_{0,T}$). Indeed, this follows from the assumption $F(x, 0) = f(x)$ and the Taylor formula

$$f^+ = f(x - \xi \cdot v) = f(x) - (f_1, \cdots, f_m) \cdot \xi \cdot v \quad \mathrm{mod}\ (\xi_{i,k})^2.$$

From these two claims we conclude that there are convergent power series $A_j, U_k, W_{i,k}$ in ξ, η, t such that $U_k \equiv \eta_k$, $W_{i,k} \equiv \xi_{i,k}$ modulo $(t)\mathcal{O} + (t, \eta, \xi)^2\mathcal{O}$, and

$$(1 + \eta \cdot v)F(x,t) \equiv (1 + \textstyle\sum_k U_k v_k)f^+ + \sum_{i,k} f_i^+ W_{i,k} \cdot v_k + \sum_j A_j g_j^+$$
$$\text{mod } J\mathcal{O}_{0,\mathbb{C}^{m+mM+n} \times T}.$$

The power series U_k and $W_{i,k}$ have independent linear terms in the $\xi_{i,k}, \eta_k$. Therefore, by the Implicit Function Theorem, the system of equations $U_k = 0, W_{i,k} = 0$ ($\forall i, k$) has a unique solution $\eta_k(t), \xi_{i,k}(t) \in (t)\mathcal{O}_{0,T}$. Set $a_j(t) = A_j(t, \eta(t), \xi(t))$ and $b = \xi(t) \cdot v$. $\square$

The second result is essentially Tougeron's Lemma as presented in [Art69a].

Lemma 4.64. *Notation as above. Let* $G(x,t) \in \mathcal{O}_{0,\mathbb{C}^m \times T}$ *be such that*

$$F(x,t) \equiv G(x,t) \quad \text{mod } (x)I^2\mathcal{O}_{0,\mathbb{C}^m \times T},$$

where $I = (F, \partial F/\partial x_1, \cdots, \partial F/\partial x_m)$. *Then there are convergent power series* $b_i(x,t), d(x,t) \in (t)\mathcal{O}_{0,\mathbb{C}^m \times T}$ *such that*

$$(1 + d(x,t)) \cdot F(x - b, t) = G(x,t).$$

Proof. Let us denote $\partial F/\partial x_i$ by F_i for simplicity. Choose $c \in (x)I$ and $c_{i,i'} \in (x)\mathcal{O}_{0,\mathbb{C}^m \times T}$ such that $F(x,t) - G(x,t) = \sum_{i,i'} c_{i,i'} F_i F_{i'} + c \cdot F$. Replacing G by $G/(1-c)$ and $c_{i,i'}$ by $c_{i,i'}/(1-c)$, we may assume that $F(x,t) - G(x,t) = \sum_{i,i'} c_{i,i'} F_i F_{i'}$. Let $\nabla F := {}^t(F_1, \cdots, F_m)$ denote the gradient of F and let $\xi := (\xi_{i,i'})$ be an $m \times m$-matrix with independent variables as entries. By Taylor's formula

$$F(x + \xi \cdot \nabla F, t) \equiv F(x,t) + \sum_{i,i'} \xi_{i,i'} F_i F_{i'} \quad \text{mod } (\xi_{i,i'})^2 (F_i)^2.$$

Thus there are convergent power series $W_{i,i'}$ in x, ξ such that

$$W_{i,i'} \equiv c_{i,i'} + \xi_{i,i'} \quad \text{mod } (\xi_{i,i'})^2 \mathcal{O}_{0,\mathbb{C}^{m+m^2} \times T}, \quad \text{and}$$
$$F(x + \xi \cdot \nabla F, t) = G(x,t) + \sum_{i,i'} W_{i,i'} F_i F_{i'}.$$

The functions $W_{i,i'}$ have independent linear terms in the variables $\xi_{i,i'}$. By the Implicit Function Theorem, the equations $W_{i,i'} = 0$ ($\forall i, i'$) have a unique solution $\xi_{i,i'}(x,t) \in (x)\mathcal{O}_{0,\mathbb{C}^m \times T}$. $\square$

4.65 (Proof of (4.62)).

Set $I = (F, \partial F/\partial x_1, \cdots, \partial F/\partial x_m)$. The $\mathbb{C}$-vector space

$$\mathcal{O}_{0,\mathbb{C}^m \times T}/I + (t) \simeq \mathbb{C}\{x\}/(f, \partial f/\partial x_1, \cdots, \partial f/\partial x_m)$$

is finite dimensional, hence $\mathcal{O}_{0,\mathbb{C}^m \times T}/I$ is finite over $\mathcal{O}_{0,T}$. Note that $\mathcal{O}_{0,\mathbb{C}^m \times T}/(x)I^2$ has a filtration of submodules $I/(x)I^2$, $I^2/(x)I^2$ whose successive quotients are finite $\mathcal{O}_{0,\mathbb{C}^m \times T}/I$-modules. Thus $\mathcal{O}_{0,\mathbb{C}^m \times T}/(x)I^2$ is also finite over $\mathcal{O}_{0,T}$. We first apply (4.63) for $J = (x)I^2$ and then (4.64) with $G(x,t) = f(x-b) + \sum_j a_j(t)g_j(x-b)$.

The uniqueness of $a_j(t) \mod (t)^2\mathcal{O}_{0,T}$ follows from the formula

$$(1 + u(x,t))F(x-b,t) \equiv F(x,t) \mod ((t)^2 + (t)I)\mathcal{O}_{0,\mathbb{C}^m \times T}. \quad \square$$

4.66 (Algebraic versions). † Although in this section we worked in the analytic setup, the arguments above work in other cases as well. First of all, there are no changes needed to use formal power series everywhere.

For us it is more important to understand how to work algebraically. If the various data (like $f(x)$ and $F(x,t)$) are regular functions on germs of algebraic varieties, then the $W_{i,i'}$ in (4.64) can be chosen to be regular functions as well (on a suitably small representative of the germ). The only difficulty is that the implicit function theorem fails for regular functions. That is, usually the solutions $\xi_{i,i'}(x,t)$ are not regular functions on any representative of the germ. (The $\xi_{i,i'}(x,t)$ are algebraic functions on $\mathbb{C}^m \times T$, according to the classical terminology, see [PS97, Sie69].)

From the point of view of algebraic spaces (cf. [Art69b, Art70]) the key point is that the equations $W_{i,i'} = 0$ define an algebraic variety T' whose projection to $\mathbb{C}^m \times T$ is étale at the origin and $\xi_{i,i'}$ are regular functions on T'.

Similarly, if f is in the local ring $\mathcal{O}_{0,\mathbb{C}^m}$ (of regular functions) then we need to pass to an étale base change $(0 \in T') \to (0 \in T)$ for (4.63). In (4.64), two hypersurfaces become algebraically isomorphic when they are pulled back to a space étale over $\mathbb{C}^m \times T'$.

Below we state an algebraic analogue (4.69) of (4.61) and show how the arguments above are actually modified. Except for (4.68), all the results below hold over an algebraically closed field of any characteristic. For a more detailed account of the algebraic theory of deformations, we refer the reader to [Art76].

Definition 4.67. Let $f : (\xi \in X) \to (\eta \in Y)$ be a morphism of germs of algebraic schemes over $\mathbb{C}$. We say that f is *étale* at ξ if we can express $(\xi \in X)$ as $\mathcal{O}_{\xi,X} = (\mathcal{O}_{\eta,Y}[x])_\xi/(g_1, \cdots, g_n)$, where $x = {}^t(x_1, \cdots, x_n)$, $\xi = (0, \eta) \in \mathbb{C}^n \times Y$ and the g_i's satisfy the conditions that $g_i(\xi) = 0$ and that the Jacobian matrix $\partial g/\partial x(\xi) := (\partial g_i/\partial x_j(\xi))$ is invertible.

---

† If you are not interested in restricting the minimal model program to projective varieties only, the rest of the section can be skipped.

If f is étale, the implicit function theorem says that $\mathcal{O}_{\xi,X}^{an} \simeq \mathcal{O}_{\eta,Y}^{an}$. In particular, the embedding dimension $\dim_{\mathbb{C}} m_{\xi,X}/m_{\xi,X}^2$ of $(\xi \in X)$ and the dimension $\dim_\xi X$ at ξ are equal to those for $(\eta \in Y)$, respectively.

Given two germs $(\xi \in X)$ and $(\eta \in Y)$, we say that they are *étale equivalent* and write $(\xi \in X) \cong_{et} (\eta \in Y)$ if there exist a germ $(\zeta \in Z)$ and two étale morphisms $(\zeta \in Z) \to (\xi \in X)$ and $(\zeta \in Z) \to (\eta \in Y)$.

Corollary 4.68 (Corollary to (4.64)). *Given two germs of polynomial hypersurfaces $\subset (0 \in \mathbb{C}^n)$ with isolated singularity at 0, they are étale equivalent iff they are biholomorphic to each other.* □

Proposition 4.69. *Notation as in (4.61). Assume that $f, g_1, \cdots, g_n$ are regular functions. Let $f : (0 \in X) \to (0 \in T)$ be a flat morphism of pointed algebraic schemes with identification $(0 \in f^{-1}(0)) \cong_{et} (0 \in X_0)$ (cf. (4.68)). Then there is an étale morphism of schemes $(0' \in T') \to (0 \in T)$ with the induced deformation $f' : (0' \in X' = X \times_T T') \to (0' \in T')$, such that $u(f') : T' \to \mathbf{U}$ (as in (4.59)) can be chosen to be a morphism of schemes.*

Proof. We show how the proof of (4.62) is modified in our case.

We may assume that $0 \in X$ is singular. Let $(\zeta \in Z)$ be a common étale cover of $(0 \in f^{-1}(0))$ and $(0 \in X_0)$. By (4.71), we have $(\zeta \in Z) \subset (0 \in \mathbb{C}^m)$ and can extend the morphism $(\zeta \in Z) \to (0 \in f^{-1}(0))$ to a morphism $(0 \in \mathbb{C}^m) \to (0 \in \mathbb{C}^m)$, which is necessarily étale (and do the same with X_0). Thus we can pull back everything to $(0 \in \mathbb{C}^m \times T) \supset (\zeta \in Z) \times \{t\}$ and assume that $(0 \in f^{-1}(0)) = (0 \in X_0)$.

Let $F(x,t) = 0$ be the equation of $X \subset \mathbb{C}^m \times T$ near 0. Set $I = (F, \partial F/\partial x_1, \cdots, \partial F/\partial x_m)$. Since $\operatorname{Spec} \mathcal{O}_{0,\mathbb{C}^m \times T}/(I + (t))$ is finite, the scheme $\operatorname{Spec} \mathcal{O}_{0,\mathbb{C}^m \times T}/I$ is quasi-finite over $0 \in T$. By (4.72) there is a pointed scheme $0' \in T'$, étale over $0 \in T$, such that the induced map $\operatorname{Spec} \mathcal{O}_{0',\mathbb{C}^m \times T'}/I\mathcal{O}_{0',\mathbb{C}^m \times T'} \to T'$ is finite. We make this base change and write T instead of T' to conform with the notation of (4.62).

Let $v(x,t), u(x)$ be polynomials such that $v(0,0)u(0) \neq 0$, $F \cdot v$, $f \cdot u, g_1 \cdot u, \cdot, g_n \cdot u$ are polynomials and $F(x,0) \cdot v(x,0) = f \cdot u$. Replacing f with $u \cdot f$ and g_j with $u \cdot g_j$ does not change the ideal $(f, \partial f/\partial x_1, \cdots \partial f/\partial x_m)$ and the g_j's remain a basis of the quotient. Replacing F with $F \cdot v$ does not change the ideal I. Thus we can assume that $F, f, g_1, \cdots, g_n$ are polynomials and $F(x,0) = f(x)$, without loss of generality.

Then we apply the argument of (4.63) for $J = (x)I^2$, and will freely use the notation in the argument. Note that $U_k, W_{i,k}, A_j$ are all polynomials. The closed subscheme $(0' \in T')$ of $\mathbb{C}^{mM+n} \times T$ defined by $U_k = W_{i,k} = 0$

for all i, k is étale at $(0' \in T')$. Then the restrictions of $\eta_k, \xi_{i,k}$ on T' satisfy $U_k = W_{i,k} = 0$. Hence there exist regular functions $a'_j \in \mathcal{O}_{0',T'}$ and $b'_j, d' \in (t)\mathcal{O}_{0',\mathbb{C}^m \times T'}$ such that

$$(1 + d') \cdot F(x, t') \equiv f(x - b') + \sum_j a'_j g_j(x - b') \mod J\mathcal{O}_{0,\mathbb{C}^m \times T'}.$$

By (4.64), $(0' \in F(x, t') = 0) \subset \mathbb{C}^m \times T'$ is biholomorphic to $(0 \in f(x - b') + \sum_j a'_j g'_j(x - b') = 0) \subset \mathbb{C}^m \times T'$ over $(0' \in T')$. Now $(a'_1, \cdots, a'_n) : T' \to U$ is a morphism of schemes. $\qquad\square$

We list below the auxiliary results used above.

Proposition 4.70. *Let $f : (x \in X) \to (t \in T)$ be a morphism of germs of algebraic schemes over $\mathbb{C}$ with $d = \dim_x f^{-1}(t)$. Then f decomposes into an open embedding $g : (x \in X) \to Y$ and a projective morphism $h : Y \to T$ such that $d = \dim h^{-1}(t)$.*

Proof. We have an open embedding of the germ $(x \in X)$ into a closed subscheme $X_1 \subset \mathbb{P}^N \times T$ for some N. Using $d + 1$ general hyperplanes $H_0, \cdots, H_d$, we make a general linear projection $\pi : \mathbb{P}^N \times T \dashrightarrow \mathbb{P}^d \times T$ which is a morphism at x. Blowing up the closed subscheme $Z = \cap_i H_i$, we get an open embedding of the germ $(x \in X)$ into $X_2 = B_Z X_1$ and a projective morphism $\phi : X_2 \to \mathbb{P}^d \times T$. Let $g : X_2 \to Y$ be the Stein factorization [Har77, III.11.5]. Since $Y \to \mathbb{P}^d \times T$ is finite, the induced morphism $Y \to \mathbb{P}^d \times T \to T$ is projective and of fiber dimension d. Note that over a neighbourhood of $g(x)$, g is finite (hence affine) and $g_* \mathcal{O}_{X_2} = \mathcal{O}_Y$. Thus g is an isomorphism near x. $\qquad\square$

Corollary 4.71. *Notation as above. Let m be the embedding dimension of $f^{-1}(t)$ at x. Then there exists an embedding $(x \in X) \subset ((0, t) \in \mathbb{C}^n \times T)$, where $n = \max\{m, d + 1\}$.*

Proof. By (4.70), take an embedding $X \subset \mathbb{P}^N \times T$ for some N such that the closure $\bar{X}$ has fiber dimension d over t. Then a general linear projection $g : \bar{X} \to \mathbb{P}^d \times T$ is a finite morphism. Let x_i be the points of $g^{-1}(g(x))$ other than x. We can choose a general linear projection $h : \bar{X} \to \mathbb{P}^n \times T$ such that g factors through h, $h(x_i) \neq h(x)$ for all i by $n > d$ and the relative tangent map $dh(x)$ at x is injective by $n \geq m$. These imply that h is an embedding near x. $\qquad\square$

Lemma 4.72. *Let $(x \in X) \to (y \in Y)$ be a quasi-finite morphism of germs of pointed algebraic schemes over $\mathbb{C}$. Then there is an étale morphism $(y' \in Y') \to (y \in Y)$ for which $X \times_Y Y'$ has an open and* *closed set $X' \ni (x, y')$ such that X' is finite over Y' and (x, y') is the only point above y'.*

Proof. By (4.70) with $d = 0$, we can decompose $(x \in X) \to (y \in Y)$ into an open embedding $(x \in X) \subset Z$ and a finite morphism $f : Z \to Y$.

Let $v_1, \cdots, v_n \in \mathcal{O}_Z$ be generators of $\mathcal{O}_Z$ as an $\mathcal{O}_Y$-module. Set $\Sigma = f^{-1}(y)$. Since $\mathcal{O}_Z/m_{y,Y}\mathcal{O}_Z = \oplus_{z \in \Sigma} \mathcal{O}_{z,Z}/m_{y,Y}\mathcal{O}_{z,Z}$, we can choose $u \in \mathcal{O}_Z$ to be such that $u \in m_{y,Y}\mathcal{O}_{x,Z}$ and $u - 1 \in m_{y,Y}\mathcal{O}_{z,Z}$ for $z \in \Sigma \setminus \{x\}$. Hence $u^2 - u \in m_{y,Y}\mathcal{O}_Z$. Let $a_i \in \mathcal{O}_{y,Y}$ be such that $u = \sum a_i v_i$. Let $\xi = (\xi_1, \cdots, \xi_n)$ be variables and consider $\phi(\xi) = \sum_i(\xi_i + a_i)v_i \in \mathcal{O}_Z[\xi]$. Since $\phi(\xi) = \sum_i \xi_i v_i + u$, we have

$$\phi^2 - \phi \equiv (\sum_i \xi_i v_i)(2u - 1) \mod m_{y,Y}\mathcal{O}_Z + (\xi)^2\mathcal{O}_Z.$$

By $(2u-1)^2 \equiv 1 \mod m_{y,Y}\mathcal{O}_Z$, $2u-1$ is a unit and we have $A_i \in \mathcal{O}_{y,Y}[\xi]$ such that $\phi^2 - \phi = (2u - 1)\sum_i A_i v_i$ and $A_i - \xi_i \in m_{y,Y}\mathcal{O}_{y,Y}[\xi] + (\xi)^2$. (The choice of $\{A_i\}$ is not unique, but we just choose one.)

Then $Y' = \operatorname{Spec} \mathcal{O}_{y,Y}[\xi]/(A_1, \cdots, A_n) \to Y$ is étale at $y' = (y, 0)$. If we denote the class of ξ_i by ξ_i', then in $\mathcal{O}_Z \otimes_{\mathcal{O}_{y,Y}} \mathcal{O}_{y',Y'}$, $\phi' = \sum_i \xi_i' v_i + u$ is an idempotent and $X' = (\phi' = 0)$ contains only (x, y') above y'. So $X' \subset X \times_Y Y'$. $\qquad\square$

5

Singularities of the Minimal Model Program

The aim of this chapter is to study the higher dimensional singularities that occur in the minimal model program.

In the first section we study properties of Cohen–Macaulay and of rational singularities. These results belong more to general algebraic geometry than to the minimal model program. We develop their theory only to the extent necessary for our purposes.

In section 2 we investigate local properties of dlt pairs. One of the most important results is that they are rational (5.22). An important consequence of this is that we can frequently work with $\mathbb{Q}$-Cartier Weil divisors as if they were Cartier (5.26, 5.27).

Section 3 is devoted to a detailed study of 3-dimensional terminal and canonical singularities. First we relate them to Du Val and elliptic surface singularities (5.34,5.35). Then we apply the results of Chapter 4 to their study. The complete classification of 3-dimensional terminal singularities is stated in (5.43), though we do not need this result in the sequel.

In section 4 we gather various results concerning adjunction and inversion of adjunction. Inversion of adjunction is a process that frequently allows one to reduce a problem concerning dlt pairs to a lower dimensional question. The main result is (5.50). Similar techniques can also be used to study the reduced part of the boundary of dlt pairs. Such results become important in Chapter 7.

Section 5 gives a quick introduction to the duality theory of CM sheaves.

Starting with this chapter, we encounter several theorems which are essentially local in nature, but their proofs are much easier for projective varieties. One of the best examples is (5.22). The difficulties can usually be traced back to duality theory, where duality on a projective variety is much easier to formulate and prove than relative duality for a proper morphism.

In such cases we state the general version of the theorem, prove the projective version and then give references for the proofs of the general algebraic and complex analytic cases.

5.1 Rational Singularities

In this section we recall the basic properties of CM sheaves and of rational singularities.

The results of this section are formulated for algebraic varieties, but they all hold in the complex analytic setting as well.

Definition 5.1. Let (R, m) be a Noetherian local ring and N a finite R-module. As usual, we set $\dim N := \dim \operatorname{Supp} N$. N is called CM (which is short for Cohen–Macaulay) if one of the following equivalent conditions holds.

(1) There is an N-regular sequence $x_1, \ldots, x_r \in m$ of length $r = \dim N$. That is, x_i is not a zero divisor on $N/(x_1, \ldots, x_{i-1})N$ for all i.
(2) If $x_1, \cdots, x_r \in m$ ($r = \dim N$) and $\dim N/(x_1, \cdots, x_r)N = 0$, then $x_1, \cdots, x_r$ is an N-regular sequence.

A coherent sheaf F on a scheme X is called CM if for every $x \in X$ the stalk F_x is CM over $\mathcal{O}_{x,X}$.

A scheme X is called CM if its structure sheaf $\mathcal{O}_X$ is CM.

The equivalence (1) $\Leftrightarrow$ (2) is easy to prove, see, for instance [Mat86, Chapter 6].

Definition 5.2. A sheaf F on a scheme X is called S_d if for every $x \in X$ the stalk F_x has a regular sequence of length at least $\min\{d, \dim \mathcal{O}_{x,X}\}$. Thus F is CM iff it is S_d for $d = \dim X$.

We use only conditions S_1 and S_2. A sheaf F is S_1 iff every associated point of F is a generic point of X. On a normal scheme a torsion free sheaf F is S_2 iff it is reflexive.

Proposition 5.3. *Let (R, m) be a local ring and N a finite R-module.*

(1) *Let $x \in m$ be a non-zero divisor on N. Then N is CM (resp. S_d) iff N/xN is CM (resp. S_{d-1}).*

(2) *Assume that R is regular, N is finite and $\dim N = \dim R$. Then N is CM iff it is free.*

Proof. These are all relatively easy. (1) is in [Mat86, 17.3]. (2) is a very special case of [Mat86, 19.1], but it is easier to prove it directly by induction on $\dim R$.   □

Proposition 5.4. *Let $f : X \to Y$ be a finite morphism of schemes and F a coherent sheaf on X such that $\operatorname{Supp} F$ is pure r-dimensional. Then F is CM (resp. S_d) iff f_*F is CM (resp. S_d).*

Proof. We may assume that Y is the spectrum of a local ring (R, m) with closed point Q. Let $\{P_1, \cdots, P_s\} = f^{-1}(Q)$. Then X is the spectrum of a semi-local ring and P_i are all the maximal ideals. Let $y_1, \cdots, y_r \in m$ be such that $\dim F/(y_1, \cdots, y_r)F = 0$. Then F is CM iff F_{P_i} is CM for all i iff $y_1, \cdots, y_r$ is F_{P_i}-regular for all i iff $y_1, \cdots, y_r$ is F-regular iff $y_1, \cdots, y_r$ is f_*F-regular.

A similar argument works for S_d.   □

Corollary 5.5. *Let $f : X \to Y$ be a finite morphism of n-dimensional schemes, and F a coherent sheaf on X. Then F is flat over Y iff f_*F is locally free. If furthermore Y is regular, then f is flat iff X is CM.*

Proof. F is flat over Y iff f_*F is locally free by [Mat69, 3.G]. If Y is regular, it is equivalent to F being CM by (5.3.2).   □

Definition 5.6. Let X be a scheme and A a locally free sheaf of $\mathcal{O}_X$-algebras. Choose an open set $U \subset X$ such that $A|_U \cong \mathcal{O}_U^r$. Left multiplication by $b \in H^0(U, A|_U)$ can be viewed as an element $\phi(b) \in \operatorname{Hom}_U(\mathcal{O}_U^r, \mathcal{O}_U^r)$. The trace (resp. the determinant) of $\phi(b)$ is independent of the isomorphism $A|_U \cong \mathcal{O}_U^r$ and is called the *trace* (resp. *norm*) of b. The trace gives an $\mathcal{O}_X$-homomorphism Trace $: A \to \mathcal{O}_X$, and the norm gives a multiplicative map Nm $: A \to \mathcal{O}_X$.

If $\mathcal{O}_X$ is S_2 and B is a sheaf of $\mathcal{O}_X$-algebras which is locally free outside a codimension at least 2 subset $Z \subset X$, then $\operatorname{Trace}_{X\backslash Z} : B|_{X\backslash Z} \to \mathcal{O}_{X\backslash Z}$ (resp. $\operatorname{Nm}_{X\backslash Z} : B|_{X\backslash Z} \to \mathcal{O}_{X\backslash Z}$) has a unique extension to Trace $: B \to \mathcal{O}_X$ (resp. Nm $: B \to \mathcal{O}_X$), which is still called the trace (resp. norm).

If $g : Y \to X$ is a finite morphism of schemes which is flat outside a codimension at least 2 subset $Z \subset X$ and X is S_2, then setting $B = g_*\mathcal{O}_Y$ we obtain $\operatorname{Trace}_{Y/X} : g_*\mathcal{O}_Y \to \mathcal{O}_X$ and $\operatorname{Nm}_{Y/X} : g_*\mathcal{O}_Y \to \mathcal{O}_X$. Note that the assumptions are satisfied if X and Y are normal and g is surjective.

(Unfortunately, the same notation is used to denote another map in (5.77), which is in some sense the dual of this trace.)

The trace can also be viewed as a bilinear pairing

$$B \times B \to \mathcal{O}_X \quad \text{given by} \quad (b, b') \mapsto \text{Trace}(bb').$$

The pairing is non-degenerate over the open set over which g is étale. This follows for instance from (5.78).

Proposition 5.7. *Let $f : X \to Y$ be a finite and surjective morphism of purely n-dimensional schemes.*

(1) *Assume that X is CM and $\mathcal{O}_Y$ is a direct summand of $f_*\mathcal{O}_X$. Then Y is also CM.*

(2) *If Y is a normal scheme over a field, X has no codimension 1 embedded points and the characteristic does not divide $\deg f$ then $\mathcal{O}_Y$ is a direct summand of $f_*\mathcal{O}_X$.*

Proof. If $\mathcal{O}_Y$ is a direct summand of $f_*\mathcal{O}_X$, then a sequence from $\mathcal{O}_Y$ is $\mathcal{O}_Y$-regular if it is $f_*\mathcal{O}_X$-regular. Thus (1) follows.

If Y is normal, then $\frac{1}{\deg f} \text{Trace}_{X/Y} : f_*\mathcal{O}_X \to \mathcal{O}_Y$ splits the injection $\mathcal{O}_Y \to f_*\mathcal{O}_X$. □

CM singularities are still too general for the purposes of higher dimensional birational geometry. A more special class is given by rational singularities. Essentially, rational singularities are those which do not affect the cohomological properties of the structure sheaf and of the dualizing sheaf.

Definition 5.8. Let Y be a variety over a field of characteristic 0 and $f : X \to Y$ a resolution of singularities. We say that $f : X \to Y$ is a *rational resolution* if

(1) $f_*\mathcal{O}_X = \mathcal{O}_Y$ (equivalently, Y is normal), and
(2) $R^i f_*\mathcal{O}_X = 0$ for $i > 0$.

We say that Y has *rational singularities* if every resolution $f : X \to Y$ is rational.

5.9. The notion of rational singularities is not so well behaved in positive characteristic. To get a good theory, one needs to assume also that $R^i f_*\omega_X = 0$ for $i > 0$. (In characteristic zero this holds by (2.68).) It is, however, not known whether smooth points in positive characteristic satisfy this property.

Theorem 5.10. *Let Y be a variety over a field of characteristic zero. The following are equivalent.*

(1) Y *has a rational resolution.*
(2) *Every resolution of Y is rational.*
(3) Y *is CM and if $f : X \to Y$ is some resolution then $f_*\omega_X = \omega_Y$ (that is, the map $\mathrm{Trace}_{X/Y}$ defined in (5.77) is an isomorphism).*

By (2.48) this immediately implies:

Corollary 5.11. *If X is a variety over $\mathbb{C}$ then X has rational singularities iff X^{an} has rational singularities.* $\qquad\square$

Proof of (5.10). Let $f : X \to Y$ be a resolution and $g : X' \to X$ a resolution of X. Then $g_*\omega_{X'} = \omega_X$ (5.77.3), thus $f_*\omega_X = (f \circ g)_*\omega_{X'}$. Since any two resolutions of Y can be dominated by a third one, this shows that $f_*\omega_X$ is independent of the resolution $f : X \to Y$. Thus condition (3) is independent of any resolution. So the theorem follows once we prove the following:

Lemma 5.12. *Let Y be a variety of dimension n over a field k of characteristic zero and $f : X \to Y$ a resolution. The following are equivalent:*

(1) f *is a rational resolution,*
(2) Y *is CM and $f_*\omega_X = \omega_Y$.*

Proof. We prove only the case when Y is projective. A relatively simple proof of the general case can be found in [Kol97, Sec. 11].

Let D be an ample Cartier divisor on Y. Then $H^i(X, \omega_X(rf^*D)) = 0$ for $i > 0, r > 0$ by (2.64), and so, by Serre duality,

$$H^{n-i}(X, \mathcal{O}_X(-rf^*D)) = 0 \text{ for } i > 0, r > 0. \tag{5.1}$$

We also use the Leray spectral sequence

$$E_2^{i,j} = H^i(Y, R^j f_* \mathcal{O}_X(-rD)) \Rightarrow H^{i+j}(X, \mathcal{O}_X(-rf^*D)). \tag{5.2}$$

We first prove (1) $\Rightarrow$ (2). By assumption $R^j f_* \mathcal{O}_X = 0$ ($j > 0$), thus $H^j(Y, \mathcal{O}_Y(-rD)) \simeq H^j(X, \mathcal{O}_X(-rf^*D))$ by (5.2). This implies that $H^j(Y, \mathcal{O}_Y(-rD)) = 0$ for $j < n, r > 0$ by (5.1). Therefore Y is CM by (5.72) and Serre duality holds for CM sheaves on Y (5.68). The isomorphism above for $j = n$ implies that

$$h^0(Y, \omega_Y(rD)) = h^0(X, \omega_X(rf^*D)) = h^0(Y, f_*\omega_X(rD)) \quad \text{for } r > 0.$$

This implies that $f_*\omega_X = \omega_Y$.

We prove (2) $\Rightarrow$ (1) by induction on n. We claim that $R^i f_* \mathcal{O}_X = 0$ outside a 0-dimensional set for all $i > 0$. To see this, let H be a general hyperplane section of Y and set $H' := f^{-1}H$. Then $f : H' \to H$ is a resolution of the CM scheme H and using (5.73) on X and on Y, we get that

$$f_* \omega_{H'} = f_*(\omega_X(H') \otimes \mathcal{O}_{H'}) = \mathcal{O}_H(H) \otimes f_* \omega_X = \mathcal{O}_H(H) \otimes \omega_Y = \omega_H.$$

By induction, $\mathcal{O}_H \otimes R^i f_* \mathcal{O}_X = R^i f_* \mathcal{O}_{H'} = 0$ for all $i > 0$, which proves the claim.

Therefore $H^p(Y, R^q f_* \mathcal{O}_X(-rD)) = 0$ if $p, q > 0$, or if $p < n$ and $q = 0$ (by the hypothesis via (5.72)). Hence by (5.2),

$$H^0(Y, R^q f_* \mathcal{O}_X \otimes_{\mathcal{O}_Y} \mathcal{O}_Y(-rD)) = 0 \quad \text{for } q < n - 1, \text{ and}$$
$$H^0(Y, R^{n-1} f_* \mathcal{O}_X(-\nu D)) \cong$$
$$\cong \ker[H^n(Y, \mathcal{O}_Y(-rD)) \xrightarrow{\alpha} H^n(X, \mathcal{O}_X(-rf^*D))].$$

$R^q f_* \mathcal{O}_X$ has zero dimensional support for $q > 0$, thus the first isomorphism implies that $R^q f_* \mathcal{O}_X = 0$ for $0 < q < n - 1$. α is dual to the map

$$H^0(Y, \omega_Y(rD)) \to H^0(X, \omega_X(rf^*D)) = H^0(Y, f_* \omega_X(rD)),$$

which is an isomorphism since $f_* \omega_X = \omega_Y$. Thus α is an isomorphism, $H^0(Y, R^{n-1} f_* \mathcal{O}_X(-rD)) = 0$ and so $R^{n-1} f_* \mathcal{O}_X = 0$. $\square$

Proposition 5.13. *Let $f : X \to Y$ be a finite morphism of n-dimensional varieties over a field of characteristic zero. Assume that X has rational singularities and Y is normal. Then Y has rational singularities.*

Proof. We already know from (5.7) that Y is CM. Let $g^Y : Y' \to Y$ be a resolution of singularities. This gives a commutative diagram

$$
\begin{array}{ccc}
X' & \xrightarrow{g^X} & X \\
f' \downarrow & & \downarrow f \\
Y' & \xrightarrow{g^Y} & Y
\end{array}
$$

where f' is finite and X' is normal. We get a diagram

$$
\begin{array}{ccccc}
f_* g_*^X \omega_{X'} & = & f_* \omega_X & \xrightarrow{\text{Trace}_{X/Y}} & \omega_Y \\
\| & & & & \| \\
g_*^Y f'_* \omega_{X'} & \xrightarrow{\text{Trace}_{X'/Y'}} & g_*^Y \omega_{Y'} & \xrightarrow{c} & \omega_Y
\end{array}
$$

$\mathcal{O}_{Y'}$ is a direct summand of $f'_* \mathcal{O}_{X'}$, thus by (5.68) and (5.77) $\text{Trace}_{X/Y}$ is surjective. Hence c is surjective and so Y has rational singularities. $\square$

Quotient singularities are interesting examples of rational singularities:

Definition 5.14. Let $x \in X$ be a germ of a complex analytic space. We say that X has a *quotient singularity* if there is a smooth germ $0 \in Y$ and a finite group G acting on $0 \in Y$ such that $(x \in X) \cong (0 \in Y)/G$.

Let X be an algebraic variety over $\mathbb{C}$. We say that X has quotient singularities if X^{an} has only quotient singularities.

Proposition 5.15. *Let X be an algebraic or analytic variety over $\mathbb{C}$ with quotient singularities only. Then X has rational singularities and X is $\mathbb{Q}$-factorial.*

Proof. Quotient singularities are rational by (5.11) and (5.13).

Let $x \in X$ be a closed point. Let $\mathcal{O}_{x,X}$ denote its local ring, $\mathcal{O}_{x,X}^{an}$ its analytic local ring and $\hat{\mathcal{O}}_{x,X}$ their common completion. If $I \subset \mathcal{O}_{x,X}$ is an ideal then I is principal iff $I\mathcal{O}_{x,X}^{an}$ is principal iff $I\hat{\mathcal{O}}_{x,X}$ is principal by [Mat69, 24.E]. Thus it is sufficient to prove that an analytic quotient singularity is $\mathbb{Q}$-factorial. This follows from the next lemma. □

Lemma 5.16. *Let $f : X \to Y$ be a finite surjective morphism of normal varieties. If X is $\mathbb{Q}$-factorial then so is Y.*

Proof. Let F be any prime divisor on X. We claim that $f(F)$ is $\mathbb{Q}$-Cartier on X. There is a positive integer a such that aF is Cartier. Given any $x \in X$, there is an open set $U \ni x$ such that aF is defined by one equation $\phi = 0$ on $V := f^{-1}(U)$. Then $\mathrm{Nm}_{V/U}(\phi) = 0$ defines $f(F)$ on U as a set by the construction of the norm. Hence the prime divisor $f(F)$ on X is $\mathbb{Q}$-Cartier. □

5.2 Log Terminal Singularities

The aim of this section is to study log terminal singularities in greater detail. We emphasize those results which hold in general.

As in the previous section, we formulate everything for algebraic varieties, but they all hold in the complex analytic setting as well.

We start with two basic methods which allow us to reduce some problems to simpler ones: taking hyperplane sections and cyclic covers.

Lemma 5.17. *Let (X, Δ) be a pair and $|H|$ a free linear system on X, $H_g \in |H|$ a general member. Then*

(1) discrep$(X, \Delta) \le$ discrep$(H_g, \Delta|_{H_g})$, *and*
(2) discrep$(X, \Delta + H_g) = \min\{0, \text{discrep}(X, \Delta)\}$.

Proof. Let $f : X' \to X$ be a log resolution of (X, Δ) and set $H'_g :=$ $f^{-1}(H_g)$. Then f is also a log resolution of $(X, \Delta + H_g)$ and $f : H'_g \to H_g$ is a log resolution of $(H_g, \Delta|_{H_g})$.

(1) follows using the adjunction formula and (2) holds by (2.32). $\square$

This can be used to describe the codimension 2 behaviour of terminal and canonical singularities:

Corollary 5.18. *If X is terminal, then X is smooth in codimension two. (That is, Sing X has codimension at least 3 in X.) If X is canonical, then K_X is Cartier in codimension 2.*

Proof: Use (5.17) $(\dim X - 2)$-times and the description of terminal and canonical surface singularities (4.5). $\square$

Next we discuss the method of index 1 covers, which frequently allows one to reduce questions to the case when K_X is Cartier.

Definition 5.19. Let X be a normal variety and D a $\mathbb{Q}$-Cartier Weil divisor on X. The smallest natural number r such that rD is Cartier is called the *index* of D on X. Thus D has index 1 iff it is Cartier.

The index of K_X is also called the *index* of X.

Assume that $\mathcal{O}_X(rD) \cong \mathcal{O}_X$ (this always holds in a suitable neighbourhood of any point of X), and choose a nowhere zero section $s \in H^0(X, \mathcal{O}_X(rD))$. Let $p : Z \to X$ be the corresponding cyclic cover (2.50). p is étale over $X \setminus \text{Sing } X$, in particular $K_Z = p^*K_X$. p^*D is a Cartier divisor, linearly equivalent to zero by (2.53).

If $D = K_X$ then we call $p : Z \to X$ an *index 1 cover* of X.

The choice of s will not be important for us, and we frequently call Z *the* index 1 cover. If the base field is algebraically closed and has characteristic zero (or at least the characteristic is relatively prime to r), then $p : Z \to X$ does not depend on the choice of s, up to isomorphism. In other cases one has to pay close attention to the role of s. (If $s' \in H^0(X, \mathcal{O}_X(rD))$ is another nowhere zero section then s/s' is a nowhere zero function on X. It is an rth power if the base field is algebraically closed and the characteristic is relatively prime to r, but not in general.)

The use of cyclic covers relies on a general principle comparing discrepancies under finite morphisms. A result of this type first appeared in [Rei80].

Proposition 5.20. *Let* $g : X' \to X$ *be a finite morphism between n-dimensional normal varieties. Let* $\Delta = \sum a_i D_i$ *be a* $\mathbb{Q}$-*divisor on* X *and* $\Delta' := \sum a'_j D'_j$ *a* $\mathbb{Q}$-*divisor on* X' *such that* $K_{X'} + \Delta' = g^*(K_X + \Delta)$. *Then*

    (1) $K_X + \Delta$ *is* $\mathbb{Q}$-*Cartier iff* $K_{X'} + \Delta'$ *is;*
    (2) $(\deg g)(\operatorname{discrep}(X, \Delta) + 1) \geq (\operatorname{discrep}(X', \Delta') + 1)$;
    (3) $\operatorname{discrep}(X', \Delta') \geq \operatorname{discrep}(X, \Delta)$;
    (4) (X, Δ) *is klt (resp. lc) iff* (X', Δ') *is.*

Proof: First we have to clarify how to pull back a Weil divisor by a finite morphism. Let $U \subset X$ be the smooth locus. $X' \setminus g^{-1}(U)$ has codimension at least 2 in X'. Thus we can take any Weil divisor B on X, restrict it to U, pull it back to $g^{-1}(U)$ and then extend uniquely to X'.

Set $U' = g^{-1}(U) \subset X'$. Then $g : U' \to U$ is finite and

$$m(K_{U'} + \Delta'|_{U'}) = g^*(m(K_U + \Delta|_U))$$

for every m. Assume that $m(K_X + \Delta)$ is Cartier. Then $g^*(m(K_X + \Delta))$ is Cartier on X' and agrees with $m(K_{X'} + \Delta')$ outside a codimension 2 set. Thus $g^*(m(K_X + \Delta)) = m(K_{X'} + \Delta')$ and so $m(K_{X'} + \Delta')$ is Cartier. A similar argument shows that if $m(K_{X'} + \Delta')$ is Cartier then $\deg g \cdot m(K_X + \Delta)$ is also Cartier.

In order to see (2) consider the fiber product diagram with exceptional divisors given below:

$$
\begin{array}{ccccc}
e' \in E' & \subset & Y' & \xrightarrow{f'} & X' \\
\downarrow & & \downarrow h & & \downarrow g \\
E & \subset & Y & \xrightarrow{f} & X.
\end{array}
$$

Let $r \leq \deg h = \deg g$ be the ramification index of h along E'. Near e' we compute that

$$
\begin{aligned}
K_{Y'} &= f'^*(K_{X'} + \Delta') + a(E', X', \Delta')E' \\
&= f'^* g^*(K_X + \Delta) + a(E', X', \Delta')E' \\
&= h^* f^*(K_X + \Delta) + a(E', X', \Delta')E', \quad \text{and} \\
K_{Y'} &= h^* K_Y + (r - 1)E' \\
&= h^* f^*(K_X + \Delta) + a(E, X, \Delta)h^* E + (r - 1)E' \\
&= h^* f^*(K_X + \Delta) + (ra(E, X, \Delta) + (r - 1))E'.
\end{aligned}
$$

This shows that $a(E', X', \Delta') + 1 = r(a(E, X, \Delta) + 1)$. This implies (3) if one of the two sides is ≥ -1. In other cases both sides are $-\infty$ by (2.31). Finally (3) implies (4). $\qquad\qquad\qquad\square$

Corollary 5.21. *Let* $x \in X$ *be a germ of a normal singularity.*

(1) $x \in X$ *is klt if and only if it is a cyclic quotient of an index 1 canonical singularity* $0 \in Y$ *by an action which is fixed point free in codimension 1.*

(2) *If* $x \in X$ *is terminal (resp. canonical) then it is a cyclic quotient of an index 1 terminal (resp. canonical) singularity* $0 \in Y$ *via an action which is fixed point free in codimension 2.*

Proof: Assume that $x \in X$ is klt. Then K_X is $\mathbb{Q}$-Cartier; let $p : Y \to X$ be the index 1 cover (5.19). By (5.20.3), discrep$(Y) \geq$ discrep(X). Since K_Y is Cartier, discrep(Y) is an integer. This shows (2) and one direction of (1). The other part of (1) follows from (5.20.4). $\qquad\square$

Remark. The converse of (2) is false. In general it is not easy to understand which quotients of a terminal (resp. canonical) singularity are again terminal (resp. canonical). The case when Y is terminal of dimension 3 is discussed in (5.43).

The most important basic result about local properties of dlt pairs is the following.

Theorem 5.22. *[Elk81, Fuj85, KMM87] Let* (X, Δ) *be a dlt pair,* Δ *effective. Then* X *has rational singularities.*

Proof: We prove the case when X is projective. The proof of the general assertion is more involved. A proof using only duality on CM schemes can be found in [Kol97, Sec. 11].

By (2.43) we may assume that (X, Δ) is klt.

Let $f : Y \to X$ be a log resolution for Δ. Write

$$K_Y \equiv f^*(K_X + \Delta) - A + B,$$

where A, B are effective $\mathbb{Q}$-divisors without common components. Then Supp $B \subset \text{Ex}(f)$, $\lfloor A \rfloor = 0$, and Supp$(A + B)$ is an snc divisor. Consider the $\mathbb{Q}$-divisor

$$\lceil B \rceil \equiv K_Y - f^*(K_X + \Delta) + A + \{-B\}.$$

Note that $f^*(K_X + \Delta)$ is numerically f-trivial, $A + \{-B\}$ is an snc divisor with $\lfloor A + \{-B\} \rfloor = 0$. Therefore $R^j f_* \mathcal{O}_Y(\lceil B \rceil) = 0$ for $j > 0$ by (2.68).

Let L be an ample Cartier divisor on X. We obtain a commutative diagram

$$
\begin{array}{ccc}
H^i(\mathcal{O}_Y(-rf^*L)) & \to & H^i(\mathcal{O}_Y(\lceil B \rceil - rf^*L)) \\
\uparrow & & \uparrow \beta \\
H^i(\mathcal{O}_X(-rL)) & = & H^i(\mathcal{O}_X(-rL)).
\end{array}
$$

We have a Leray spectral sequence

$$
H^i(X, \mathcal{O}_X(-rL) \otimes R^j f_* \mathcal{O}_X(\lceil B \rceil)) \Rightarrow H^{i+j}(Y, \mathcal{O}_Y(\lceil B \rceil - rf^*L)).
$$

This gives that β is an isomorphism since $R^j f_* \mathcal{O}_X(\lceil B \rceil) = 0$ for $j > 0$.

Furthermore, $H^i(\mathcal{O}_Y(-rf^*L)) = 0$ for $i < n$ and $r > 0$ by (2.64). Thus $H^i(\mathcal{O}_X(-rL)) = 0$ for $i < n$ and so X is CM by (5.72).

For $i = n$ we obtain an injection

$$
H^n(\mathcal{O}_X(-rL)) \hookrightarrow H^n(\mathcal{O}_Y(-rf^*L)).
$$

Since X is CM, by duality (5.71) this gives a surjection

$$
H^0(Y, \omega_Y(rf^*L)) = H^0(X, f_*\omega_Y \otimes \mathcal{O}_X(rL)) \twoheadrightarrow H^0(X, \omega_X \otimes \mathcal{O}_X(rL)).
$$

Thus $f_*\omega_Y \to \omega_X$ is surjective and so X has rational singularities by (5.10).   □

Note. There are two quite short published proofs of (5.22). One is in [Fle81, p.36] and the other in [MP97, p.141]. We have been unable to follow these arguments.

Example 5.23. A cone over an Abelian variety of dimension at least 2 is log canonical but not CM.

Corollary 5.24. *Let X be a normal variety such that K_X is Cartier. Then X has rational singularities iff $(X, 0)$ is canonical.*

Proof. If $(X, 0)$ is canonical then X has rational singularities by (5.22).

Conversely, assume that X has rational singularities. Let $f : Y \to X$ be any resolution. We can write $K_Y = f^*K_X + E$ where E is f-exceptional. $f_*\mathcal{O}_Y(K_Y) = \mathcal{O}_X(K_X)$, thus E is effective. This shows that every exceptional divisor over X has non-negative discrepancy, hence $(X, 0)$ is canonical.   □

In several cases we can work with $\mathbb{Q}$-Cartier Weil divisors as if they were Cartier:

Corollary 5.25. *Let (X, Δ) be a dlt pair and D a $\mathbb{Q}$-Cartier Weil divisor on X. Then $\mathcal{O}_X(D)$ is a CM sheaf. If D is effective then $\mathcal{O}_D$ is CM.*

Proof. By (2.43) we may assume that (X, Δ) is klt. Choose $m > 0$ such that mD is Cartier and let L be a Cartier divisor such that $mL - mD$ is very ample. Let $s \in H^0(X, \mathcal{O}(mL - mD))$ be a general section with divisor E and $p : X' \to X$ the corresponding cyclic cover.

$(X, \Delta + (1 - \frac{1}{m})E)$ is klt by (5.17.2), hence $(X', p^*\Delta)$ is klt by (5.20.4). So $\mathcal{O}_{X'}$ is CM by (5.22). $\mathcal{O}_X(D - L)$ is a direct summand of $p_*\mathcal{O}_{X'}$, so it is CM by (5.4). $\mathcal{O}_X(D)$ and $\mathcal{O}_X(D - L)$ are locally isomorphic, hence $\mathcal{O}_X(D)$ is also CM.

Assume that D is effective and let $D' := p^*D$. Then $D' \subset X$ is a Cartier divisor, thus $\mathcal{O}_{D'}$ is CM. Tensoring $\frac{1}{\deg p} \operatorname{Trace}_{X'/X} : p_*\mathcal{O}_{X'} \to \mathcal{O}_X$ with $\mathcal{O}_D$ gives a splitting of $\mathcal{O}_D \to p_*\mathcal{O}_{D'}$, thus $\mathcal{O}_D$ is CM. $\qquad \square$

Proposition 5.26. *Let (X, Δ) be a projective dlt pair and S, D $\mathbb{Q}$-Cartier Weil divisors. Assume that S is effective and D is Cartier in codimension 2 on X. Then the restriction sequence*

$$0 \to \mathcal{O}_X(D - S) \to \mathcal{O}_X(D) \to \mathcal{O}_S(D|_S) \to 0 \quad \text{is exact.}$$

Proof. Take $p : X' \to X$ as in the previous proof and set $S' := p^*S$. $D' := p^*D$ is Cartier, thus we have an exact sequence

$$0 \to \mathcal{O}_{X'}(D' - S') \to \mathcal{O}_{X'}(D') \to \mathcal{O}_{S'}(D'|_{S'}) \to 0.$$

Pushing this forward to X and taking a suitable direct summand we obtain

$$0 \to \mathcal{O}_X(D - S - L) \to \mathcal{O}_X(D - L) \to F \to 0,$$

where F is a direct summand of $p_*\mathcal{O}_{S'}(D'|_{S'})$. We are done if we can identify F and $\mathcal{O}_S(D - L|_S)$.

The problem is that our construction of $X' \to X$ does not identify $p_*\mathcal{O}_{S'}$ with $p_*\mathcal{O}_{X'} \otimes \mathcal{O}_S$; but this holds over the set where p is étale. Let $U \subset X$ be the largest open set where D is Cartier. We see that $F|_{U \setminus E} \cong \mathcal{O}_S(D - L|_S)|_{U \setminus E}$. D' is Cartier thus $p_*\mathcal{O}_{S'}(D'|_{S'})$ is CM by (5.25). $\mathcal{O}_S(D - L|_S)$ is S_2 and $S \cap (X \setminus U)$ has codimension at least 2 in S. Thus $F|_{X \setminus E} \cong \mathcal{O}_S(D - L|_S)|_{X \setminus E}$.

We can move E, so in fact $F \cong \mathcal{O}_S(D - L|_S)$. $\qquad \square$

Corollary 5.27 (Serre duality). *Let (X, Δ) be a projective dlt pair of pure dimension n over a field of characteristic zero and D a $\mathbb{Q}$-Cartier Weil divisor on X. Then*

$$H^i(X, \mathcal{O}_X(D)) \quad \text{is dual to} \quad H^{n-i}(X, \omega_X(-D)).$$

Proof. This follows from (5.25) and (5.71).   □

Finally we mention, without proof, a useful property of canonical singularities.

Theorem 5.28. *[Kol83] Let $f : X \to S$ be a flat morphism whose fibers have canonical singularities. Then*

(1) *The formation of $\omega_{X/S}^{[q]}$ (that is, the double dual of $\omega_{X/S}^{\otimes q}$) commutes with base change.*

(2) *If f is proper, then $s \mapsto \chi(X_s, \omega_{X_s}^{[q]})$ is locally constant on S.*

5.3 Canonical and Terminal Threefold Singularities

The aim of this section is to study terminal and canonical threefold singularities in greater detail. We prove a structure theorem for index 1 terminal and canonical threefold singularities in (5.34) and (5.35). Arbitrary terminal threefold singularities are classified based on this result (5.43). The fine classification, sometimes referred to as the 'terminal lemma', is not used in this book.

Notation 5.29. The arguments in this section work in the algebraic and analytic settings as well. Accordingly, $(0 \in X)$ denotes either a normal affine variety over $\mathbb{C}$, or a normal Stein space, with a marked closed point. 0 need not be an isolated singular point of X.

We think of X as a neighbourhood of 0. In the course of the proofs it is sometimes necessary to replace X with a smaller neighbourhood; we do this without special mention.

The following result, due to [Rei80], is the first step toward the classification.

Lemma 5.30. *Let $(0 \in X)$ be an index 1 canonical threefold singularity and $0 \in H \subset X$ a general hypersurface section. Then $(0 \in H)$ is either a Du Val or an elliptic singularity.*

Proof: X is CM by (5.22), thus H is also CM. Since H is a general hypersurface section, H is smooth in codimension one. Thus H is a normal surface. Let $g : H' \to H$ be any resolution. We need to check that either $g_* \omega_{H'} = \omega_H$ (thus H is Du Val), or $g_* \omega_{H'} = m_{0,H} \omega_H$ (thus H is elliptic).

Let $B_0 X \to X$ be the blow up of the maximal ideal $m_{0,X} \subset \mathcal{O}_X$ and $f : Y \to X$ a resolution dominating $B_0 X$. $f^* m_{0,X} \subset \mathcal{O}_Y$ is an ideal sheaf which defines a Cartier divisor $E \subset Y$. Let $H' = f_*^{-1}H \subset Y$ be the birational transform of H by f. Then $H' + E = f^*H$ and H' is smooth since the linear system $|H'|$ is free. Since $(0 \in X)$ is canonical, $\omega_Y = f^*\omega_X(F)$, where F is effective. Thus, by adjunction,

$$\omega_{H'} = \omega_Y(H')|_{H'} = f^*(\omega_X(H))(F - E)|_{H'} = f^*\omega_H((F - E)|_{H'}).$$

Therefore,

$$\begin{aligned} f_*\omega_{H'} &\supset \omega_H \otimes f_*\mathcal{O}_{H'}(-E|_{H'}) \\ &= \omega_H \otimes (f_*\mathcal{O}_Y(-E)|_H) = m_{0,H} \cdot \omega_H. \end{aligned}$$

This implies that $(0 \in H)$ is either Du Val or elliptic. $\qquad\square$

Remark 5.31. The above proof works in all dimensions and it shows that the general hypersurface section $0 \in H \subset X$ of an index 1 canonical singularity is either rational or elliptic.

Definition 5.32. Let $(0 \in X)$ be a threefold singularity. We say that it is a *compound Du Val* or *cDV* singularity if a general hypersurface section $0 \in H \subset X$ is a Du Val singularity.

Definition 5.33. Let $(0 \in X)$ be a complex analytic singularity. Let $\text{Aut}(0 \in X)$ denote the set of all isomorphisms $\phi : (0 \in U_1) \cong (0 \in U_2)$ where $0 \in U_i \subset X$ are open Euclidean neighbourhoods. Two such isomorphisms ϕ_1, ϕ_2 are identified if they agree on some neighbourhood of 0. We treat $\text{Aut}(0 \in X)$ as a discrete group though it is easy to endow it with a topology.

The two cases in (5.30) do behave very differently, and both classes have several equivalent characterizations:

Theorem 5.34. *Let $(0 \in X)$ be a threefold singularity with a general hypersurface section $0 \in H \subset X$. The following are equivalent:*

(1) $0 \in H$ *is a Du Val singularity.*
(2) (X, H) *is a canonical pair.*
(3) X *is canonical of index 1 and if $f : Y \to X$ is any resolution and $E \subset f^{-1}(0)$ an exceptional divisor then $a(E, X) \geq 1$.*

Theorem 5.35. *Let $0 \in X$ be an index 1 canonical threefold singularity. The following are equivalent:*

(1) *The general hypersurface section $0 \in H \subset X$ is elliptic.*

    (2) *There exists a birational projective morphism $f : Y \to X$ such that $f^{-1}(0)$ is non-empty of pure codimension 1, $f : Y \setminus f^{-1}(0) \to X \setminus \{0\}$ is an isomorphism and $K_Y = f^* K_X$. Moreover, we can choose Y such that the $\mathrm{Aut}(0 \in X)$-action on X lifts to an action on Y.*

    (3) *If $f : Y \to X$ is any resolution of singularities then there is a divisor $E \subset f^{-1}(0)$ such that $a(E, X) = 0$.*

5.36 (Plan of the proofs of (5.34) and (5.35)).
The two theorems are proved together as follows.

    First we prove that $(5.34.1) \Rightarrow (5.34.2) \Rightarrow (5.34.3)$. $(5.35.2) \Rightarrow (5.35.3)$ turns out to be easy, and then we prove $(5.35.1) \Rightarrow (5.35.2)$.

    Using (5.30) we observe that $(5.34.1) \Rightarrow (5.34.3)$ is equivalent to $(5.35.3) \Rightarrow (5.35.1)$ and $(5.34.3) \Rightarrow (5.34.1)$ is equivalent to $(5.35.1) \Rightarrow (5.35.3)$. This completes the proof of both theorems.

    Now to the proofs.

    We start with $(5.34.1) \Rightarrow (5.34.2)$.

    H has embedding dimension ≤ 3, so X has embedding dimension ≤ 4. Thus $0 \in X$ is a hypersurface singularity, in particular it is CM and ω_X is locally free. We have an exact sequence

$$0 \to \omega_X \to \omega_X(H) \to \omega_H \to 0$$

by (5.73).

    Let $f : Y \to X$ be a log resolution of (X, H) and set $H' = f_*^{-1}H$. On Y we have the sequence

$$0 \to \omega_Y \to \omega_Y(H') \to \omega_{H'} \to 0.$$

Push this forward by f_*. Since $R^1 f_* \omega_Y = 0$ (cf. (2.68)), we obtain the following commutative diagram with exact rows:

$$
\begin{array}{ccccccccc}
0 & \to & f_*\omega_Y & \to & f_*\omega_Y(H') & \to & f_*\omega_{H'} & \to & 0 \\
 & & \downarrow \delta & & \downarrow \beta & & \downarrow \gamma & & \\
0 & \to & \omega_X & \to & \omega_X(H) & \xrightarrow{\alpha} & \omega_H & \to & 0.
\end{array}
\tag{5.3}
$$

    Since H is canonical, γ is an isomorphism and by Nakayama's Lemma we conclude that

$$f_*\omega_Y(H') = \omega_X(H).$$

As in (2.26) we can write

$$\omega_Y(H') = f^*\omega_X(H) \otimes \mathcal{O}_Y\big(\textstyle\sum a_i E_i\big),$$

and $\sum a_i E_i$ is effective iff (X, H) is canonical. From this we get that

$$f_* \omega_Y(H') = \omega_X(H) \otimes f_* \mathcal{O}_Y\left(\sum a_i E_i\right).$$

Comparing this with the above equality, we conclude that $\sum a_i E_i$ is effective and (X, H) is canonical.

Assume next that (5.34.2) holds. Using the above notation, let $f^* H = H' + \sum b_i E_i$. By assumption $K_X + H$ is $\mathbb{Q}$-Cartier, hence so is K_X. Thus we can write

$$K_Y \equiv f^* K_X + \sum (a_i + b_i) E_i,$$

which shows that $(X, \emptyset)$ is canonical, and $a(E_i, X) = a_i + b_i \geq b_i$ is positive if $f(E_i) \subset H$. In order to get (5.34.3) we still need to establish that K_X is Cartier.

Restricting $K_Y + H' \equiv f^*(K_X + H) + \sum a_i E_i$ to H' we obtain that $K_{H'} \equiv f^* K_H + \sum a_i(E_i|_{H'})$, which shows that H is Du Val by (4.20). As in the beginning of the proof we see that X is a hypersurface singularity, in particular K_X is Cartier.

Assume (5.35.2). Then there is a divisor E over X such that $a(E, X) = 0$ and $\text{center}_X(E) = \{0\}$. By (2.32.1), such divisors appear on any resolution, which shows (5.35.3).

Next we prove that (5.35.1) $\Rightarrow$ (5.35.2).

Let $0 \in H \subset X$ be a general hypersurface section. By assumption, $(0 \in H)$ is an elliptic singularity with invariant $k(\geq 1)$ (4.57). In (4.57) we identified a specific birational morphism $B_0^w H \to H$ for every elliptic surface singularity. Our aim is to construct a morphism $f : Y \to X$ which is compatible with these $B_0^w H \to H$ for every general $H \subset X$. As in (4.57), we have to consider three separate cases. For now let us assume that we succeeded and we have the following:

Assumption: There is a birational morphism $f : Y \to X$, satisfying the following properties:

(1) f^{-1} is an isomorphism outside $\{0\}$.
(2) Let $0 \in H \subset X$ be a general hypersurface. Write $f^* H = H' + E$ where E is f-exceptional. Then $f : H' \to H$ is isomorphic to the (weighted) blow up specified in (4.57) and $\mathcal{O}_Y(H')$ is f-ample.
(3) The $\text{Aut}(0 \in X)$ action on X lifts to Y.

Let us show how to finish the proof, assuming that $f : Y \to X$ exists.

By (4.57), H' has only Du Val singularities and $K_{H'} = -E|_{H'}$. Thus Y is normal in a neighbourhood of H'. Since H' is f-ample, any neighbourhood of H' contains a subset of the form $E \setminus \{\text{finite set of points}\}$.

We replace Y with its normalization; this does not change the neighbourhood of H'. (A more careful argument shows that Y is in fact normal, but we do not need it.)

$K_Y \equiv f^*K_X + F$ for some effective exceptional divisor F. By adjunction we obtain that

$$-E|_{H'} \ = K_{H'} = (K_Y + H')|_{H'} = (f^*(K_X + H) + F - E)|_{H'}$$
$$= (F - E)|_{H'}.$$

In particular, $F|_{H'} = 0$. H' is f-ample, thus it intersects every exceptional divisor, hence if $F|_{H'} = 0$ then $F = 0$ since F is effective and $F \subset f^{-1}(0)$. This shows (5.35.2).

It remains to establish the existence of $f : Y \to X$ satisfying the above assumptions. Let $0 \in H \subset X$ be a general hypersurface section with invariant $k = k(H)$.

If $k \geq 3$ then $Y = B_0 X$ satisfies the assumptions.

Assume next that $k = 1$ or $k = 2$. In these cases H has embedding dimension 3, hence X has embedding dimension 4. We may replace X by a small analytic neighbourhood of 0 and view X as a hypersurface in $\mathbb{C}^4$ defined by an equation $f(x, y, z, t) = 0$.

We use the equations of H described in (4.57). They show that X has a double point, thus its equation can be written as $x^2 + g(y, z, t) = 0$. If $\mathrm{mult}_0\, g = 2$ then the equation can be further transformed into $x^2 + y^2 + h(z, t)$ and a hyperplane section $t = \lambda z$ gives a Du Val singularity of type A, which is not our case. Thus $\mathrm{mult}_0\, g \geq 3$.

If $\mathrm{mult}_0\, g \geq 4$, then assign weights $\mathrm{wt}(x, y, z, t) = (2, 1, 1, 1)$ and let $f : Y \to X$ be the corresponding weighted blow up.

If $\mathrm{mult}_0\, g = 3$, then, as in (4.25), we can write our equation as $x^2 + y^3 + yq_4(z, t) + q_6(z, t)$. If $\mathrm{mult}_0\, q_4 < 4$ or $\mathrm{mult}_0\, q_6 < 6$ then a hyperplane section $t = \lambda z$ gives a Du Val singularity by (4.25. Step 6). Thus $\mathrm{mult}_0\, q_4 \geq 4$ and $\mathrm{mult}_0\, q_6 \geq 6$. Assign weights $\mathrm{wt}(x, y, z, t) = (3, 2, 1, 1)$ and let $f : Y \to X$ be the corresponding weighted blow up.

In both of these cases, the assumptions (1–2) are easy to check. By definition, $Y = \mathrm{Proj}_X \oplus_n m^w(n)$ (where the ideals $m^w(n)$ are defined in (4.56)), thus (3) is implied by the following:

Claim 5.37. Let $f(x_1, x_2, x_3, x_4) \in \mathbb{C}[[x_1, x_2, x_3, x_4]]$ be a power series. Assume that either

(1) $f = (\text{unit}) \cdot (x_1^2 + g(x_2, x_3, x_4))$, $\mathrm{mult}_0\, g \geq 4$ and we have weights $\mathrm{wt}(x_1, x_2, x_3, x_4) = (2, 1, 1, 1)$; or

(2) $f = (\text{unit}) \cdot (x_1^2 + x_2^3 + x_2 q_4(x_3, x_4) + q_6(x_3, x_4))$ where $\text{mult}_0 \, q_r \geq r$ and we have weights $\text{wt}(x_1, x_2, x_3, x_4) = (3, 2, 1, 1)$.

Let ϕ be an automorphism of $\mathbb{C}[[x_1, x_2, x_3, x_4]]$ such that $f \circ \phi$ again has the above form. Then $\phi(m^w(n)) = m^w(n)$ for every $n \geq 0$.

Proof. ϕ is given by coordinate functions $x_i \mapsto \phi_i(x_1, x_2, x_3, x_4)$. We need to prove that if $\sum a_i w(x_i) \geq n$ then every term in

$$\prod_i \phi_i(x_1, x_2, x_3, x_4)^{a_i}$$

has weight at least n. Equivalently: in $\phi_i(x_1, x_2, x_3, x_4)$ every term has weight at least $w(x_i)$.

In the first case the only non-obvious assertion is that $\phi_1 = (\text{const}) x_1 +$ (degree ≥ 2 terms). Assume the contrary; say x_2 appears in ϕ_1 with non-zero coefficient. Then x_2^2 appears in ϕ_1^2 with non-zero coefficient and it does not appear in $g(\phi_2, \phi_3, \phi_4)$ since $\text{mult } g \geq 4$. Thus x_2^2 appears in $\phi_1^2 + g(\phi_2, \phi_3, \phi_4)$ with non-zero coefficient. This contradicts the assumption that $f \circ \phi$ is of the form $(\text{unit}) \cdot (x_1^2 + g(x_2, x_3, x_4))$.

A similar argument settles the second case; this is left to the reader. □

Several consequences of these results are worth mentioning:

Corollary 5.38. *A 3 dimensional normal singularity $0 \in X$ is terminal of index 1 iff it is an isolated cDV singularity.* □

For higher index terminal singularities, we can immediately combine (5.38) and (5.21) to obtain the following.

Corollary 5.39. *If $0 \in X$ is a terminal threefold singularity, then it is a cyclic quotient of an isolated cDV singularity (or of a smooth point) $0 \in Y$ via an action which is fixed point free outside the origin.* □

Corollary 5.40. *Let X be a canonical 3-fold. Then X has only finitely many non-cDV points.*

Proof. By (5.18) there are only finitely many points where K_X is not Cartier. If K_X is Cartier at x but $x \in X$ is not cDV, then by (5.35) there is an exceptional divisor E over X such that $\text{center}_X E = \{x\}$ and $a(E, X) = 0$. By (2.36) there are only finitely many divisors E over X such that $a(E, X) = 0$. Thus there are only finitely many non-cDV points. □

Corollary 5.41. *Let* $0 \in X'$ *be a 3 dimensional canonical singularity of index* r. *Let* $p : X \to X'$ *be the index 1 cover. Assume that* X *is not a cDV singularity. Then there is an* $m_{0,X'}$-*primary ideal* $I \subset \mathcal{O}_{X'}$ *such that* $Y' := B_I X'$ *is normal,* $g : Y' \to X'$ *is birational with non-empty exceptional divisor* $E \subset Y'$ *and* $K_{Y'} \equiv p^* K_{X'}$.

Proof. Let $f : Y \to X$ be the weighted blow up constructed in (5.35). Let N be a common multiple of the weights and $E \subset Y$ the unique effective f-exceptional Cartier divisor linearly equivalent to $\mathcal{O}_Y(-N)$. Let G denote the Galois group of X/X'. By (5.35), the ideals $m^w(n)$ are G-invariant. Thus we have a G-action on Y and E is G-invariant.

Since we have not proved that Y is normal, let $\bar{f} : \bar{Y} \to Y \to X$ denote its normalization and $\bar{E}$ the pull back of E to $\bar{Y}$. Then $-\bar{E}$ is $\bar{f}$-ample, thus we can choose $s > 0$ such that $J := \bar{f}_* \mathcal{O}_{\bar{Y}}(-s\bar{E})$ generates $\oplus_n \bar{f}_* \mathcal{O}_{\bar{Y}}(-ns\bar{E})$. J is naturally an $m_{0,X}$-primary ideal in $\mathcal{O}_X$ and $Y = B_J X$.

Set $I := J^G \subset \mathcal{O}_{X'}$. Then $B_I X'$ is the quotient of Y by G and I is $m_{0,X'}$-primary. Y' is normal since it is a quotient of the normal variety Y by a group action.

The only remaining question is to show that $K_{Y'} \equiv p^* K_{X'}$. In any case, $K_{Y'} \equiv p^* K_{X'} + \sum a(E'_i, X') E'_i$ where E'_i are the exceptional divisors. Let $E_i \subset Y$ be an exceptional divisor lying over E'_i. We proved in (5.20) that $a(E_i, X) + 1 = m(a(E'_i, X') + 1)$ where m is the ramification index along E_i. $a(E'_i, X') \geq 0$ since X' is canonical and $a(E_i, X) = 0$ by (5.35). This implies that $a(E'_i, X') = 0$. $\square$

As an aside we note that the proof of (5.34) also shows the following theorem of [Elk78].

Theorem 5.42. *Let* $0 \in X$ *be a singularity with a Cartier divisor* $0 \in H \subset X$. *If* $0 \in H$ *is a rational singularity then* $0 \in X$ *is also a rational singularity.*

Sketch of the proof. X is CM since H is CM (5.10). In the diagram (5.3) we established that β and γ are isomorphisms, thus $\delta : f_* \omega_Y \to \omega_X$ is also an isomorphism. X has rational singularities by (5.10). $\square$

In the rest of this section we state the fine classification of 3–dimensional terminal singularities. These results are not used in the sequel.

A classification by explicit equations was given by [Mor85]. The sufficiency of the conditions was checked in [KSB88] and this was simplified in [Ste88]. The connection with covers of Du Val singularities was observed in [Rei87b]. We refer for the details to [Mor85] and [Rei87b].

Theorem 5.43. *Let* $(0 \in X)$ *be a normal isolated threefold singularity. Assume that* K_X *is* $\mathbb{Q}$-*Cartier of index* r *and let* $\pi : (\tilde{0} \in \tilde{X}) \to (0 \in X)$ *be the index 1 cover. The group* μ_r *of* r^{th}-*roots of unity acts on* $\tilde{X}$.

(1) $(0 \in X)$ *is terminal iff a general member* $H \in |-K_X|$ *containing* 0 *is Du Val.*
(2) *The following is a complete list of all possible* $\tilde{H} := \pi^*(H)$, H *and the action of* μ_r *on* $\mathbb{C}^4$.

name	Type of $\tilde{H} \to H$	r	Type of action
cA/r	$A_{k-1} \to A_{kr-1}$	r	$1/r(a, -a, 1, 0; 0)$
$cAx/2$	$A_{2k-1} \to D_{k+2}$	2	$1/2(0, 1, 1, 1; 0)$
$cAx/4$	$A_{2k-2} \to D_{2k+1}$	4	$1/4(1, 1, 3, 2; 2)$
$cD/2$	$D_{k+1} \to D_{2k}$	2	$1/2(1, 0, 1, 1; 0)$
$cD/3$	$D_4 \to E_6$	3	$1/3(0, 2, 1, 1; 0)$
$cE/2$	$E_6 \to E_7$	2	$1/2(1, 0, 1, 1; 0)$

In the list, $1/r(a_1, \cdots, a_4; b)$ means that the generator ξ of μ_r acts on the coordinates $x_1, \cdots, x_4$ and on the equation f of $\tilde{X}$ as $(x_1, \cdots, x_4; f) \mapsto (\xi^{a_1} x_1, \cdots, \xi^{a_4} x_4; \xi^b f)$.

The column $\tilde{H} \to H$ is a complete list of all possible cyclic covers between Du Val singularities, which are unramified outside the singular point.

We note that by (5.28), a flat deformation of a terminal singularity $0 \in X$ of index r can be obtained by taking the quotient of a flat deformation of the index 1 cover $\tilde{X}$ by μ_r. Deformations of hypersurface singularities can be described explicitly (4.61). Looking at the equations given in [Mor85], we can read off the following:

Corollary 5.44. *Let* $f : X \to S$ *be a flat morphism of relative dimension three. Pick closed points* $s_0 \in S$ *and* $0 \in X_{s_0}$. *Assume that* $0 \in X_{s_0}$ *is terminal of index* r. *Then there is a neighbourhood* $0 \in X^0 \subset X$ *such that the following holds for every* $s \in S$ *such that* $X_s^0 \neq \emptyset$.

(1) X_s^0 *has terminal singularities.*
(2) *The index of every singularity of* X_s^0 *divides* r *and* X_s^0 *has at least one terminal singularity of index* r.

5.4 Inversion of Adjunction

Let $(X, S + B)$ be an n-dimensional pair, where S is a Weil divisor and B a $\mathbb{Q}$-divisor. The aim of this section is to prove several results which compare the discrepancies of $(X, S + B)$ with the discrepancies of $(S, B|_S)$. Theorems of this type are crucial in the inductive treatment of many questions. At the end of the section we apply these results to study the case when $(X, S + B)$ is dlt.

As usual, we formulate everything algebraically but all the results hold for complex analytic spaces as well.

We need the following refined version of (2.28).

Definition 5.45. Let (X, Δ) be a pair and $Z \subset S \subset X$ closed subschemes. Define

$$\text{discrep}(\text{center} \subset Z, X, \Delta) :=$$
$$\inf\{a(E, X, \Delta) : E \text{ is exceptional and center}_X\, E \subset Z\}, \quad \text{and}$$
$$\text{discrep}(\text{center} \cap S \subset Z, X, \Delta) :=$$
$$\inf\{a(E, X, \Delta) : E \text{ is exceptional and center}_X\, E \cap S \subset Z\},$$

where E runs through the set of all exceptional divisors over X. Both of these have a totaldiscrep version where we allow E to be a divisor on X.

Proposition 5.46. *[K$^+$92, 17.2] Let X be a normal variety, S a normal Weil divisor which is Cartier in codimension 2, $Z \subset S$ a closed subvariety and $B = \sum b_i B_i$ a $\mathbb{Q}$-divisor. Assume that $K_X + S + B$ is $\mathbb{Q}$-Cartier. Then*

$$\begin{aligned} \text{totaldiscrep}(\text{center} \subset Z, S, B|_S) &\geq \text{discrep}(\text{center} \subset Z, X, S + B) \\ &\geq \text{discrep}(\text{center} \cap S \subset Z, X, S + B). \end{aligned}$$

Proof. Let $f : Y \to X$ be a log resolution of $(X, S + B)$ and set $S' := f_*^{-1}S$. By further blowing up we may assume that $f_*^{-1}(S + B)$ is smooth and if E_i is an exceptional divisor of f which intersects S' then center$_X\, E_i \subset S$.

Write $K_Y + S' \equiv f^*(K_X + S + B) + \sum e_i E_i$. By the usual adjunction formula,

$$K_{S'} = K_Y + S'|_{S'}, \quad \text{and} \quad K_X + S + B|_S = K_S + B|_S.$$

This gives that

$$K_{S'} \equiv f^*(K_S + B|_S) + \sum e_i (E_i \cap S').$$

S' is disjoint from $f_*^{-1}B$, thus if $E_i \cap S' \neq \emptyset$ then E_i is f-exceptional and center$_X$ $E_i \subset S$. This shows that every discrepancy which occurs in $S' \to S$ also occurs among the exceptional divisors of $Y \to X$ whose center on X is in S. (It may happen that E_i is f-exceptional but $E_i \cap S'$ is not $f|_{S'}$-exceptional. This is why we have totaldiscrep on the left hand side.) □

In general there are exceptional divisors E_j of $f : Y \to X$ which do not intersect S', and there is no obvious connection between the discrepancies of such divisors and the discrepancies occurring in $S' \to S$. Despite this, [Sho92, 3.3] and [K$^+$92, 17.3] conjectured that equality holds in (5.46).

The conjecture (or similar results and conjectures) is frequently referred to as *adjunction* if we assume something about X and obtain conclusions about S, or *inversion of adjunction* if we assume something about S and obtain conclusions about X.

Remark 5.47. A recurring assumption in this section is that $S \subset X$ is a Weil divisor which is Cartier in codimension 2. Under this assumption $K_S = (K_X + S)|_S$. If S is not Cartier in codimension 2, then this formula needs a correction term. With this correction term, the assumption about being Cartier in codimension 2 is not necessary. For some applications this is crucial, but we do not need it. See [K$^+$92], especially Chapters 16–17 for details.

For many applications of inversion of adjunction the important case is when one of the two sides is klt or lc. The proof of these cases relies on the following connectedness result which is of interest in itself.

Theorem 5.48. *[K$^+$92, 17.4] Let $g : Y \to X$ be a proper and birational morphism, Y smooth, X normal. Let $D = \sum d_i D_i$ be a snc $\mathbb{Q}$-divisor on Y such that g_*D is effective and $-(K_Y + D)$ is g-nef. Write*

$$A = \sum_{i:d_i<1} d_i D_i, \quad and \quad F = \sum_{i:d_i\geq1} d_i D_i.$$

Then $\operatorname{Supp} F = \operatorname{Supp}\lfloor F \rfloor$ *is connected in a neighbourhood of any fiber of g.*

Proof. By definition

$$\lceil -A \rceil - \lfloor F \rfloor = K_Y - (K_X + D) + \{A\} + \{F\},$$

and therefore by (2.68)

$$R^1 f_* \mathcal{O}_Y(\lceil -A \rceil - \lfloor F \rfloor) = 0.$$

Applying g_* to the exact sequence

$$0 \to \mathcal{O}_Y(\lceil -A \rceil - \lfloor F \rfloor) \to \mathcal{O}_Y(\lceil -A \rceil) \to \mathcal{O}_{\lfloor F \rfloor}(\lceil -A \rceil) \to 0$$

we obtain that

$$g_* \mathcal{O}_Y(\lceil -A \rceil) \to g_* \mathcal{O}_{\lfloor F \rfloor}(\lceil -A \rceil) \quad \text{is surjective.}$$

Let D_i be an irreducible component of A. Then either D_i is g-exceptional or $d_i > 0$ since $g_* D$ is effective. Thus $\lceil -A \rceil$ is g-exceptional, effective and $g_* \mathcal{O}_Y(\lceil -A \rceil) = \mathcal{O}_X$. Assume that $\lfloor F \rfloor$ has at least two connected components $\lfloor F \rfloor = F_1 \cup F_2$ in a neighbourhood of $g^{-1}(x)$ for some $x \in X$. Then

$$g_* \mathcal{O}_{\lfloor F \rfloor}(\lceil -A \rceil)_{(x)} \cong g_* \mathcal{O}_{F_1}(\lceil -A \rceil)_{(x)} + g_* \mathcal{O}_{F_2}(\lceil -A \rceil)_{(x)},$$

and neither of these summands is zero. Thus $g_* \mathcal{O}_{\lfloor F \rfloor}(\lceil -A \rceil)_{(x)}$ cannot be the quotient of $\mathcal{O}_{x,X} \cong g_* \mathcal{O}_Y(\lceil -A \rceil)_{(x)}$. $\square$

Corollary 5.49. *Let $g : Y \to X$ be a proper and birational morphism and $D = \sum d_i D_i$ a $\mathbb{Q}$-divisor on Y such that $g_* D$ is effective and $-(K_Y + D)$ is g-nef.*

Let $Z \subset Y$ be the set of points where (Y, D) is not klt. Then Z is connected in a neighbourhood of any fiber of g.

Proof. Apply (5.48) to a log resolution of (Y, D). $\square$

As a corollary we obtain the following results which were proved by [Sho92] in dimension 3 and by [K⁺92, 17.6–7] in general.

Theorem 5.50 (Inversion of adjunction). *Let X be normal and $S \subset X$ a normal Weil divisor which is Cartier in codimension 2. Let B be an effective $\mathbb{Q}$-divisor and assume that $K_X + S + B$ is $\mathbb{Q}$-Cartier. Then*

(1) *$(X, S + B)$ is plt near S iff $(S, B|_S)$ is klt.*
(2) *Assume in addition that B is $\mathbb{Q}$-Cartier and S is klt. Then $(X, S + B)$ is lc near S iff $(S, B|_S)$ is lc.*

Proof. In both cases the only if part follows from (5.46).

In order to see (1), let $g : Y \to X$ be a log resolution of $(X, S + B)$ and write $K_Y + D \equiv g^*(K_X + S + B)$ as in (2.26).

Let $S' = g_*^{-1} S$ and $F = S' + F'$. By adjunction $K_{S'} = g^*(K_S + B|_S) + (A - F')|_{S'}$. $(X, S + B)$ is plt near S iff $F' \cap g^{-1}(S) = \emptyset$, and $(S, B|_S)$ is klt iff $F' \cap S' = \emptyset$. Assume that $(S, B|_S)$ is klt. By (5.48) every $x \in S$ has an open neighbourhood $U_x \subset X$ such that $(S' \cup F') \cap g^{-1}(U_x)$ is connected, hence $F' \cap g^{-1}(U_x) = \emptyset$. Moving $x \in S$, we obtain (1).

By (2.43), $(X, S + B)$ is lc iff $(X, S + cB)$ is plt for every $c < 1$, and $(S, B|_S)$ is lc iff $(S, cB|_S)$ is klt for $c < 1$. Thus (1) implies (2). ☐

The above ideas are very useful in understanding the reduced boundary of dlt pairs:

Proposition 5.51. *Let (X, Δ) be a dlt pair, Δ effective. The following are equivalent:*

(1) (X, Δ) *is plt,*
(2) $\lfloor \Delta \rfloor$ *is normal,*
(3) $\lfloor \Delta \rfloor$ *is the disjoint union of its irreducible components.*

Proof. Assume (1) and let $g : Y \to X$ be a log resolution such that $\Delta' := g_*^{-1}\Delta$ is smooth. Write

$$K_Y + \Delta' \equiv g^*(K_X + \Delta) + E,$$

where $\lceil E \rceil$ is effective and g-exceptional since (X, Δ) is plt. We have an exact sequence

$$0 \to \mathcal{O}_Y(-\lfloor \Delta' \rfloor + \lceil E \rceil) \to \mathcal{O}_Y(\lceil E \rceil) \to \mathcal{O}_{\lfloor \Delta' \rfloor}(\lceil E \rceil|_{\lfloor \Delta' \rfloor}) \to 0.$$

By definition $-\lfloor \Delta' \rfloor + \lceil E \rceil \equiv_g K_Y + \{\Delta'\} + (\lceil E \rceil - E)$, and therefore by (2.68) $R^1 f_* \mathcal{O}_Y(-\lfloor \Delta' \rfloor + \lceil E \rceil) = 0$. Thus we get a surjection

$$g_* \mathcal{O}_Y(\lceil E \rceil) \twoheadrightarrow g_* \mathcal{O}_{\lfloor \Delta' \rfloor}(\lceil E \rceil|_{\lfloor \Delta' \rfloor}).$$

$\lceil E \rceil$ is effective and g-exceptional, hence $\mathcal{O}_X = g_* \mathcal{O}_Y(\lceil E \rceil)$. Let $\lfloor \Delta \rfloor^n$ denote the normalization of $\lfloor \Delta \rfloor$. We have maps and inclusions

$$g_* \mathcal{O}_Y(\lceil E \rceil) = \mathcal{O}_X \to \mathcal{O}_{\lfloor \Delta \rfloor} \subset \mathcal{O}_{\lfloor \Delta \rfloor^n} \subset g_* \mathcal{O}_{\lfloor \Delta' \rfloor}(\lceil E \rceil|_{\lfloor \Delta' \rfloor}),$$

and the composite is surjective. Thus $\mathcal{O}_{\lfloor \Delta \rfloor} = \mathcal{O}_{\lfloor \Delta \rfloor^n}$ and so $\lfloor \Delta \rfloor$ is normal.

(2) $\Rightarrow$ (3) is clear.

Assume (3) and let $Z \subset X$ be the smallest subset such that $X \setminus Z$ is smooth and $\Delta|_{X \setminus Z}$ is snc. If E is an exceptional divisor over X and $\mathrm{center}_X E \subset Z$ then $a(E, X, \Delta) > -1$ by definition. If $\mathrm{center}_X E \not\subset Z$ then $a(E, X, \Delta) > -1$ by (2.31.3). ☐

Corollary 5.52. *Let (X, Δ) be a dlt pair, Δ effective. Then every irreducible component of $\lfloor \Delta \rfloor$ is normal.*

Proof. Let $S \subset \lfloor \Delta \rfloor$ be an irreducible component and set $\Delta_1 := \Delta - S$. By (2.43) there is an effective $\mathbb{Q}$-divisor D such that $(X, \Delta - \epsilon\Delta_1 + \epsilon D)$

is dlt for all rational numbers $0 < \epsilon \ll 1$. $\lfloor \Delta - \epsilon\Delta_1 + \epsilon D \rfloor = S$ and S is normal by (5.51). □

Corollary 5.53. *Let $(X, S + \Delta)$ be a dlt pair, Δ effective. Assume that S is irreducible, Cartier in codimension 2 and $\lfloor \Delta \rfloor = 0$. Then $\lfloor \Delta|_S \rfloor = 0$.*

Proof. $(X, S + \Delta)$ is plt and S is normal (5.51), hence $(S, \Delta|_S)$ is klt by (5.50). Thus $\lfloor \Delta|_S \rfloor = 0$. □

Remark 5.54. (5.53) fails for $(X, S + \Delta)$ lc. Take for instance $X = \mathbb{C}^2$, $S = (x = 0)$ and $\Delta = (1/2)(x + y = 0) + (1/2)(x - y = 0)$.

Corollary 5.55. *Let $(0 \in X, \Delta)$ be a dlt surface pair, Δ effective. Then*

(1) *either X is smooth and $\Delta = \lfloor \Delta \rfloor$ has two irreducible components intersecting transversally at 0,*
(2) *or $(0 \in X, \Delta)$ is plt and $\lfloor \Delta \rfloor$ is smooth.*

Proof. Assume that X is not smooth or Δ is not a snc divisor near 0. If $f : Y \to X$ is any birational morphism and $E \subset Y$ an exceptional divisor with $f(E) = 0$ then $a(E, X, \Delta) > -1$ by definition. Thus $(0 \in X, \Delta)$ is plt and so $\lfloor \Delta \rfloor$ is normal and hence smooth by (5.51). □

Corollary 5.56. *Let $(S, D + \Delta)$ be a dlt surface pair where D is a Cartier divisor, Δ effective. Then S is smooth near D.*
If $(X, D + \Delta)$ is a dlt pair (of any dimension) where D is a Cartier divisor and Δ is effective then the irreducible components of D are Cartier at every codimension 2 point of X.

Proof. By (5.55), either S or D is smooth. In the latter case S is also smooth since D is Cartier.

Let $x \in X$ be a codimension 2 point. If $x \notin \operatorname{Supp} D$ then there is nothing to prove. If $x \in \operatorname{Supp} D$ then X is smooth at x by the first part (using either localization or taking hyperplane sections), thus every Weil divisor is Cartier at x. □

Corollary 5.57. *Let $(0 \in X, D + \Delta)$ be a surface pair such that $0 \in X$ and $0 \in D$ are smooth. Then $(0 \in X, D + \Delta)$ is plt (resp. lc) near 0 iff $(D \cdot \Delta)_0 < 1$ (resp. ≤ 1), where $(D \cdot \Delta)_0$ denotes the intersection number at 0.* □

The following is a rather general version of inversion of adjunction for lc surfaces.

Proposition 5.58. *Let* $f : S \to T$ *be a resolution of a surface germ* $(0 \in T)$. *Let* $D \subset S$ *be an irreducible and smooth non-exceptional curve and* Δ *a* $\mathbb{Q}$-*divisor on* S *such that* $f_* \Delta$ *is effective and* $D \not\subset \operatorname{Supp} \Delta$. *Assume that*

(1) $K_S + D + \Delta \equiv 0$, *and*

(2) $(D \cdot \Delta) \leq 1$.

Then $(S, D + \Delta)$ *is lc.*

Proof. We run the $(K + D)$-MMP over T. All the steps exist by (3.47) and the program stops with a surface

$$f : S \xrightarrow{g} S' \xrightarrow{h} T. \quad \text{Set} \quad D' := g_* D, \Delta' := g_* \Delta.$$

We claim that S' and D' are smooth, $(D' \cdot \Delta') \leq 1$, Δ' is effective and $-\Delta'$ is h-nef.

We consider these conditions in one extremal contraction, contracting a curve E. Then E is exceptional over T, so $E \neq D$. Thus $(K \cdot E) \leq ((K + D) \cdot E) < 0$. Hence E is a (-1)-curve and $D \cap E = \emptyset$. Thus after contraction, S and D remain smooth and $(D \cdot \Delta)$ remains unchanged.

If Δ' is not effective (or $-\Delta'$ is not nef), then by (3.41) there is an exceptional curve E' such that $(E' \cdot \Delta') > 0$. But then $(E' \cdot (K + D')) < 0$, a contradiction.

Since $K_S + D + \Delta \equiv 0$, we have that $K_S + D + \Delta \equiv g^*(K_{S'} + D' + \Delta')$ hence it is sufficient to show that $(S', D' + \Delta')$ is lc.

Let $Z \subset h^{-1}(0)$ be the set of points where $(S', D' + \Delta')$ is not klt. Z is connected by (5.49) and $Z \supset D' \cap h^{-1}(0)$. Thus either $Z = D' \cap h^{-1}(0)$ is a point or Z is the union of those curves in $h^{-1}(0)$ whose coefficient in Δ' is at least 1. Set $D_0 := D'$ and note that $(S', D' + \Delta')$ is lc at $D' \cap h^{-1}(0)$ by (5.57).

Since Δ' is effective and $(D' \cdot \Delta') \leq 1$, we have two possibilities:

(1) All curves in Δ' intersecting D' have coefficient < 1. Thus $Z = D' \cap h^{-1}(0)$ and $(S', D' + \Delta')$ is lc.

(2) There is a unique curve $D_1 \subset \operatorname{Supp} \Delta'$ intersecting D' and it has coefficient 1 in Δ'.

We continue by induction. Assume that we have already found $D_1, \ldots, D_i \subset h^{-1}(0)$ such that they have coefficient 1 in Δ', $(D_i \cdot D_{i-1}) =$

1 and $(S', D' + \Delta')$ is lc along $D_1 \cup \cdots \cup D_{i-1}$. We have the equation

$$
\begin{aligned}
0 &= (D_i \cdot (K_{S'} + D' + \Delta')) \\
&= (D_i \cdot (K_{S'} + D_i)) + (D_i \cdot D_{i-1}) + (D_i \cdot (D' + \Delta' - D_i - D_{i-1})) \\
&= 2p_a(D_i) - 2 + 1 + (D_i \cdot (D' + \Delta' - D_i - D_{i-1})).
\end{aligned}
$$

This is only possible if $D_i \cong \mathbb{P}^1$ and $(D_i \cdot (D' + \Delta' - D_i - D_{i-1}) = 1$. Thus $(S', D' + \Delta')$ is lc along D_i by (5.57). Again we get two cases:

(1) All curves in $D' + \Delta' - D_i - D_{i-1}$ intersecting D_i have coefficient < 1 in Δ'. Thus $Z = D_1 \cup \cdots \cup D_i$ and $(S', D' + \Delta')$ is lc.
(2) There is a unique curve $D_{i+1} \subset \operatorname{Supp}(D' + \Delta' - D_i - D_{i-1})$ intersecting D_i and it has coefficient one in Δ'. D_{i+1} is different from the curves $D_0, \ldots, D_i$.

Eventually we end up with case (1) which completes the proof.   □

Adjunction holds for dlt pairs:

Proposition 5.59. *Let X be a normal variety, S an irreducible Weil divisor which is Cartier in codimension 2, and $B = \sum b_i B_i$ an effective $\mathbb{Q}$-divisor. If $(X, S + B)$ is dlt then $(S, B|_S)$ is also dlt.*

Proof. S is normal by (5.52). Since $(X, S + B)$ is dlt, there is a closed subscheme $Z \subset X$ such that $X \setminus Z$ is smooth, $S + B|_{X \setminus Z}$ is a snc divisor and discrep(center $\subset Z, X, S + B) > -1$. Set $Z' := Z \cap S$. Then $S \setminus Z'$ is smooth, $B|_{S \setminus Z'}$ is a snc divisor and

$$
\operatorname{discrep}(\text{center} \subset Z', S, B|_S) \geq \operatorname{discrep}(\text{center} \subset Z, X, S + B) > -1
$$

by (5.46).   □

Remark 5.60. The converse of (5.59) fails as shown by the example given in (5.54).

Proposition 5.61. *Let X be a normal variety, S an irreducible Weil divisor which is Cartier in codimension 2, and $B = \sum b_i B_i$ an effective $\mathbb{Q}$-divisor. If $(X, S + B)$ is dlt and $\lfloor B \rfloor$ is the disjoint union of its irreducible components, then $(S, B|_S)$ is plt.*

Proof. Pick $Z \subset X$ as in the proof of (5.59) and let $f : Y \to X$ be a log resolution of $(X, S + B)$. Set $S' := f_*^{-1}S$. If E is an f-exceptional divisor such that $a(E, X, S + B) = -1$ then $f(E)$ is an irreducible component of $S \cap \lfloor B \rfloor$ by (5.53). Thus either $E \cap S' = \emptyset$ or $E \cap S'$ maps birationally to an irreducible component of $S \cap \lfloor B \rfloor$. Thus $E \cap S'$ is not

$f|_{S'}$-exceptional. By (5.46), there exist no exceptional divisors F over S such that $a(F, S, B|_S) \leq -1$. □

Corollary 5.62. *Let X be a normal variety, S an irreducible Weil divisor which is Cartier in codimension 2, S' an irreducible Weil divisor and B an effective $\mathbb{Q}$-Cartier $\mathbb{Q}$-divisor. If $(X, S + S' + B)$ is dlt then $S \cap S'$ is the disjoint union of its irreducible components.*

Proof. We may assume that $\lfloor B \rfloor = 0$, thus $\lfloor S' + B \rfloor$ is irreducible. Hence $(S, S'|_S + B|_S)$ is plt by (5.61). Thus $S'|_S = \lfloor S'|_S \rfloor$ is the disjoint union of its irreducible components by (5.51). It is clear that $\mathrm{Supp}(S \cap S') = \mathrm{Supp}(S'|_S)$, hence we are done. □

Proposition 5.63. *Let X be a normal variety and S_i $\mathbb{Q}$-Cartier Weil divisors passing through a point $0 \in X$. If $(X, \sum_{i=1}^k S_i)$ is dlt then $k \leq \dim X$.*

Proof. The proof is by induction on $\dim X$. The assertion is clear if $\dim X = 1$. Assume that S_1 is Cartier. Let T be one of its irreducible components. Then T is normal (5.52), Cartier in codimension 2 (5.56) and $(T, (S_1 - T)|_T + \sum_{i=2}^k S_i|_T)$ is dlt by (5.59), hence $k \leq \dim X$ by induction.

Next we reduce to the case when S_1 is Cartier. Let $m > 0$ be the smallest integer such that mS_1 is Cartier. Let $L = \mathcal{O}_X(S_1)$ and $p : X' \to X$ the corresponding cyclic cover (2.50). p is étale over the open set where S_1 is Cartier. Set $S_i' := p^{-1}(S_i)$. S_1' is Cartier.

Pick $Z \subset X$ as in the proof of (5.59) and set $Z' := p^{-1}(Z)$. $X \setminus Z$ is smooth, so $(X' \setminus Z') \to (X \setminus Z)$ is étale. Together with (5.20) this shows that $(X', \sum_{i=1}^k S_i')$ is dlt, thus $k \leq \dim X$ as we proved above. □

Remark 5.64. (5.63) is a special case of the following more general result of [K+92, 18.2]:
If $(X, \sum b_i S_i)$ is lc, then $\sum b_i \leq \dim X$.

5.5 Duality Theory

The aim of this section is to discuss results from duality theory that are used elsewhere in this book. Grothendieck's general duality theory (cf. [Har66]) contains all these results as a very special case, but it is frequently quite hard to disentangle the simpler results from the complications caused by the great generality considered in [Har66]. Much of what we need can also be derived from the more elementary treatment given in [Har77, III.5–7], but this also needs some work.

Therefore we decided to develop duality for CM sheaves. One of the main advantages of CM sheaves is that Serre duality works for them without passing to derived categories.

The proofs can be done very efficiently for projective varieties. Unfortunately, this approach needs substantial changes to handle the case of non-projective varieties or of complex spaces. The main problem is that the definition of the dualizing sheaf (5.66) does not make sense for non-proper varieties. One can adopt (5.68) as the local definition of the dualizing sheaf in general. It is unfortunately not easy to check that this is independent of the choices made.

Therefore we consider only the projective case. The general algebraic case follows from [Har66] and the complex analytic setting is discussed in [BS76].

For ease of reference we state the only result we assume from duality theory:

5.65 (Serre duality on $\mathbb{P}^n$). Let F be a coherent sheaf on $\mathbb{P}^n$ (over a field k). Then $H^n(\mathbb{P}^n, F)$ and $\mathrm{Hom}_{\mathbb{P}^n}(F, \mathcal{O}_{\mathbb{P}^n}(K_{\mathbb{P}^n}))$ are dual k-vector spaces. If F is locally free then $H^i(\mathbb{P}^n, F)$ and $H^{n-i}(\mathbb{P}^n, \mathcal{O}_{\mathbb{P}^n}(K_{\mathbb{P}^n}) \otimes F^*)$ are dual k-vector spaces for every i.

Definition 5.66. Let X be a proper scheme of dimension n over a field k. A *dualizing sheaf* is a coherent sheaf ω_X and a surjection $\mathrm{Trace}_X : H^n(X, \omega_X) \to k$ such that for an arbitrary coherent sheaf F, Trace_X induces a natural k-isomorphism

$$\mathrm{Hom}_X(F, \omega_X) \cong \mathrm{Hom}_k(H^n(X, F), k).$$

It is easy to see that $(\omega_X, \mathrm{Trace}_X)$ is unique if it exists. Usually we suppress Trace_X for simplicity of notation.

The following general result is useful for computing the dualizing sheaves.

Proposition 5.67. *Let $f : X \to Y$ be a finite morphism, F, G coherent sheaves on X and on Y, respectively. Set $f^! G := \mathcal{H}om_{\mathcal{O}_Y}(f_* \mathcal{O}_X, G)$ with the natural $\mathcal{O}_X$-module structure.*

(1) *There is a natural $f_* \mathcal{O}_X$-isomorphism*

$$f_* \mathcal{H}om_{\mathcal{O}_X}(F, f^! G) = \mathcal{H}om_{\mathcal{O}_Y}(f_* F, G).$$

(2) *There is a natural k-isomorphism*

$$\mathrm{Hom}_X(F, f^!G) = \mathrm{Hom}_Y(f_*F, G).$$

Proof. (1) is local, so let $X = \mathrm{Spec}\,A, Y = \mathrm{Spec}\,B, M = \Gamma(X, F)$ and $N = \Gamma(Y, G)$. Then proving (1) amounts to checking

$$\mathrm{Hom}_A(M, \mathrm{Hom}_B(A, N)) = \mathrm{Hom}_B(M, N),$$

where the elements ψ on the left side and ϕ on the right are related by $\psi(m) : a \mapsto \phi(am)$ for $a \in A$, $m \in M$. This gives a one-to-one correspondence.

(2) follows from (1) by taking global sections. $\qquad\qquad\square$

Proposition 5.68. *Let $f : X \to Y$ be a finite morphism of proper schemes both of pure dimension n. If ω_Y exists then ω_X exists and $\omega_X \cong f^!\omega_Y$.*

Proof. By (5.67.2) we have a natural isomorphism $\mathrm{Hom}_X(F, f^!\omega_Y) = \mathrm{Hom}_Y(f_*F, \omega_Y)$. By duality on Y, the latter is dual to $H^n(Y, f_*F) = H^n(X, F)$. $\qquad\qquad\square$

Corollary 5.69. *ω_X exists and is S_2 for any projective scheme over k.*

Proof. Any projective scheme of dimension n over k has a finite morphism $f : X \to \mathbb{P}_k^n$. We know that $\omega_{\mathbb{P}_k^n}$ exists (cf. (5.65)). Thus ω_X exists by (5.68).

By (5.4), ω_X is S_2 iff $f_*\omega_X$ is. The latter is S_2 since it is obtained by taking $\mathcal{H}om$ to a locally free sheaf. $\qquad\qquad\square$

Corollary 5.70. *Let X be a projective scheme of pure dimension n over k and F a coherent sheaf on X such that $\mathrm{Supp}\,F$ is of pure dimension n.*

(1) *If F is CM then $\mathcal{H}om_{\mathcal{O}_X}(F, \omega_X)$ is also CM, and the converse also holds if F is S_2.*

(2) *If X is S_2 then $\mathcal{O}_X$ is CM iff ω_X is CM.*

Proof. Let $f : X \to Y \cong \mathbb{P}_k^n$ be a finite morphism. F is CM iff f_*F is locally free by (5.5), and $\mathcal{H}om_{\mathcal{O}_X}(F, \omega_X)$ is CM iff $f_*\mathcal{H}om_{\mathcal{O}_X}(F, \omega_X) = \mathcal{H}om_{\mathcal{O}_Y}(f_*F, \omega_Y)$ is locally free.

If f_*F is locally free then $\mathcal{H}om_{\mathcal{O}_Y}(f_*F, \omega_Y)$ is locally free and the converse also holds if f_*F is known to be S_2. The latter holds exactly when F is S_2, proving (1).

(2) is a special case of (1). $\qquad\qquad\square$

Theorem 5.71 (Serre duality for CM sheaves). *Let X be a projective scheme of pure dimension n over a field k. Let F be a CM sheaf on X such that $\operatorname{Supp} F$ is of pure dimension n. Then*

$$H^i(X, F) \quad \text{is dual to} \quad H^{n-i}(X, \mathcal{H}om_{\mathcal{O}_X}(F, \omega_X)).$$

Proof. There is a finite morphism $f : X \to P = \mathbb{P}^n$. Then $H^i(X, F) = H^i(P, f_*F)$. f_*F is locally free by (5.5), so the latter group is dual to $H^{n-i}(P, \mathcal{H}om_{\mathcal{O}_P}(f_*F, \omega_P))$ by the Serre duality on $\mathbb{P}^n$. By (5.67), the latter group is isomorphic to $H^{n-i}(P, f_*\mathcal{H}om_{\mathcal{O}_X}(F, \omega_X))$, which is equal to $H^{n-i}(X, \mathcal{H}om_{\mathcal{O}_X}(F, \omega_X))$. $\qquad\square$

The first consequence is a cohomological characterization of CM sheaves:

Corollary 5.72. *Let X be a projective scheme over a field k of pure dimension n with ample Cartier divisor D. Let F be a coherent sheaf on X such that $\operatorname{Supp} F$ is of pure dimension n. The following are equivalent.*

(1) *F is CM.*
(2) *$H^i(X, F(-rD)) = 0$ for every $i < n$ and $r \gg 0$.*

Proof. Assume (1). By (5.71),

$$H^i(X, F(-rD)) \quad \text{is dual to} \quad H^{n-i}(X, \mathcal{H}om_{\mathcal{O}_X}(F, \omega_X)(rD))$$

which vanishes for $r \gg 0$ and $i < n$ by the Serre vanishing. This proves (2).

We prove (2) $\Rightarrow$ (1) by induction on n. There is nothing to prove if $n = 0$. Take any $x \in X$. Since $H^0(X, F(-rD)) = 0$, F does not contain any subsheaf whose support is $\{x\}$. Thus, for $r' \gg 0$, there is an $s \in H^0(X, \mathcal{O}(r'D))$ such that $s(x) = 0$ and s does not vanish at the associated points of F. Thus $s : F \to F(r'D)$ is an injection. Set $Y := (s = 0)$. We have an exact sequence

$$0 \to F(-(r+r')D) \to F(-rD) \to F_Y(-rD) \to 0.$$

$H^i(Y, F_Y(-rD)) = 0$ for $i < n-1$ and $r \gg 0$ from the long cohomology sequence and (2). Thus F_Y is CM by induction and F is CM at x by (5.3.1). $\qquad\square$

Proposition 5.73 (Adjunction formula). *Let X be a projective CM scheme of pure dimension n over a field k and $D \subset X$ an effective Cartier divisor. Then $\omega_D \cong \omega_X(D) \otimes \mathcal{O}_D$.*

Proof. Let $f : X \to P = \mathbb{P}^n$ be a finite morphism, $L \subset P$ a hyperplane and $H = f^{-1}(L)$. Then $\omega_L \cong \omega_P(L) \otimes \mathcal{O}_L$ since we know the dualizing sheaves of the projective spaces P and L. $f_* \mathcal{O}_X$ is locally free by (5.5), hence

$$
\begin{aligned}
f_* \omega_H &\cong \mathcal{H}om(f_* \mathcal{O}_H, \omega_L) = \mathcal{H}om(f_* \mathcal{O}_X, \omega_P(L) \otimes \mathcal{O}_L) \\
&= \mathcal{H}om(f_* \mathcal{O}_X, \omega_P) \otimes \mathcal{O}_P(L) \otimes \mathcal{O}_L = f_*(\omega_X(H) \otimes \mathcal{O}_H).
\end{aligned}
$$

If $D \subset X$ is an arbitrary Cartier divisor, then there is a finite morphism $f : X \to \mathbb{P}^n$ and a hyperplane $L \subset \mathbb{P}^n$ such that $H = D + E$ where Supp D and Supp E have no common irreducible components.

Let $j : D \to H$ be the injection. Then $j_* \omega_D = \mathcal{H}om_H(j_* \mathcal{O}_D, \omega_H)$ and $\omega_H \cong \omega_X(H) \otimes \mathcal{O}_H$. We need to prove that

$$
\mathcal{H}om_X(\mathcal{O}_D, \omega_H) \cong \omega_X(D) \otimes \mathcal{O}_D.
$$

We have an exact sequence

$$
0 \to \mathcal{O}_X(D) \to \mathcal{O}_X(H) \to \mathcal{O}_E(H) \to 0.
$$

Tensoring by ω_X we obtain

$$
\omega_X(D) \to \omega_X(H) \to \omega_X \otimes \mathcal{O}_E(H) \to 0,
$$

which is also left exact since ω_X is S_2. Furthermore, $\omega_X \otimes \mathcal{O}_E(H)$ is S_1 as an $\mathcal{O}_E$-sheaf by (5.3) since ω_X is S_2 (5.69).

Consider the exact sequence

$$
0 \to \omega_X(D)/\omega_X \to \omega_X(H)/\omega_X \to \omega_X(H)/\omega_X(D) \to 0.
$$

We have proved that its middle term is ω_H and we have just identified the right hand side. This gives the exact sequence

$$
0 \to \omega_X(D) \otimes \mathcal{O}_D \to \omega_H \to \omega_X \otimes \mathcal{O}_E(H) \to 0.
$$

Since $\omega_X \otimes \mathcal{O}_E(H)$ is S_1 as an $\mathcal{O}_E$-sheaf and Supp D and Supp E have no common components, we obtain that

$$
\begin{aligned}
j_* \omega_D &= \mathcal{H}om_H(j_* \mathcal{O}_D, \omega_H) = \mathcal{H}om_H(j_* \mathcal{O}_D, \omega_X(D) \otimes \mathcal{O}_D) \\
&= \mathcal{H}om_D(\mathcal{O}_D, \omega_X(D) \otimes \mathcal{O}_D) = \omega_X(D) \otimes \mathcal{O}_D. \quad \square
\end{aligned}
$$

Remark 5.74. If X is not CM then from the above proof we see that there is an injection $\omega_X(D) \otimes \mathcal{O}_D \to \omega_D$, but it is not necessarily an isomorphism.

For instance, let X be a cone over an Abelian surface $A \subset \mathbb{P}^n$ with vertex $0 \in X$. Let ω_X be the dualizing sheaf. One can see that $\omega_X \cong \mathcal{O}_X$ (and X is lc). Let $0 \in D \subset X$ be a general hyperplane section. D is not normal at 0; let $\bar{D}$ denote its normalization. Then $\omega_D \cong \mathcal{O}_X(D)|_D \otimes \mathcal{O}_{\bar{D}}$ is not isomorphic to $\omega_X(D)|_D$. X is not CM at 0 and ω_X does not coincide with the dualizing complex of [Har66].

Proposition 5.75. *Let X be a normal projective variety of dimension n over k. Then $\omega_X \cong \mathcal{O}_X(K_X)$.*

Proof. We use the fact that $\omega_{\mathbb{P}^n} \cong \mathcal{O}_{\mathbb{P}^n}(K_{\mathbb{P}^n})$. Embed X into $\mathbb{P}^N$ and fix a general projection $\pi : \mathbb{P}^N \dashrightarrow \mathbb{P}^n$ such that the induced morphism $f : X \to P := \mathbb{P}^n$ is finite and separable. Write $K_X = f^*K_P + R$, where R is supported on the ramification locus. The trace map (5.6) gives a bilinear pairing $f_*\mathcal{O}_X \times f_*\mathcal{O}_X \to \mathcal{O}_P$. Tensoring with $\mathcal{O}_P(K_P)$ gives a bilinear pairing $f_*\mathcal{O}_X \times f_*\mathcal{O}_X(f^*K_P) \to \mathcal{O}_P(K_P)$, which in turn gives

$$f_*\mathcal{O}_X \times f_*(\mathcal{O}_X(K_X - R)) \to \mathcal{O}_P(K_P).$$

This extends to

$$\Phi : f_*\mathcal{O}_X \times f_*(\mathcal{O}_X(K_X)) \to (\text{sheaf of rational sections of } \mathcal{O}_P(K_P)),$$

which, in terms of local coordinates on P, is defined as

$$(g, h \cdot f^*(dx_1 \wedge \cdots \wedge dx_n)) \mapsto \text{Trace}(gh)dx_1 \wedge \cdots \wedge dx_n.$$

Claim 5.76. Φ gives a pairing

$$\Phi : f_*\mathcal{O}_X \times f_*(\mathcal{O}_X(K_X)) \to \mathcal{O}_P(K_P),$$

which is non-degenerate at the points over which f is flat.

Proof. At such points, $f_*\mathcal{O}_X$ is free and $f_*(\mathcal{O}_X(K_X))$ is reflexive. Thus it is sufficient to prove the claim outside a subset of P of codimension at least 2. So let $p \in P$ be a point such that X is smooth at all points of $f^{-1}(p)$.

We can factor $\pi : \mathbb{P}^N \dashrightarrow \mathbb{P}^n$ as

$$\pi : \mathbb{P}^N \xrightarrow{\pi'} \mathbb{P}^{n+1} \xrightarrow{\pi''} \mathbb{P}^n,$$

such that $\pi' : X \to Y := \pi'(X)$ is an isomorphism over $g^{-1}(p)$. On $\mathbb{P}^{n+1}$ choose coordinates $(x_0 : \cdots : x_n : y)$ such that π'' is the last coordinate projection and $p = (1 : 0 : \cdots : 0)$. Let $(F(x_1, \ldots, x_n, y) = 0)$ be an affine equation of Y. F is a separable polynomial in y of degree $r = \deg f$.

A local free basis of $f_*(\mathcal{O}_Y(K_Y))$ is given by $y^j \sigma$ for $j = 0, \cdots, r - 1$, where

$$\sigma := \frac{1}{\partial F/\partial y} f^*(dx_1 \wedge \cdots \wedge dx_n).$$

$1, y, \ldots, y^{r-1}$ gives a free basis of $f_*\mathcal{O}_X$. By (5.78),

$$\text{Trace}(y^j\sigma) = \delta_{j,r-1}dx_1 \wedge \cdots \wedge dx_n \quad \text{for} \quad j = 0, \ldots, r-1.$$

By the definition of Φ, im Φ is generated by the $\text{Trace}(y^j\sigma)$ for $j = 0, \cdots, r-1$. Hence im $\Phi \subset \mathcal{O}_P(K_P)$.

$$(y^i, y^j\sigma) = \text{Trace}(y^{i+j}\sigma) = \delta_{i+j,r-1}dx_1 \wedge \cdots \wedge dx_n \quad \text{for} \quad i + j \leq r-1.$$

Hence the matrix of the pairing has only 1's on the skew diagonal and 0's above it. $\square$

Another proof can also be obtained along the lines of [Har77, III.7.12].

Proposition 5.77 (Relative trace map). *Let X and Y be projective schemes of pure dimension n over a field k and $f : X \to Y$ a generically finite morphism.*

(1) *There is a natural (non-zero) map $\text{Trace}_{X/Y} : f_*\omega_X \to \omega_Y$.*
(2) *If f is birational and X, Y normal then $\text{Trace}_{X/Y}$ is an isomorphism over the points where f^{-1} is an isomorphism.*
(3) *If f is birational and X, Y smooth then $\text{Trace}_{X/Y}$ is an isomorphism.*

Proof. Let F be any coherent sheaf on X. Supp $R^j f_* F$ has dimension at most $n - j - 1$ for $j > 0$, thus $H^i(Y, R^j f_* F) = 0$ for $i + j \geq n$ and $j > 0$. By the Leray spectral sequence we get a surjection

$$H^n(Y, f_*F) \twoheadrightarrow H^n(X, F).$$

Applying this to $F = \omega_X$ we obtain an injection

$$H^0(X, \mathcal{O}_X) = H^0(X, \mathcal{H}om(\omega_X, \omega_X)) \hookrightarrow H^0(Y, \mathcal{H}om(f_*\omega_X, \omega_Y)).$$

The image of $1 \in H^0(X, \mathcal{O}_X)$ gives the map $f_*\omega_X \to \omega_Y$.

If f is birational then $\text{Trace}_{X/Y}$ is an injection. If X and Y are normal, then $\text{Trace}_{X/Y}$ gives a non-zero map $f_*\mathcal{O}_X(K_X) \to \mathcal{O}_Y(K_Y)$. $f_*\mathcal{O}_X(K_X)$ is naturally a subsheaf of $\mathcal{O}_Y(f_*(K_X)) = \mathcal{O}_Y(K_Y)$, and the two are isomorphic over the points where f^{-1} is an isomorphism.

Finally, (3) is easy (cf. [Har77, II.8.19]). $\square$

Lemma 5.78. *Let K be a field and $f \in K[x]$ a monic polynomial of degree r without multiple roots. Set $V := K[x]/(f)$ as a free K-module. Then the multiplication by $\partial f/\partial x$ is invertible on V and*

$$\text{Trace}\left(\frac{x^s}{\partial f/\partial x}\right) = \delta_{s,r-1} \text{ for } 0 \leq s \leq r-1.$$

Proof. Both claims are invariant under base extension, thus we may assume that K is an algebraically closed field. We can write $f(x) = \prod_i (x - a_i)$ and then $K[x]/(f) \simeq \oplus_i K[x]/(x - a_i)$. The summands of this decomposition are generated by eigenvectors of x and $\partial f / \partial x$, thus we obtain that

$$\text{Trace}\left(\frac{x^s}{\partial f / \partial x}\right) = \sum_i a_i^s \prod_{j \neq i} \frac{1}{a_i - a_j}.$$

The lemma follows from the identity

$$g_s(A_1, \cdots, A_r) := \sum_i A_i^s \prod_{j \neq i} \frac{1}{A_i - A_j} = \delta_{s,r-1} \text{ for } s = 0, \ldots, r - 1$$

in $\mathbb{Z}[A_1, \cdots, A_r, \Delta^{-1}]$, where $\Delta = \prod_{j < i}(A_j - A_i)$.

Let us check the identity. It is obvious that g_s is a symmetric function in the A_i's and that $g \cdot \Delta$ is an alternating polynomial. Every alternating polynomial is divisible by Δ. Since $\deg g_s \cdot \Delta = \deg \Delta - (r - 1) + s$, we obtain that $g_s = 0$ for $s < r - 1$ and g_{r-1} is a constant. To see $g_{r-1} = 1$, we note that $g_{r-1} \to 1$ as a $\mathbb{Q}$-valued function as $A_1 \to \infty$ while $A_2, \cdots, A_r$ are fixed. Thus $g_{r-1} = 1$. $\square$

6

Three-dimensional Flops

In this chapter we begin the detailed study of 3–dimensional birational geometry. Compared with algebraic surfaces, the main new feature of 3–fold geometry is the appearance of flips and flops. This chapter is devoted to the study of flops.

Section 1 is a general introduction to flips and flops. These results hold in all dimensions.

Flops of 3–dimensional varieties with terminal singularities are discussed in section 2. The existence of flops is proved using the classification of index 1 terminal 3–fold singularities. This is one of the main reasons why the existence of higher dimensional flops is unknown. We also consider sequences of flops. For the applications it is important to establish that our procedures to improve a variety do not result in an infinite sequence of flips or flops.

Section 3 is a continuation of our earlier studies of 3–fold canonical singularities. The main theorems (6.23) and (6.25) show how to simplify canonical singularities by explicit blow ups until we reach a variety with only Q–factorial terminal singularities. The proof uses the classification of index 1 canonical singularities established in section 5.3. The existence of these Q–factorial terminalizations follows from the 3–dimensional MMP, but we will use it to establish parts of the MMP.

In section 4 we use Q–factorial terminalizations to construct flops of varieties with canonical singularities. The method, called 'crepant descent', reduces the existence of canonical flops to terminal flops. This method works in many different situations and also in higher dimensions.

6.1 Flips and Flops

In this section we discuss general results concerning flips and flops. We mainly consider uniqueness and the connection of flips with the finite generation of certain algebras. Much less is known about the existence of flips and flops, which is the main topic of subsequent sections.

Remark 6.1. Throughout this section, all definitions and results are formulated for normal varieties. It is, however, important to note that the definitions make sense and the results are valid in the following settings:

(1) schemes of finite type (even in mixed characteristic), and
(2) complex analytic spaces.

In the latter case it is sometimes convenient to work in a neighbourhood of a compact set.

A general definition of $(K + D)$-flips is given in (3.33). The following lemma of [Kaw88] connects flips with the finite generation of certain rings, and it also implies that flips are unique. See [Kol91b, 2.1] for another approach to uniqueness.

Lemma 6.2. *Let Y be a normal algebraic (resp. analytic) variety and B a Weil divisor on Y. The following are equivalent.*

(1) *$R(Y, B) := \oplus_{m \geq 0} \mathcal{O}_Y(mB)$ is a sheaf of finitely generated $\mathcal{O}_Y$-algebras.*
(2) *There exists a projective (resp. proper) birational morphism $g : Z \to Y$ such that Z is normal, $\mathrm{Ex}(g)$ has codimension at least 2, $B' := g_*^{-1}B$ is $\mathbb{Q}$-Cartier and g-ample over Y (resp. over a suitable neighbourhood of any compact subset of Y).*

$g : Z \to Y$ is unique with the above properties.

Proof. In the analytic case, we work near a compact subset of Y.

Assume (2). First we claim that $\mathcal{O}_Y(mB) = g_*\mathcal{O}_Z(mB')$. There is an injection $g_*\mathcal{O}_Z(mB') \to \mathcal{O}_Y(mB)$. Let $U \subset Y$ be an open set and $\sigma \in H^0(U, \mathcal{O}_Y(mB))$. σ lifts to $\sigma' \in H^0(g^{-1}(U) \setminus \mathrm{Ex}(g), \mathcal{O}_Z(mB'))$. Since $\mathrm{Ex}(g)$ has codimension at least 2, σ' can be extended to a section $g^*\sigma \in H^0(g^{-1}(U), \mathcal{O}_Z(mB'))$. Thus $g_*\mathcal{O}_Z(mB') \to \mathcal{O}_Y(mB)$ is also surjective. Since B' is g-ample, $\oplus_{m \geq 0} \mathcal{O}_Y(mB) = \oplus_{m \geq 0} g_*\mathcal{O}_Z(mB')$ is finitely generated.

Also, $Z = \mathrm{Proj}_Y \oplus_{m \geq 0} \mathcal{O}_Y(mB)$, thus Z is unique.

Let us assume (1). Replacing B with a suitable multiple, we may assume that $\mathcal{O}_Y(B)$ generates the $\mathcal{O}_Y$-algebra $R(Y, B)$. We set $Z = \text{Proj}_Y R(Y, B)$. Then $\mathcal{O}_Z(1)$ is very ample and $g_* \mathcal{O}_Z(m) = \mathcal{O}_Y(mB)$. Assume that $g : Z \to Y$ has an exceptional divisor E. Then $\mathcal{O}_Z \subsetneq \mathcal{O}_Z(E)$ and we get an injection

$$\tau : \mathcal{O}_Y(mB) = f_* \mathcal{O}_Z(m) \subsetneq f_*(\mathcal{O}_Z(m)(E)) \quad \text{for } m \gg 0.$$

This is a contradiction since $\mathcal{O}_Y(mB)$ is reflexive and τ is an isomorphism outside the codimension 2 set $g(\text{Ex}(g))$. □

Remark 6.3. Assume that $Y = \text{Spec } S$ is affine and B is an effective divisor corresponding to an ideal $I \subset S$. Then $R(Y, -B) = \oplus_{m \geq 0} I^{(m)}$ where $I^{(m)}$ denotes the mth symbolic power of I. $R(Y, -B)$ is also called the *symbolic power algebra* of I. There are very few cases when such algebras are known to be finitely generated. [Ree58] and [Cut88b] contain examples when they are not.

Corollary 6.4. *Let* $f : X \to Y$ *be a* $(K + D)$-*flipping contraction. That is,* X, Y *are normal varieties,* $f : X \to Y$ *is a proper birational morphism such that* $\text{Ex}(f)$ *has codimension at least 2 in* X *and* D *is a* $\mathbb{Q}$-*divisor on* X *such that* $-K_X - D$ *is* $\mathbb{Q}$-*Cartier and* f-*ample. Choose* $r \in \mathbb{Z}_{>0}$ *such that* rD *is an integral divisor. Then:*

(1) *The* $(K + D)$-*flip of* f *coincides with the canonical model of* (X, D) *over* Y *(3.50).*

(2) *The* $(K + D)$-*flip of* f *exists iff* $\oplus_{m \geq 0} \mathcal{O}_Y(m f_*(r K_X + rD))$ *is a sheaf of finitely generated* $\mathcal{O}_Y$-*algebras.*

(3) *The* $(K + D)$-*flip of* f *is unique.*

(4) *Let* D' *be another* $\mathbb{Q}$-*divisor on* X *such that* $K_X + D'$ *is* $\mathbb{Q}$-*Cartier and there are* $a, a' \in \mathbb{Z}_{>0}$ *such that* $a(K_X + D)$ *is linearly* f-*equivalent to* $a'(K_X + D')$. *Then the* $(K + D)$-*flip is also the* $(K + D')$-*flip.*

Proof. Comparing the definitions gives (1), which implies (3) by (3.52). Uniqueness also follows from (6.2) by setting $B = f_*(K_X + D)$.

(2) follows from (3.52) and also from (6.2.2).

If $a(K_X + D)$ is linearly f-equivalent to $a'(K_X + D')$ then $a(K_{X^+} + D^+)$ is linearly f^+-equivalent to $a'(K_{X^+} + D'^+)$, thus $K_{X^+} + D'^+$ is also f^+-ample. □

Definition 6.5. If $f : X \to Y$ is an extremal flipping contraction (3.34), then by (6.4.3) the $(K + D)$-flip does not depend on the choice of D. In this case we call $f^+ : X^+ \to Y$ the *flip* of f.

Proposition 6.6. *Let Y be a normal variety and B a Weil divisor on Y. For a closed point $0 \in Y$ let $\hat{\mathcal{O}}_{0,Y}$ denote the completion of the local ring $\mathcal{O}_{0,Y}$ at the maximal ideal and set $\hat{Y} = \operatorname{Spec} \hat{\mathcal{O}}_{0,Y}$. The following are equivalent*

(1) $\oplus_{m \geq 0} \hat{\mathcal{O}}_{0,Y}(mB|_{\hat{Y}})$ *is a finitely generated $\hat{\mathcal{O}}_{0,Y}$-algebra.*
(2) *There is a Zariski open neighbourhood $0 \in U \subset Y$ such that $\oplus_{m \geq 0} \mathcal{O}_U(mB|_U)$ is a finitely generated $\mathcal{O}_U$-algebra.*

Note: $\hat{\mathcal{O}}_{0,Y}$ is also normal, but we do not need this.

Proof. First, (2) implies (1) by

$$\oplus_{m \geq 0} \hat{\mathcal{O}}_{0,Y}(mB|_{\hat{Y}}) = \hat{\mathcal{O}}_{0,Y} \otimes (\oplus_{m \geq 0} \mathcal{O}_U(mB|_U)).$$

Conversely, assume that $\oplus_{m \geq 0} \hat{\mathcal{O}}_{0,Y}(mB|_{\hat{Y}})$ is generated by the summands of degree at most n, and let $R = \oplus R_m \subset \oplus_{m \geq 0} \mathcal{O}_Y(mB)$ be the subalgebra generated by $\oplus_{m=0}^n \mathcal{O}_Y(mB)$. Set $Z = \operatorname{Proj}_Y R$ with projection $g : Z \to Y$. By construction,

$$\operatorname{Spec} \hat{\mathcal{O}}_{0,Y} \times_Y Z = \operatorname{Proj}_{\hat{Y}} \left(\oplus_{m \geq 0} \hat{\mathcal{O}}_{0,Y}(mB|_{\hat{Y}}) \right).$$

In particular, no exceptional divisor of g intersects $g^{-1}(0)$. Let $F \subset \operatorname{Ex}(g)$ be the union of all exceptional divisors. Then $0 \notin g(F)$ and so $U := Y \setminus g(F)$ is a Zariski open neighbourhood of 0.
(6.2) shows that U satisfies (2).   □

This implies that the existence of flips is a local problem in the Euclidean topology:

Corollary 6.7. *Let X, Y be normal varieties or complex spaces over $\mathbb{C}$ and $f : X \to Y$ a $(K_X + D)$-flipping contraction. The $(K + D)$-flip of f exists iff the following holds:*

Every point $y \in Y$ has a (Zariski or Euclidean) open neighbourhood $y \in U_y \subset Y$ such that the $(K + D)$-flip of $f_y : f^{-1}(U_y) \to U_y$ exists.

Proof. Set $B = f_*D$. If the $(K + D)$-flip of $f_y : f^{-1}(U_y) \to U_y$ exists then $\oplus_{m \geq 0} \mathcal{O}_{U_y}(mB|_{U_y})$ is finitely generated, and by (6.6) so is $\oplus_{m \geq 0} \hat{\mathcal{O}}_{y,Y}(mB|_{\hat{Y}})$. Thus again by (6.6), $\oplus_{m \geq 0} \mathcal{O}_Y(mB)$ is finitely generated in a neighbourhood of y. We can do this for every y, thus $\oplus_{m \geq 0} \mathcal{O}_Y(mB)$ is a sheaf of finitely generated algebras.   □

The next result is used repeatedly in later sections. (The pull back of Weil divisors by finite morphisms is discussed in the proof of (5.20).)

Proposition 6.8. *[Kaw88, 3.2] Let Y and Y' be normal, irreducible varieties and $h : Y' \to Y$ a finite and surjective morphism. Let B be a Weil divisor on Y and set $B' = h^*B$. Then $\oplus_{m \geq 0} \mathcal{O}_Y(mB)$ is finitely generated iff $\oplus_{m \geq 0} \mathcal{O}_{Y'}(mB')$ is.*

Proof. Suppose that $\oplus_{m \geq 0} \mathcal{O}_Y(mB)$ is finitely generated. By (6.2) there is $g : Z \to Y$ satisfying the properties (6.2.2). Let $g' : Z' \to Y'$ be the normalization of $Y' \times_Y Z$. It has all the properties required in (6.2.2), thus $\oplus_{m \geq 0} \mathcal{O}_{Y'}(mB')$ is finitely generated by (6.2).

Conversely, assume that $\oplus_{m \geq 0} \mathcal{O}_{Y'}(mB')$ is finitely generated. For simplicity we consider only the case when h is separable (this always holds in characteristic zero). Let $q : Y'' \to Y' \to Y$ be the Galois closure of $Y' \to Y$, G the Galois group of Y''/Y and $B'' := q^*B$. By the already proved direction, $\oplus_{m \geq 0} \mathcal{O}_{Y''}(mB'')$ is finitely generated, and by (6.2) we have $g'' : Z'' \to Y''$. Z'' is unique, thus G acts on Z''. Set $Z := Z''/G$. $\qquad \square$

Corollary 6.9. *Let $f : X \to Y$ be a flipping contraction with respect to $K + D$. Let $g : X' \to X$ be a finite morphism and $X' \xrightarrow{f'} Y' \xrightarrow{h} Y$ the Stein factorization. Define D' by the formula $g^*(K_X + D) = K_{X'} + D'$. Then the $(K + D)$-flip of f exists iff the $(K + D')$-flip of f' exists.*

Proof. This follows from (6.8) by setting $B := f_*(K_X + D)$ and $B := f'_*(K_{X'} + D')$. $\qquad \square$

The above corollary is used in section 7.3 to reduce the existence of certain flips to the existence of flops:

Definition 6.10. Let X be a normal variety. A *flopping contraction* is a proper birational morphism $f : X \to Y$ to a normal variety Y such that $\mathrm{Ex}(f)$ has codimension at least two in X and K_X is numerically f-trivial.

If D is a $\mathbb{Q}$-Cartier $\mathbb{Q}$-divisor on X such that $-(K_X + D)$ is f-ample, then the $(K + D)$-flip of f is also called the *D-flop*.

If $f : X \to Y$ is extremal, then by (6.4.3) the D-flop does not depend on the choice of D, and we call $f^+ : X^+ \to Y$ or $\phi : X \dashrightarrow X^+$ the *flop* of f.

A flop $\phi : X \dashrightarrow X^+$ is called *canonical* resp. *terminal* if $(X, 0)$ is canonical resp. terminal.

If $(X, 0)$ is terminal and D is $\mathbb{Q}$-Cartier then $(X, \epsilon D)$ is also terminal for $0 < \epsilon \ll 1$ by (2.35.2). Thus a terminal flop can be viewed is a terminal flip. If $(X, 0)$ is canonical and D is $\mathbb{Q}$-Cartier, then frequently $(X, \epsilon D)$ is not canonical for any $0 < \epsilon$, thus a canonical flop is not a special case of canonical flips.

Lemma 6.11. *Let $f : X \to Y$ be a flipping contraction. Let $B \sim_f -2K_X$ be a reduced divisor and $g : X' \to X$ the corresponding double cover (2.50). Then*

(1) $f' : X' \to Y'$ *(as in (6.9)) is a flopping contraction, and*
(2) *if $(X, (1/2)B)$ is canonical then $(X', 0)$ is canonical.*

Proof. There is a smooth open subset $U \subset X$ such that $B|_U$ is also smooth and $X \setminus U$ has codimension 2. Then $U' := g^{-1}(U)$ is smooth by (2.51). Thus X' is normal and g is ramified along B. Therefore $K_{X'} = g^*(K_X + (1/2)B)$ is f'-trivial.

The second part follows from (5.20.3).   □

In sections 6.4 and 7.4 we use the MMP to construct certain flips. This method relies on the following special case of (3.53).

Proposition 6.12. *Let $f : X \to Y$ be a flipping contraction with respect to $K + D$. Let $g : X' \to X$ be a proper birational morphism and (X', D') an lc pair such that $K_{X'} + D' - g^*(K_X + D)$ is effective and g-exceptional. Then the canonical model of (X', D') over Y (3.50) is also the flip of $f : X \to Y$ with respect to $K + D$.*   □

6.2 Terminal Flops

The purpose of this section is to prove the existence of terminal flops and the termination of extremal canonical flips in case of dimension 3.

Remark 6.13. The arguments of this section work in the algebraic and analytic settings as well. (Note, however, that they are based on the partial classification of terminal threefold singularities (5.39), which is fully proved in this book only for projective threefolds, because of (5.22).)

Theorem 6.14. *Let $f : X \to Y$ be a D-flopping contraction of a three-fold X with terminal singularities only. Then the D-flop exists.*

Proof. We see first that $(Y, 0)$ is terminal. For $0 < \epsilon \ll 1$, $(X, \epsilon D)$ is terminal (6.10). $\overline{NE}(X/Y)$ is a $(K_X + \epsilon D)$-negative extremal face and f can be viewed as the contraction of $\overline{NE}(X/Y)$. Then by (3.25.4), K_Y is $\mathbb{Q}$-Cartier and $K_X = f^*K_Y$. Thus $(Y, 0)$ is terminal by (2.30).

Let $Q \in Y$ such that $f^{-1}(Q)$ is not a point. We treat Y as the analytic germ $(Q \in Y)$. In view of (6.7), it is enough to prove that $R(Y, K_Y + f_* D)$ is a finitely generated $\mathcal{O}_{Q,Y}$-algebra.

By (5.39), $Q \in Y$ has a finite covering $\pi : (\tilde{Q} \in \tilde{Y}) \to (Q \in Y)$ where $(\tilde{Q} \in \tilde{Y})$ is a hypersurface singularity $u^2 = f(x, y, z)$ in $(\mathbb{C}^4, 0)$. $\tilde{Y}$ is also a double cover $g : (\tilde{Q} \in \tilde{Y}) \to (0 \in \mathbb{C}^3)$ by $(x, y, z, u) \mapsto (x, y, z)$. Let $\iota : \tilde{Y} \to \tilde{Y}$ be the involution $(x, y, z, u) \mapsto (x, y, z, -u)$. For a Weil divisor F on $\tilde{Y}$, one sees that $F + \iota^* F = g^* g_* F \sim 0$, that is $\iota^* F \sim -F$. Thus $\iota^* R(\tilde{Y}, F) \simeq R(\tilde{Y}, -F)$.

$R(Y, -K_Y - f_* D)$ is finitely generated since $-K_X - D$ is f-ample, hence $R(\tilde{Y}, -\pi^* f_* D)$ is finitely generated by (6.8). Thus $R(\tilde{Y}, \pi^* f_* D) \simeq \iota^* R(\tilde{Y}, -\pi^* f_* D)$ is also finitely generated. Hence $R(Y, K_Y + f_* D)$ is finitely generated, again by (6.8). □

The above argument actually proves the following theorem when K_X is Cartier. When K_X is not Cartier, it can be proved by examining the cyclic group action on the index-1 cover of the flopping contraction.

Theorem 6.15. *[Kol89] Three dimensional terminal flops preserve the analytic singularity type. To be precise, if X is a threefold with only terminal singularities $\{x_1, \cdots, x_n\}$, then its flop X' has the same number of singular points $\{x_1', \cdots, x_n'\}$ and (after renumbering them) $(X, x_i) \simeq (X', x_i')$, as analytic germs, for all i.*

Next, we treat the termination of flops. To make it explicit, we introduce the following setup.

Definition 6.16. A *sequence of $(K_X + \Delta)$-flips* is a sequence of pairs (X^i, Δ^i) and maps $\phi^i : X^i \dashrightarrow X^{i+1}$, such that $(X^1, \Delta^1) = (X, \Delta)$, $\phi^i : X^i \dashrightarrow X^{i+1}$ is a $(K_{X^i} + \Delta^i)$-flip and $\Delta^{i+1} = \phi_*^i \Delta^i$. If ϕ^i is defined for all $i \geq 1$, it is called an infinite sequence.

Theorem 6.17. *An arbitrary sequence of 3-dimensional canonical K_X-flips is finite. Furthermore an arbitrary sequence of 3-dimensional extremal canonical $(K_X + \Delta)$-flips is finite.*

Remark 6.18. In the above theorem (6.17) and in (6.19), we need to assume that X is a suitably small neighbourhood of a compact set if X is an analytic space, in order to avoid a sequence of flips taking place on curves moving out of any compact set.

Since a flop is a special kind of flip, we obtain:

Corollary 6.19. *An arbitrary sequence of 3-dimensional extremal terminal D-flops is finite.* □

The proof of (6.17) uses induction on the following invariant.

Definition 6.20. [Kol89, Kaw92b, K+92] Let $(X, \Delta = \sum a_i D_i)$ be a canonical pair, where the D_i are distinct prime divisors. Set $a := \max\{a_i\}$ and $S := \sum a_i \mathbb{Z}_{\geq 0} \subset \mathbb{Q}$. ($a = 0$ and $S = \{0\}$ if $\Delta = 0$.) We set

$$d(X, \Delta) := \sum_{\xi \in S, \, \xi \geq a} \# \left\{ \begin{array}{l} \text{Exceptional divisors } E \text{ over } X \\ \text{with } a(E, X, \Delta) < 1 - \xi \end{array} \right\}.$$

Note that totaldiscrep$(X, \Delta) = -a$, hence $d(X, \Delta) < \infty$ by (2.36.2), and $d(X, \Delta)$ does not increase under $(K_X + \Delta)$-flips by (3.38).

Lemma 6.21. *Let $\phi : X \dashrightarrow X'$ be the $(K_X + \Delta)$-flip of a three dimensional canonical pair $(X, \Delta = \sum_{i=1}^k a_i D_i)$. Let $C' \subset X'$ be a flipped curve (3.33), and $E_{C'}$ the exceptional divisor obtained by blowing up C' near a general point of C'. Then X' is smooth along C' and*

$$0 \leq a(E_{C'}, X, \Delta) < a(E_{C'}, X', \Delta') = 1 - \sum a_i \cdot \text{mult}_{C'}(D_i'),$$

where $\text{mult}_{C'}(D')$ is the multiplicity of D' along C'.

Proof. (X', Δ') is terminal along C' by (3.38). So X' is smooth along C', and the rest is an obvious computation (cf. (2.29)). □

Proof of (6.17). Write $\Delta = \sum_{i=1}^k a_i D_i$ with $a_1 \leq \cdots \leq a_k$. We use the notation of (6.16). The birational transform of an object $A \subset X$ on X^j is denoted by A^j. The proof is by induction on k.

If $k = 0$, then $d(X^{j-1}, 0) > d(X^j, 0)$ for all j by (3.38) and (6.21). This settles the case $k = 0$.

Assume that $k > 0$. By (6.20), $d(X^j, \Delta^j)$ is non-increasing. If a flipped curve C^j for ϕ^{j-1} is contained in D_k^j, then $a_k < 1$ and $d(X^{j-1}, \Delta^{j-1}) > d(X^j, \Delta^j)$ by (3.38) and (6.21). Thus for $j \gg 0$, D_k^j contains no flipped curve. Let $\bar{D}_k^j$ be the normalization of D_k^j. Then ϕ^{j-1} induces a birational morphism $\bar{D}_k^{j-1} \to \bar{D}_k^j$. Since the exceptional curves for $\bar{D}_k^j \to \bar{D}_k^l$ for $l > j$ are numerically independent (3.40), we have $\bar{D}_k^j \simeq \bar{D}_k^l$ for all $l \geq j \gg 0$. That is, D_k^j does not contain a flipping curve (3.33) for $j \gg 0$. If $f^j : X^j \to Y^j$ is the extremal flipping contraction, the $\mathbb{Q}$-Cartier divisor D_k^j is nef over Y^j. Thus $-(K_{X^j} + \sum_{i=1}^{k-1} a_i D_i^j)$ is f^j-ample and $\phi^j : X^j \dashrightarrow X^{j+1}$ is an extremal $(K_{X^j} + \sum_{i=1}^{k-1} a_i D_i^j)$-flip. This is in the case $k - 1$. □

6.3 Terminalization and $\mathbb{Q}$-factorialization

In this section we prove two theorems using extremal terminal flops studied in the previous section.

Definition 6.22. [Rei83b] Let (X, Δ) be a pair and $f : Y \to X$ a birational morphism. An f-exceptional irreducible divisor $E \subset Y$ is called *crepant* (with respect to (X, Δ)) if $a(E, X, \Delta) = 0$. f is called *crepant* (with respect to (X, Δ)) iff all f-exceptional irreducible divisors $E \subset Y$ are crepant.

The first main result is the terminalization theorem of [Rei83b].

Theorem 6.23 (Terminalization). *Let X be an algebraic (resp. analytic) threefold with only canonical singularities. Then there is a crepant birational morphism $f : Y \to X$ such that Y has only terminal singularities and f is projective (resp. projective over a suitable neighbourhood of a compact subset of Y).*

Remark 6.24. The terminalization constructed in the proof of (6.23), temporarily denoted by $\pi_X (= f) : X^{ter}(= Y) \to X$, has functorial properties:

(1) Any open embedding $g : X_1 \to X_2$ lifts uniquely to an open embedding $g^{ter} : X_1^{ter} \to X_2^{ter}$ so that $g \circ \pi_{X_1} = \pi_{X_2} \circ g^{ter}$,

(2) $(\pi_X)^{an} = \pi_{X^{an}}$ (compatibility with passing to X^{an}).

We call X^{ter} a *standard terminalization*.

The second main result is the $\mathbb{Q}$-factorialization theorem of [Kaw88].

Theorem 6.25 ($\mathbb{Q}$-factorialization). *Let X be an algebraic (resp. analytic) threefold with only terminal singularities. Then there is a birational morphism $f : Y \to X$ such that Y is terminal and $\mathbb{Q}$-factorial (i.e. every global Weil divisor on Y is $\mathbb{Q}$-Cartier), f is an isomorphism in codimension 1 and projective (resp. projective over a suitable neighbourhood of an arbitrary compact subset of Y).*

Remark 6.26. We emphasize that if Y is a $\mathbb{Q}$-factorialization of a non-proper algebraic threefold X in (6.25) then the associated analytic space Y^{an} might not be a $\mathbb{Q}$-factorialization of X^{an}. Similarly if $U \subset X$ is an open subset of an analytic X, then $f^{-1}(U)$ might not be a $\mathbb{Q}$-factorialization of U, cf. (2.17).

We use the following result to prove (6.23).

Theorem 6.27. *[Rei83b] Let X be a threefold with only canonical singularities such that at each point $P \in X$ the index one cover of $(P \in X)$ is a cDV point. Let C be the 1-dimensional part of the singular locus of X with its reduced structure, and I the defining ideal of C in X. Then:*

(1) *$R = R(X) := \oplus_{\nu \geq 0} I^{(\nu)}$ is a sheaf of normal and finitely generated $\mathcal{O}_X$-algebras, where $I^{(\nu)}$ denotes the ν-th symbolic power of I. Furthermore, the fibers of $f : Y = \mathrm{Proj}_X R \to X$ are of dimension ≤ 1.*

(2) *Y is canonical, $K_Y \equiv f^*(K_X)$, and $\dim f^{-1}(x) = 1$ for every $x \in C$.*

Proof. First we prove (2) assuming (1).

Set $R' := \oplus_{\nu \geq 0} I^\nu$. For each ν, $I^{(\nu)}/I^\nu$ is a sheaf of finite length. To see this, note that there are a finite number of primary ideals Q_i of $\mathcal{O}_X$ (depending on ν) such that $I^\nu = I^{(\nu)} \cap (\cap_i Q_i)$ and $\sqrt{Q_i}$ is a maximal ideal defining a point $x_i \in X$. So $\mathcal{O}_X/Q_i$ and

$$I^{(\nu)}/I^\nu \subset \oplus_i I^{(\nu)}/(I^{(\nu)} \cap Q_i) \subset \oplus_i (I^{(\nu)} + Q_i)/Q_i$$

are of finite length as claimed.

Since R is finitely generated, there is a finite set $S \subset X$ such that $R|_{X \setminus S} = R'|_{X \setminus S}$. Set $f' : Y' := \mathrm{Proj}_X R' \to X$. We have $f|_{X \setminus S} = f'|_{X \setminus S}$. We may enlarge S so that it contains all non-cDV points of X (5.40).

Pick any $x \in C \setminus S$ and let $x \in H \subset U_x \subset X$ be a general hypersurface section (where U_x is a suitable open neighbourhood such that $H \cap C = \{x\}$). $x \in H$ is a Du Val singularity and $(f')^{-1}(H) = B_x H$. $B_x H$ is normal and $K_{B_x H} = (f')^* K_H$ (cf. (1) in the proof of (4.20)). Thus $K_{Y' \setminus (f')^{-1}(S)} = (f')^* K_{X \setminus S}$ and $K_Y \equiv f^* K_X$ since $f^{-1}(S)$ has codimension 2. Y is canonical by (2.30). This proves (2).

Now we treat (1). We can check it locally analytically. Let $(P \in X)$ be an analytic germ. We may assume $P \in C$, otherwise there is nothing to prove. Then (1) for $P \in C \subset X$ follows from (1) for the index 1 cover by taking the cyclic group quotient (cf. (6.8)). So we may assume that $(P \in X)$ is a cDV singularity. Let $\pi : (P \in X) \to (0 \in V)$ be a morphism to a smooth curve such that $X_0 := \pi^{-1}(0)$ has only Du Val singularities and $\dim X_0 \cap C = 0$.

Note that we can take a finite Galois base change V' of the base curve V with Galois group G, because $X \times_V V'$ remains cDV and if $R(X \times_V V')$ is normal and finitely generated then so is $R(X)$ which is the G-invariant part of $R(X \times_V V')$.

So we may assume that π has a projective simultaneous minimal resolution (4.28) $h : Z \to X$ such that $Z_v \to X_v$ is a minimal resolution for all $v \in V$. Let F be the divisor part of the closed subspace of Z defined by $I\mathcal{O}_Z$. By (6.28) there is an effective divisor G such that $G \sim -F$. We note that G is h-nef over general points ξ of V, $\dim h^{-1}(P) \leq 1$ and $K_Z = h^*(K_X)$.

Since Z is smooth, $(Z, \epsilon G)$ is terminal for a small $\epsilon > 0$. Run the $(K_Z + \epsilon G)$-MMP over X. Since $K_Z \sim 0$, the canonical divisor remains trivial during the MMP. Since $\dim h^{-1}(P) \leq 1$ and G is nef over $X \setminus \{P\}$, we see only G-flops and no divisorial contractions as long as the birational transform G' is not nef. Hence by the existence and termination of extremal terminal flops, we get $h' : Z' \to X$ isomorphic to h in codimension 1 and a nef G'. Since $G' - K_{Z'}$ is nef and big over X, $\nu G'$ is basepoint–free for all $\nu \gg 0$ by the Relative Basepoint-free Theorem (3.24) and thus $\oplus_{\nu \geq 0} h'_* \mathcal{O}_{Z'}(\nu G')$ is finitely generated. If Z' is any normal variety and G' a Weil divisor on Z' then $\oplus_{\nu \geq 0} H^0(Z', \mathcal{O}_{Z'}(\nu G'))$ is a normal ring (though it may not be Noetherian) (cf. [Har77, Ex.II.5.14] for the global case).

We have $h'_* \mathcal{O}_{Z'}(\nu G') = I^{(\nu)} (= I^\nu)$ on $X \setminus \{P\}$ by the earlier argument for (1) $\Rightarrow$ (2) and (4.14). Note that, for $s \in \mathcal{O}_{P,X}$, $s \in h'_* \mathcal{O}_{Z'}(\nu G')$ iff $s|_{X \setminus \{P\}} \in h'_* \mathcal{O}_{Z'}(\nu G')|_{X \setminus \{P\}}$ because h' has fiber dimension ≤ 1 over P. The primary ideals $I^{(\nu)}$ have a similar property. Thus $h'_* \mathcal{O}_{Z'}(\nu G') = I^{(\nu)}$ and so R given in (1) is normal and finitely generated. Since h' has fiber dimension ≤ 1, so does f given in (1). $\qquad\square$

Lemma 6.28. *Let* $f : U \to V$ *be a proper birational morphism and* D *a Cartier divisor on* U. *Assume that* V *is quasi-projective in the algebraic case and Stein in the analytic case. Then there is an effective Cartier divisor* F *on* U *such that* $F \sim_f D$.

Proof. Let H be a Cartier divisor on V such that $f_* \mathcal{O}_U(D) \otimes \mathcal{O}_V(H)$ has a non-zero global section s. Then $0 \neq f^* s \in H^0(U, \mathcal{O}_U(D + f^* H))$ and $F := (f^* s = 0) \sim D + f^* H$. $\qquad\square$

Remark 6.29. If X in (6.27) is an algebraic variety, then one can manage with only the MMP for projective varieties during the proof.

First $\pi : (P \in X) \to (0 \in V^\circ)$ in the above proof is chosen to be a morphism onto an open set V° of a non-singular projective curve V.

Then we analyse how (4.39) is used in the proof of the simultaneous resolution (4.28). For $(0 \in V^\circ)$ we can find an étale morphism $(0 \in V^\#) \to (0 \in V^\circ)$ of schemes such that $(0 \in V^\#)$ has an induced morphism of schemes to the deformation space $\mathbf{U}$ of the Du Val singularity by (4.69). An étale morphism $(C^n)^\# \to C^n$ with similar properties exists for C^n near the reference point in (4.39). So find a finite covering $q : W \to V$ by a non-singular W such that $W^\circ = q^{-1}(V^\circ) \to V^\circ$ factors through $V^\circ \times_\mathbf{U} (C^n)^\#$. Then on W° we have two algebraic families of Du Val singularities: $X \times_{V^\circ} W^\circ$ for which we need to prove the assertion (1), and the other $\bar{Y}_n \times_{C^n} W^\circ$ which is a part of a projective family $\bar{Y}_n \times_{C^n} W$ of projective surfaces with only Du Val singularities. Notice that $\bar{Y}_n \times_{C^n} W$ has a global simultaneous projective minimal resolution $Y_n \times_{C^n} W$.

Unfortunately we do not know if $X \times_{V^\circ} W^\circ$ is algebraically isomorphic to $\bar{Y}_n \times_{C^n} W^\circ$, but they are analytically isomorphic locally at the singular point, which was how (4.28) was proved. Notice that to prove (1) for $X \times_{V^\circ} W^\circ$ we may instead work on the locally analytically isomorphic $\bar{Y}_n \times_{C^n} W^\circ$ which has a simultaneous projective minimal resolution. So if we replace $h : Z \to X$ in the proof with our $Y_n \times_{C^n} W \to \bar{Y}_n \times_{C^n} W$, we need only the MMP for projective varieties.

Definition 6.30. Let X be a three-fold with only canonical singularities. Let $e(X)$ denote the number of exceptional divisors E over X such that $a(E, X) = 0$. $e(X) < \infty$ by (2.36), (where we restrict X to a suitably small neighbourhood of a compact subset K if X is analytic).

Proof of (6.23). We use induction on $e(X)$. We note that X is terminal if $e(X) = 0$. Assume that X has a point P such that the index 1 cover of $(P \in X)$ is not a cDV point. Then by (5.35), there exists a projective birational morphism $f : Y \to X$ such that $K_Y = f^* K_X$ and f contracts at least one divisor to P. Hence $e(Y) < e(X)$ and we are done by induction in this case.

Assume next that the index 1 cover of an arbitrary $(P \in X)$ is a cDV point and $\dim \mathrm{Sing}(X) = 1$. By (6.27) there exists a projective birational morphism $f : Y \to X$ such that $K_Y = f^* K_X$ and f contracts at least one divisor to a curve. $e(Y) < e(X)$ and we are again done.

Finally, if the index 1 cover of an arbitrary $(P \in X)$ is a cDV point and $\dim \mathrm{Sing}(X) = 0$, then X is already terminal by (6.31) and we are done.

We note that our construction has the properties in (6.24) by (5.35).

If X is analytic, we work on a suitably small neighbourhood of a compact set K and move K around. These glue by (6.24) and the blow up construction shows that the result is projective over a suitable neighbourhood of any compact subset of X. $\qquad\square$

Lemma 6.31. *[Rei83b] Let $\pi : (\tilde{P} \in \tilde{X}) \to (P \in X)$ be an index 1 cover of a canonical singularity of any dimension (5.19). If $(\tilde{P} \in \tilde{X})$ is terminal, then so is $(P \in X)$.*

Proof. Let ω be a generator of $\mathcal{O}_X(rK_X)$. Then π is the cyclic covering obtained by taking the r-th root of ω (5.19). Let $g : W \to X$ be a resolution, E an exceptional divisor and w a general point of E. Let Ω be a local generator of $\mathcal{O}_W(K_W)$ at w. Let $\sigma : \tilde{W} \to W$ be the normalization of W in the function field of $\tilde{X}$. We thus have a diagram

$$
\begin{array}{ccc}
\tilde{W} & \xrightarrow{\sigma} & W \\
\downarrow \tilde{g} & & \downarrow g \\
\tilde{X} & \xrightarrow{\pi} & X
\end{array}
$$

Assume that $a(E, X) = 0$. This means that $g^*\omega = u \cdot \Omega^{\otimes r}$ for some unit u in a neighbourhood of w. Since σ is obtained by taking the r-th root of u (and then the normalization), we see that σ is étale at $\tilde{w}$ lying over w. Let $\tilde{E}$ ($\subset \sigma^{-1}(E)$) be the prime divisor containing $\tilde{w}$. All these mean that $rK_{\tilde{W}} = \sigma^*g^*rK_X = \tilde{g}^*rK_{\tilde{X}}$ in a neighbourhood of $\tilde{w}$, and $a(\tilde{E}, \tilde{X}) = 0$. This contradicts the assumption that $\tilde{X}$ is terminal and we are done. $\qquad\square$

For the proof of (6.25), we need to prepare two results.

Theorem 6.32. *[Kaw88] Let X be an algebraic (resp. analytic) three-fold with only terminal singularities and D a Weil divisor. Then*

(1) *$R(X, D) = \oplus_{\nu \geq 0}\mathcal{O}_X(\nu D)$ is a sheaf of finitely generated $\mathcal{O}_X$-algebras.*

(2) *There exists a projective (resp. proper) birational morphism $g : Y \to X$ such that Y is normal, $\mathrm{Ex}(g)$ has codimension at least 2, $D' := g_*^{-1}D$ is $\mathbb{Q}$-Cartier and g-ample over X (resp. a suitable neighbourhood of any compact subset of X).*

Remark 6.33. It is obvious that $g : Y \to X$ in (6.32) (called the *symbolic blow up* of X by $-D$) has the same functorial properties as the standard terminalization in (6.24).

Proof. The proof goes along the same lines as the one for (6.27).

(2) and (1) are equivalent by (6.2). By (6.7), it is sufficient to prove (1) for a germ of an analytic terminal singularity of index 1, that is, for an isolated cDV singularity.

Let $\pi : (P \in X) \to (0 \in V)$ be a morphism to a smooth analytic curve such that X_0 is Du Val and X_ξ is smooth for a general $\xi \in V$.

Furthermore, by (6.8), (6.32) follows from (6.32) $\times_V W$ for an arbitrary base change by a smooth curve W. This allows us to assume that π has a simultaneous minimal resolution $h : Z \to X$. Unlike in (6.27), h is an isomorphism in codimension 1 in our case and we run the $(K_Z + \epsilon G)$-MMP, where $G = h_*^{-1}D$. We get $h' : Z' \to X$ and a nef G' on Z'. Then G' is basepoint–free and $h'_* \mathcal{O}_{Z'}(\nu G') = \mathcal{O}_X(\nu D)$ for all $\nu \geq 0$. This implies (6.32) as in the proof of (6.27). $\qquad\square$

Remark 6.34. If X in (6.32) is an algebraic variety, then one can manage with only the MMP for projective varieties during the proof just like (6.29). Since the difference is minor, we explain what extra is needed in our case under the notation of (6.29).

What we need is a global Weil divisor D' on $\bar{Y}_n \times_{C^n} W$ whose restriction is linearly equivalent to the image of D by the local analytic isomorphism $X \times_{V^\circ} W^\circ \to \bar{Y}_n \times_{C^n} W^\circ$. Note that $\bar{Y}_n \times_{C^n} W^\circ$ has only rational singularities since it has only cDV singularities (5.34) and (5.22). Note also that by the construction of Y_n (4.39), $\operatorname{Pic} Y_n$ restricts surjectively to the Picard group of the fiber of $Y_n \to \bar{Y}_n$ at the image of our singular point. Thus the D' we need is provided by the following (6.35), and we can do a MMP of projective varieties.

Lemma 6.35. *Let $h : Z \to X$ be a resolution of an algebraic variety over $\mathbb{C}$ with only rational singularities and $x \in X$ a point such that the restriction map $\operatorname{Pic} Z \to \operatorname{Pic} h^{-1}(x)$ is a surjection. Let $x \in U \subset X^{an}$ be an open neighbourhood so that $H^1(h^{-1}(U), \mathcal{O}_{h^{-1}(U)}) = 0$ and D a Weil divisor on U.*

Then there is an algebraic Weil divisor D' on X such that D and $D'|_U$ are linearly equivalent.

Proof. Note that X^{an} has only rational singularities (2.48). By (4.13), $h_*^{-1}D|_{h^{-1}(U)}$ is characterized by its image in $\operatorname{Pic} h^{-1}(x)$. Hence by the hypothesis, there is a Cartier divisor F on Z with $F|_{h^{-1}(U)} \sim h_*^{-1}D|_{h^{-1}(U)}$. Hence $h_*F|_U \sim D|_U$. $\qquad\square$

Lemma 6.36. *Let X be a terminal singularity and $Z \subset X$ a closed subset of codimension ≥ 2. Then*

$$\Sigma(X, Z) = \{ \text{divisors } E \text{ such that } \emptyset \neq \operatorname{center}_X E \subset Z, a(E, X, \Delta) \leq 1 \}$$

is a finite set.

Proof. Let $f : Y \to X$ be a resolution such that $\operatorname{Ex}(f) \cup f^{-1}(Z) = \cup_{i=1}^m E_i$ is an snc divisor. Write $f^*K_X = K_Y - \sum_i a(E_i, X)E_i$. We claim that $\Sigma(X, Z) \subset \{E_1, \cdots, E_m\}$. To see this, assume that $E \in \Sigma(X, Z)$ is exceptional over Y. Since Y is non-singular we have $a(E, Y) \geq 1$. Hence

$$a(E, X) = a(E, Y, -\sum_i a(E_i, X)E_i)$$

$$= a(E, Y) + \sum_i a(E_i, X) \cdot \text{mult}_E E_i > 1.$$

This is a contradiction and the claim is proved. □

Proof of (6.25). Let $\text{Sing}(X)$ be the singular locus of X. We note that $\text{codim}_X \text{Sing}(X) \geq 3$. We use induction on $|\Sigma(X, \text{Sing}(X))|$ (6.36). If X is $\mathbb{Q}$-factorial, there is nothing to prove. Otherwise let D be a Weil divisor which is not $\mathbb{Q}$-Cartier. Then $f : Y = \text{Proj}_X R(X, D) \to X$ has no exceptional divisors and $f_*^{-1}D$ is $\mathbb{Q}$-Cartier by (6.2) and (6.32). Thus $K_Y = f^*K_X$ and Y is terminal. Hence $\Sigma(X, \text{Sing}(X)) \supset \Sigma(Y, \text{Sing}(Y))$. On the other hand, $\dim \text{Ex}(f) = 1$ and blowing up a curve in $\text{Ex}(f)$ gives a divisor E such that $a(E, X) = 1$ and $\text{center}_X E \subset \text{Sing}(X)$, in particular $E \in \Sigma(X, \text{Sing}(X)) \setminus \Sigma(Y, \text{Sing}(Y))$. This shows that $\Sigma(X, \text{Sing}(X)) \neq \Sigma(Y, \text{Sing}(Y))$. Thus we are done by induction.

If X is a non-proper analytic threefold with $\text{Sing}(X) = \infty$, the above construction gives a proper birational morphism $f : Y \to X$ as a limit of symbolic blow ups σ_i of X: on any compact set K of X, the symbolic blow ups become isomorphisms for $i \gg 0$. □

6.4 Canonical Flops

The main result of this section is the existence of 3-dimensional canonical flops (6.44) and (6.45) and a special case of the termination of 3-dimensional canonical flops (6.43). Although the full termination of canonical flops in dimension 3 is known [Kol89], we limit ourselves to the current version for simplicity of exposition.

Remark 6.37. The arguments of this section work in projective, open algebraic and analytic settings as well (if the relevant cone theorem and MMP, etc. are used.) In the analytic case, we work on a suitable neighbourhood of a compact set as usual, though we may not mention it explicitly.

Theorem 6.38. *[Kaw88, Kol89] Let $f_i : X_i \to Z$ $(i = 1, 2)$ be projective morphisms with a birational map $\pi : X_1 \dashrightarrow X_2$ over Z. Assume that X_i is a $\mathbb{Q}$-factorial threefold with only terminal singularities and K_{X_i} is f_i-nef $(i = 1, 2)$. Let D_2 be an f_2-ample effective divisor on X_2 and*

$D_1 := \pi_*^{-1}(D_2)$. Then π is a composition of finitely many D_1-flops. (In the analytic case, see (6.37).)

Proof of (6.38). We run a $(K_X + \epsilon D_1)$-MMP. D_1 is ample outside a codimension 2 subset of X_1 by (3.54). Hence we see only flops and no divisorial contractions during the MMP. At the end we get X_3, a nef D_3 and a birational map $X_3 \dashrightarrow X_2$ isomorphic in codimension 1. The theorem follows from (6.39). $\qquad\square$

Lemma 6.39. *Let $\pi : X_1 \dashrightarrow X_2$ be a birational map over Z such that π is isomorphic in codimension 1. Assume that there exist*

(1) *an f_2-ample Cartier divisor D_2 on X_2 such that $D_1 := \pi_*^{-1} D_2$ is an f_1-nef Cartier divisor, and*
(2) *an f_1-ample Cartier divisor E_1 on X_1 such that $E_2 := \pi_* E_1$ is Cartier.*

Then π is an isomorphism.

Proof. For $n \gg 0$, $L_1 = E_1 + nD_1$ is f_1-ample on X_1 (1.18) and $L_2 = E_2 + nD_2$ is f_2-ample on X_2. Since $H^0(X, \mathcal{O}(mL_1)) = H^0(X_2, \mathcal{O}(mL_2))$, $|mL_1|$ and $|mL_2|$ embed X_1 and X_2 into the same image over Z. Thus π is an isomorphism. $\qquad\square$

We formulate a special case of the Relative Cone Theorem (3.25).

Proposition 6.40. *Let $f : X \to Z$ be a projective morphism with an effective divisor D on X such that $-D$ is f-ample. Let Δ be an effective $\mathbb{Q}$-divisor on X such that (X, Δ) is klt and $K_X + \Delta$ is numerically f-trivial. Then the cone $NE(X/Z)$ ($\subset N(X/Z) \simeq \mathbb{R}^{\rho(X/Z)}$) of effective 1-cycles is spanned by a finite number of extremal rays. (In the analytic case, see (6.37).)*

Proof. $(X, \Delta + \epsilon D)$ is klt for small enough $\epsilon > 0$ (2.35.2). The Relative Cone Theorem (3.25.2) applied to $K_X + \Delta + \epsilon D + (\epsilon/2)(-D) \equiv_f (\epsilon/2)D$ says that $\overline{NE}(X/Z)$ is generated by a finite number of extremal rays and $\overline{NE}(X/Z)_{D \geq 0}$. Since $-D$ is f-ample, $\overline{NE}(X/Z)_{D \geq 0} = 0$ (1.44). Hence we have (6.40). $\qquad\square$

Suppose that two birational morphisms $f_i : Y_i \to Z$ ($i = 1, 2$) are given. Then there exists a unique birational map

$$\sigma = f_2^{-1} \circ f_1 : Y_1 \dashrightarrow Y_2$$

over Z. We identify f_1 and f_2 if σ is an isomorphism.

Corollary 6.41. *Let Z be a quasi-projective or Stein variety with only canonical singularities and $f : X \to Z$ a crepant birational projective morphism. Then $NE(X/Z)$ is generated by a finite number of extremal rays, and there are only finitely many projective birational morphisms $h : Y \to Z$ with Y normal and dominated by X.*

Proof. By (6.28), there is an effective divisor D on X such that $-D$ is f-ample. Then $NE(X/Z)$ is finitely generated (6.40). The Y's as in (6.41) are in 1-1 correspondence with the faces $NE(X/Y)$ of $NE(X/Z)$. □

Theorem 6.42. *[KM87] Let Z be a three-fold with only canonical singularities. Then up to Z-isomorphisms, there are only finitely many projective birational crepant morphisms $h : X \to Z$ from normal varieties X. (In the analytic case, see (6.37).)*

Proof (due to [Kol89]). To prove the finiteness, we may work on quasi-projective or Stein open sets of Z. So we may assume Z is quasi-projective or Stein. Since every X as in (6.42) has a $\mathbb{Q}$-factorial terminalization, it is enough to prove the finiteness of $\mathbb{Q}$-factorial X with only terminal singularities by (6.41). Fix one such $f^1 : X^1 \to Z$ with an ample effective divisor D^1 and an effective divisor $E^1 \sim_{f^1} -D^1$ (6.28).

Then starting from any other X, one can reach X^1 by finitely many extremal D^1-flops. Since the inverse of a D^1-flop is an E^1-flop, all other X are obtained from X^1 by finitely many E^1-flops. By (6.41) there are only finitely many ways to do an E^1-flop at each step. Now suppose there are infinitely many X's. Then there is one E^1-flop $\phi^1 : X^1 \dashrightarrow X^2$ through which infinitely many X's are obtained by further E^2-flops, where $E^2 = \phi^1_* E^1$. Similarly one can construct an E^2-flop $\phi^2 : X^2 \dashrightarrow X^3$, and so on. This leads to an infinite sequence of extremal terminal E^1-flops. This contradicts the termination of terminal flops (6.19). □

Theorem 6.43 (Easy Termination Theorem). *Let Z be a threefold with only canonical singularities and $f : X \to Z$ a projective birational crepant morphism from a normal variety. Let D be an effective $\mathbb{Q}$-Cartier $\mathbb{Q}$-divisor. Then any sequence of D-flops over Z is finite. (In the analytic case, see (6.37).)*

Proof. Let $X^1 \dashrightarrow X^2 \dashrightarrow \cdots$ be an infinite sequence of D-flops such that all X^i dominate Z. By (6.42), there are $a < b$ such that $X^a = X^b$ and $D^a = D^b$. However by (3.38) the discrepancy $a(E, X^i, \epsilon D^i)$ is a non-decreasing function for each E, and there exists an E such that

$$a(E, X^a, \epsilon D^a) < a(E, X^{a+1}, \epsilon D^{a+1}) \le a(E, X^b, \epsilon D^b).$$

This is impossible. $\square$

Theorem 6.44. *[Kaw88, Kol89] Let $(X, 0)$ be a threefold canonical pair.*

(1) *There is a birational projective morphism $f : \overline{X} \to X$ from a $\mathbb{Q}$-factorial three-fold such that f is isomorphic in codimension 1.*

(2) *If $e(X) > 0$ (6.30) and X is $\mathbb{Q}$-factorial, then there is a crepant morphism $g : X' \to X$ such that X' is $\mathbb{Q}$-factorial, $\mathrm{Ex}(g)$ is an irreducible divisor, $\rho(X'/X) = 1$ and $e(X') = e(X) - 1$.*

(3) *If X is $\mathbb{Q}$-factorial then extremal D-flops exist for any effective divisor D. (In the analytic case, see (6.37).)*

Proof. We prove the assertions (1)–(3) by induction on $e(X)$. $(j)_k$ stands for: (j) is valid if $e(X) \le k$.

When $e(X) = 0$, then X is terminal and $(1)_0$ is the $\mathbb{Q}$-factorialization (6.25). $(2)_0$ is vacuous, and $(3)_0$ is the existence of terminal flops (6.14).

Let $e(X) = k > 0$ and assume that $(1)_{k-1}, (2)_{k-1}$ and $(3)_{k-1}$ have been proved. Let $\pi : Z \to X$ be the $\mathbb{Q}$-factorial terminalization of X (6.23, 6.25). We note that π is crepant and Z is terminal and $\mathbb{Q}$-factorial. Let E be the sum of all the exceptional divisors of π. We run the $(K_Z + \epsilon E)$-MMP over X. As long as $\pi^i : Z^i \to X$ has an exceptional divisor, $K_{Z^i} + \epsilon E^i \equiv \epsilon E^i$ is not nef (3.39). Thus by $e(Z^i) < k$, we can apply $(3)_{k-1}$ and the easy termination (6.43). Hence we end up with $\pi^n : Z^n \to X$ such that Z^n is $\mathbb{Q}$-factorial and π^n is isomorphic in codimension 1. This proves $(1)_k$.

Assume furthermore that X is $\mathbb{Q}$-factorial. The exceptional set of a morphism to a $\mathbb{Q}$-factorial variety is of pure codimension 1 (2.63). Thus, in the above MMP π^n is an isomorphism and $\pi^{n-1} : Z^{n-1} \to X$ is the divisorial contraction required for $(2)_k$.

Assume that $e(X) = k$ and let $h : X \to Y$ be an extremal small contraction such that K_X is h-trivial and $-D$ is h-ample. By $(2)_k$, we get a crepant $g : X' \to X$. Let $D' := g^*D$, and run the $(K_{X'} + \epsilon D')$-MMP on X' over Y.

Since $-\mathrm{Ex}(g) - mg^*D$ is $h \circ g$-ample for $m \gg 0$, X' has exactly one extremal ray R not corresponding to g (6.40). $(R \cdot D') < 0$ since $-D$ is

h-ample. We start the MMP with this ray R to obtain

$$
\begin{array}{ccccc}
X' = X'^1 & \dashrightarrow & X'^2 \cdots & X'^{i-1} & \\
\downarrow g & & & \downarrow \phi^{i-1} & \\
X & & & X^+ = X'^i & \\
& \searrow h & & \nearrow h^+ & \\
& & Y & &
\end{array}
$$

Since $e(X') = k - 1$, we can use $(3)_{k-1}$ as long as we have D'-flops. By (6.43), this stops and we must have a divisorial contraction ϕ^{i-1} : $X'^{i-1} \to X^+ = X'^i$. Then $h^+ : X^+ \to Y$ is an isomorphism in codimension 1. Since we encountered an extremal ray R during the MMP, $a(E, X, \epsilon D) = a(E, X', \epsilon D') < a(E, X^+, \epsilon D^+)$ for some E (3.38). Hence $X^+ \not\simeq X$ over Y. We note that $\operatorname{Pic}(X/Y) \otimes \mathbb{Q} \simeq \operatorname{Pic}(X^+/Y) \otimes \mathbb{Q} \simeq \mathbb{Q}$ is generated by D and hence by D^+. If $-D^+$ is h^+-nef then $X \simeq X^+$ (6.39), which is a contradiction. Thus D^+ is h^+-ample and X^+ is the flop of X. This proves $(3)_k$. $\qquad\square$

Corollary 6.45. *Let $f : X \to Y$ be a (not necessarily extremal) D-flopping contraction of a threefold with canonical singularities only. Let $g : \bar{X} \to X$ be a $\mathbb{Q}$-factorialization. Then the D-flop $f' : X' \to Y$ of f exists. Furthermore, there is a $\mathbb{Q}$-factorialization $g' : \bar{X}' \to Y$ such that the birational map $\bar{X} \dashrightarrow \bar{X}'$ is a composition of a finite number of extremal $\bar{D}$-flops, where $\bar{D} = g^*D$.*

Proof. We run the $(K_{\bar{X}} + \epsilon\bar{D})$-MMP over Y. Since $\bar{X} \to Y$ has no exceptional divisors, we have only extremal $\bar{D}$-flops during the MMP. Hence we get $\bar{X}' \to Y$ such that $\bar{D}'$ is nef over Y. By the Relative Basepoint-free Theorem (3.24), $\phi_{|n\bar{D}'|/Y}$ is a morphism over Y for some $n > 0$. Let $\bar{X}' \to X'$ be the Stein factorization of $\phi_{|n\bar{D}'|/Y}$. Then X' is the D-flop of f. $\qquad\square$

Corollary 6.46. *An arbitrary sequence of 3-dimensional terminal flops is finite.*

Proof. If there is an infinite sequence, we get an infinite sequence of extremal terminal flops of dimension 3 by (6.45). This is impossible by (6.19). $\qquad\square$

Although we do not need it in this book, the full termination of canonical flops is known:

Theorem 6.47. *[Kol89] An arbitrary sequence of 3-dimensional canonical flops is finite. (In the analytic case, see (6.37).)*

The following is an approach to (6.47). First the termination of extremal canonical flops in dimension 3 is proved in the same way as (6.44) though the argument is more involved. We can then drop the adjective 'extremal' as in (6.46).

7

Semi-stable Minimal Models

This chapter is devoted to a special case of the MMP, called the semi-stable MMP. Instead of dealing with a threefold in itself, we view it as a family of surfaces over a curve. Semi-stability is a somewhat technical assumption requiring that the surfaces be not too complicated. Under this assumption we prove that 3–dimensional flips exist and so the corresponding MMP works.

The original proof of the existence of 3–dimensional flips [Mor88] and the more general approach of [Sho92] are both long and involved. While semi-stable flips are rather special, their study shows many of the interesting features of the general case. Moreover, the semi-stable MMP has some very interesting applications. As a consequence of the classification of 3–dimensional flips [KM92] we know that almost all flips are semi-stable, but this may be very hard to prove directly.

Section 1 establishes the general setting of the semi-stable MMP.

Section 2 contains a proof of the semi-stable reduction theorem of [KKMSD73] in dimension 3.

Sections 3 and 4 are devoted to semi-stable flips. First we consider the so-called special semi-stable flips. Then we show that the general case can be reduced to this one. Starting with any semi-stable flipping contraction, an auxiliary construction leads to another semi-stable MMP which involves only special semi-stable flips. This method was first used by [Sho92] in a somewhat different setting. Our approach is based on some ideas of Corti.

Three applications of the semi-stable MMP are considered in section 5. In all three cases the semi-stable MMP provides the solution to a crucial step of the problem. These points are explained in detail. The remaining parts of the proofs depend on well established methods of

207

moduli spaces or of singularity theory. For these the reader is referred to the original papers.

7.1 Semi-stable MMP

The MMP takes slightly different forms depending on the settings it is applied to. In this section, we state the semi-stable MMP. We note that the full proof is given here: the existence of semi-stable flips (7.42) is proved in Section 7.4 and the termination of semi-stable flips (7.7) is proved at the end of this section.

We begin with introducing lc and dlt morphisms, which help us to define the semi-stable MMP more efficiently.

Definition 7.1. Let X be a normal variety, $B = \sum b_i B_i$ an effective $\mathbb{Q}$-divisor and $f : X \to C$ a non-constant morphism to a smooth curve C.

We say that $f : (X, B) \to C$ is *log canonical* (resp. *divisorial log terminal* (abbreviated as *lc* (resp. *dlt*)) if $(X, B + f^{-1}(c))$ is lc (resp. dlt) for every closed point $c \in C$.

Lemma 7.2. *Assume that $f : (X, B) \to C$ is lc. Then*

(1) *Every fiber of f is reduced.*

(2) *B is horizontal, that is, none of the irreducible components of B is contained in a fiber of f.*

(3) *(X, B) is lc.*

(4) *If E is an exceptional divisor over X such that $\mathrm{center}_X E$ is contained in a fiber then $a(E, X, B) \geq 0$.*

(5) *(X, B) is canonical (resp. klt) iff $(X_{gen}, B|_{X_{gen}})$ is canonical (resp. klt) where X_{gen} is a general fiber of f.*

Proof. Pick $c \in C$. Every irreducible component of $B + f^{-1}(c)$ has coefficient ≤ 1 by (2.34). This implies that $f^{-1}(c)$ is reduced and B and $f^{-1}(c)$ have no irreducible components in common. (X, B) is lc by (2.35).

Assume that $\mathrm{center}_X E \subset f^{-1}(c)$. Then

$$a(E, X, B) \geq a(E, X, B + f^{-1}(c)) + 1 \geq 0.$$

If E is an exceptional divisor over X such that $\mathrm{center}_X E$ is not contained in any fiber, let E_{gen} denote a general fiber of $E \to C$. Any irreducible component E'_{gen} of E_{gen} is a prime divisor of X_{gen}. We claim that $a(E, X, B) = a(E'_{gen}, X_{gen}, B|_{X_{gen}})$, which shows (5). Indeed if E is a divisor on a birational morphism $h : Z \to X$, then we obtain $K_{Z_{gen}} = h^*_{gen}(K_{X_{gen}} + B_{gen}) + a(E, X, B)E'_{gen} + \cdots$ by restricting $K_Z = h^*(K_X + B) + a(E, X, B)E + \cdots$ to Z_{gen}. $\qquad\square$

Remark 7.3. Although (7.2.4) suggests it, (X, B) is not canonical in general. For instance if S is any lc surface then $X = C \times S \to C$ is lc, and X is canonical iff S is. Still, in many respects, (X, B) behaves like a canonical pair.

If we want to use a theorem which holds for canonical pairs but not for lc pairs, it is worthwhile to go through the proof. Small changes will frequently make it work for lc morphisms.

Lemma 7.4. *Assume that* $f : (X, B) \to C$ *is lc and let* $c \in C$ *and* $x \in f^{-1}(c)$ *be closed points.*

(1) *If* $\dim X = 2$ *then either*

    (a) $f^{-1}(c)$ *has an ordinary node at* x *and* $x \notin \operatorname{Supp} B$, *or*

    (b) $f^{-1}(c)$ *is smooth at* x *and* $\operatorname{mult}_x B \leq 1$.

(2) *If* $f^{-1}(c)$ *is normal at* x *then* $(f^{-1}(c), B|_{f^{-1}(c)})$ *is lc at* x.

Proof. If $\dim X = 2$, then (X, B) is canonical by (7.2.4) and (4.5) implies the rest of (1). (2) follows from (5.46). $\qquad\square$

Remark 7.5. For many applications it is important to understand the non-normal singularities of $f^{-1}(c)$ as well. It is precisely this question that led to the non-normal version of the notion of a log canonical pair. The theory of these so-called semi-log canonical (slc for short) pairs is not very much different from the lc case but it needs some foundational work. See [KSB88, K$^+$92] for details.

Lemma 7.6. *Assume that* $f : (X, B) \to C$ *is lc. Let* $g : C' \to C$ *be a non-constant morphism from a smooth curve* C', $X' := X \times_C C'$ *with projections* $h : X' \to X$ *and* $f' : X' \to C'$. *Set* $B' := h^* B$ *(cf. (5.20)).*
Then $f' : (X', B') \to C'$ *is also lc.*

Proof. Pick a finite set $R \subset C$ which contains all branch points of g and let $R' := \operatorname{red}(g^{-1}(R))$. We claim that X' is normal and

$$K_{X'} + f'^{-1}(R') + B' = h^*(K_X + f^{-1}(R) + B).$$

Since $C' \to C$ is flat, so is $X' \to X$. Since X is S_2, this implies that X' is also S_2 (5.4). So X' is normal iff it is normal at all codimension 1 points.

Since all fibers of f are reduced, f is smooth outside a codimension 2 subset $W \subset X$. Thus f' is also smooth outside the codimension 2 subset $W' := h^{-1}(W)$. Hence X' is normal at all codimension 1 points and all fibers of f' are reduced.

This also shows that $K_{X'/C'} = h^* K_{X/C}$. By the Hurwitz formula $K_{C'} + R' = g^*(K_C + R)$. These imply that $K_{X'} + f'^{-1}(R') = h^*(K_X + f^{-1}(R))$ and the claim follows.

$(X, f^{-1}(R) + B)$ is lc by assumption, hence $(X', f^{-1}(R') + B')$ is lc by the claim above and by (5.20). For any given $c' \in C'$ we can choose R such that $c' \in R'$, thus $f' : (X', B') \to C'$ is lc. $\square$

The following two theorems form the core of the semi-stable minimal model program. The termination is proved later in this section, and the existence is proved in section 7.4.

Theorem 7.7 (Termination of semi-stable flips). *Let (X, B) be a $\mathbb{Q}$-factorial 3-dimensional dlt pair, B effective, $f : X \to Y$ a projective morphism and $g : Y \to C$ a flat morphism to a smooth curve such that $g \circ f : (X, B) \to C$ is lc. Then an arbitrary sequence of extremal $(K_X + B)$-flips over Y is finite.*

Theorem 7.8 (Existence of semi-stable flips (7.42)). *Assume that (X, B) is a 3-dimensional $\mathbb{Q}$-factorial dlt pair with B effective. Let $f : X \to Y$ be an extremal flipping contraction with respect to $K_X + B$. Assume that there exists a flat morphism $g : Y \to C$ to a smooth curve such that $g \circ f$ is lc. Then f has a flip.*

Based on these, we have the main theorems of this chapter.

Theorem 7.9 (Semi-stable Minimal Model Theorem). *Take a 3-dimensional $\mathbb{Q}$-factorial pair (X, B) and a smooth curve C. Let $f : X \to Y$ be a projective morphism and $g : Y \to C$ a flat morphism such that $g \circ f : (X, B) \to C$ is dlt.*

Then all the steps of the MMP for (X, B) over Y exist, and the program stops with (X^m, B^m) and $f^m : X^m \to Y$ such that

(1) *$g \circ f^m : (X^m, B^m) \to C$ is dlt;*

(2) *Exactly one of the following alternatives holds:*

   (a) *$K_{X^m} + B^m$ is f^m-nef, or*

   (b) *there is a Fano fiber space structure $X^m \to Z^m \to Y$.*

The above theorem still holds if we only assume that (X, B) is dlt and $g \circ f$ is lc, in which case $g \circ f^m : (X^m, B^m) \to C$ is only lc.

Proof. Section 3.7 explains how to run the MMP. The dlt condition is preserved by (3.44). The flips exist by (7.42) and the program stops by (7.7). This shows (1), while (2) is a special case of (3.31. Step 3). ☐

Theorem 7.10 (Semi-stable canonical model theorem). *Notation and assumptions as in (7.9). Assume in addition that $K_X + B$ is f-big.*

Then (X, B) has a canonical model over Y, denoted by (X^c, B^c) and $f^c : X^c \to Y$, such that $g \circ f^c : (X^c, B^c) \to C$ is lc.

Proof. We give a proof only if (X, B) is klt; the general case is discussed at the end.

Let (X^m, B^m) and $f^m : X^m \to Y$ be the minimal model (7.9). (X^m, B^m) is klt by (3.42) and (3.43).

Since $K_{X^m} + B^m$ is f^m-nef and f^m-big, $r(K_{X^m} + B^m)$ is f^m-free for $r \gg 0$ by the Relative Basepoint-free Theorem (3.24). This gives a morphism whose Stein factorization is denoted by

$$f^m : X^m \xrightarrow{h} X^c \xrightarrow{f^c} Y.$$

Set $B^c := h_* B^m$. Then

$$K_{X^m} + B^m \equiv h^*(K_{X^c} + B^c),$$

thus (X^c, B^c) is klt and $g \circ f^c : (X^c, B^c) \to C$ is lc by (2.30).

If (X, B) is only dlt then the general base point free theorems do not apply and one needs to pay very careful attention to $\lfloor B^m \rfloor$. This is worked out in [K$^+$92, 8.4]). The proof relies on a rather detailed knowledge of the surface $\lfloor B^m \rfloor$, and so currently it works only in dimension 3. ☐

The rest of this section is devoted to the proof of the termination of semi-stable flips (7.7). The following (7.11) is the core local version of it, and (7.11) $\Rightarrow$ (7.7) is obvious by (7.12). After stating the two results, we prove them.

Theorem 7.11. *Let (X, B) be a 3-dimensional $\mathbb{Q}$-factorial dlt pair, $f : X \to Y$ be a projective morphism, and $g : Y \to C$ a flat morphism to a smooth curve. Let $T \subset Y$ be a g-trivial Cartier divisor and $S = f^*T$. Assume that $(X, S + B)$ is lc. Then an arbitrary sequence of $(K_X + B)$-flips over Y, which are isomorphisms over $Y \setminus T$, is finite.*

Lemma 7.12. *Notations as in (7.7) without $\mathbb{Q}$-factoriality of X. Then* *there is a Zariski open set $C^\circ \subset C$ such that every $(K_X + B)$-flipping contraction $h : X \to X^*$ over Y is an isomorphism on $(g \circ f)^{-1}(C^\circ)$.*

Proof of (7.12). We give two sets of arguments. Case 1 is straightforward and covers many cases. Case 2 gives a sketch of an argument for the general case involving deformation theory. The arguments work in the analytic setting with small changes since we do not use $\mathbb{Q}$-factoriality.

Case 1. We assume that $f_*\mathcal{O}_X(m(K_X + B)) \neq 0$ for some $m > 0$. This holds if f is birational or if $K_X + B$ is big on the generic fiber of f.

To check (7.12), it is enough to work on $f^{-1}(U)$ for each affine open $U \subset Y$ since Y is covered by finitely many such U's. Hence we may assume Y is affine. So we have an effective Cartier divisor V on X with $V \sim_f m(K_X + B)$. Let V_i be the irreducible components of V, $\bar{V}_i$ the normalization of V_i and $\bar{V}_i \to W_i$ the Stein factorization of $\bar{V}_i \to Y$.

We show that (7.12) is satisfied by any Zariski open set $C^\circ \subset C$ such that $\bar{V}_i|_{C^\circ} \to W_i|_{C^\circ}$ is smooth of fiber dimension ≤ 1 for every i. Such C° exists since $\bar{V}_i$ is a normal surface.

Assume that an irreducible curve $\ell \subset \text{Ex}(h)$ satisfies $P := (g \circ f)(\ell) \in C^\circ$. By $(V \cdot \ell) = m(K_X + B \cdot \ell) < 0$, we have $\ell \subset V_a$ for some a. Since $f(\ell)$ is a point, $\bar{V}_a|_{C^\circ} \to W_a|_{C^\circ}$ has 1-dimensional fibers and ℓ is the image of a whole fiber of $\bar{V}_a|_{C^\circ} \to W_a|_{C^\circ}$ as a set.

By the Rigidity Lemma (1.6), all the fibers of $\bar{V}_a|_{C^\circ} \to W_a|_{C^\circ}$ are sent to points by $\bar{V}_a \to X \to X^*$. Thus h contracts V_a, hence it is not a flip.

Case 2. (Sketch of the general case). We will show that (7.12) is satisfied by any Zariski open set C° over which $X^\circ = (g \circ f)^{-1}(C^\circ)$ has a resolution $\pi : Z \to X^\circ$ such that K_Z is h-nef and $Z \to C^\circ$ is smooth.

Let $\ell \subset \text{Ex}(h)$ be an irreducible curve with $P := (g \circ f)(\ell) \in C^\circ$. We will derive a contradiction. Arguing as in Case 1, we can assume that $(\ell \cdot B) \geq 0$.

We have $\pi^* K_{X^\circ} = K_Z + E$ for some effective π-exceptional $\mathbb{Q}$-divisor E (3.39). For $\ell' := \pi_*^{-1}(\ell)$ on the smooth surface Z_P, we have

$$(\ell' \cdot K_{Z_P}) = (\ell' \cdot K_Z) \leq (\ell' \cdot \pi^* K_{X^\circ}) = (\ell \cdot K_X) \leq (\ell \cdot K_X + B) < 0.$$

Thus ℓ' is a (-1)-curve on Z_P because $h|_{Z_P} : Z_P \to h(Z_P)$ contracts ℓ. It is known that the (-1)-curve ℓ' deforms as $P \in C^\circ$ moves. (For instance one can apply the method of Step 2 in the proof of (1.10) to $\ell' \to Z$ and see that the deformation space of $\ell' \to Z$ has dimension ≥ 4, whence ℓ' moves and hence $\ell \subset X$ moves.) This is a contradiction as in case 1. $\qquad\square$

7.13 (Proof of (7.11)). The proof of the termination of canonical flips (6.17) works for (7.11) with small changes. So we only provide the necessary lemmas and definitions involved.

Lemma 7.14. *Let X be a normal n-fold, S a Cartier divisor and $\Delta = \sum a_i D_i$, where the D_i are distinct prime divisors and $a_i \geq 0$. Assume that $(X, S + \Delta)$ is lc. Set*

$$\Sigma_S(c) = \left\{ \begin{array}{l} \text{Exceptional divisors } E \text{ over } X \text{ such that} \\ \text{center}_X(E) \subset S, a(E, X, \Delta) < c \end{array} \right\}.$$

Then $\Sigma_S(0)$ is empty. If (X, Δ) is klt then $\Sigma_S(1)$ is a finite set.

Proof. $a(E, X, \Delta) \geq a(E, X, S + \Delta) + 1 \geq 0$ shows the first claim.

If (X, Δ) is klt, then $(X, (1 - \epsilon)S + \Delta)$ is also klt by (2.35). Hence there are only finitely many exceptional divisors E over X such that $a(E, X, (1 - \epsilon)S + \Delta) < 0$ (2.36). If center$_X(E) \subset S$ and $a(E, X, (1 - \epsilon)S + \Delta) \geq 0$, then $a(E, X, \Delta) \geq a(E, X, (1 - \epsilon)S + \Delta) + (1 - \epsilon) \geq 1 - \epsilon$. If $m(K_X + \Delta)$ is Cartier then $a(E, X, \Delta)$ is a rational number whose denominator divides m. Thus if $0 < \epsilon < 1/m$, then $a(E, X, \Delta) \geq 1 - \epsilon$ is equivalent to $a(E, X, \Delta) \geq 1$. □

Definition 7.15. (cf. (6.20)) Let $(X, S + \Delta)$ be a lc pair with (X, Δ) dlt, where S is a Cartier divisor, $\Delta = \sum a_i D_i$ and the D_i are distinct prime divisors. Set $a = \max\{a_i\}$. Let $M = \sum a_i \mathbb{Z}_{\geq 0} \subset \mathbb{Q}$. We set

$$d_S(X, \Delta) = \sum_{\xi \in M, \ \xi \geq a} \# \left\{ \begin{array}{l} \text{Exceptional divisors } E \text{ over } X \text{ such that} \\ \text{center}_X(E) \subset S \text{ and } a(E, X, \Delta) < 1 - \xi \end{array} \right\}.$$

We have $d_S(X, \Delta) < \infty$ by (7.14), and $d_S(X, \Delta)$ does not increase under the $(K_X + S + \Delta)$-flips as in (7.11) by (3.38).

Lemma 7.16. *(cf. (6.21)) Let $\phi : X \dashrightarrow X'$ be a $(K_X + S + \Delta)$-flip as in (7.11), where $\Delta = \sum a_i D_i$. Let $C' \subset X'$ be a flipped curve, and let E_C be the exceptional divisor obtained by blowing up C near a general point of C. Then X is smooth along C and*

$$0 \leq a(E_C, X, \Delta) < a(E_C, X', \phi_* \Delta) = 1 - \sum a_i \cdot \text{mult}_C(D_i),$$

where $\text{mult}_C(D)$ is the multiplicity of D along C.

Indeed, $a(E, X', \phi_* \Delta) > 0$ for any exceptional divisor E such that center$_{X'}(E) \subset C'$ by (3.38). So X' is smooth at a general point of C', and the rest is an obvious computation. □

With these modified lemmas and definitions, the proof of (6.17) now works for (7.11).

7.2 Semi-stable Reduction Theorem

In this section, we take the first step toward proving (7.42). This step, called semi-stable reduction, is known in all dimensions.

Let X be a normal algebraic variety over $\mathbb{C}$ or a suitable neighbourhood of a compact set in a normal analytic space. Let $f : X \to C$ be a flat morphism to a smooth curve C over $\mathbb{C}$ and $B \subset X$ a closed subset. For any morphism $C' \to C$, set $X' := X \times_C C'$ and let X'_n be the normalization of X'. For a morphism $X'' \to X'_n$, let $B'' \subset X''$ be the pull back of B by the induced morphism $X'' \to X$.

Theorem 7.17 (Semi-stable reduction). *[KKMSD73] Notation as above. There exists a finite morphism $C' \to C$ from a non-singular curve C' and a projective resolution $g : X'' \to X'_n$ such that the induced morphism $f'' : X'' \to C'$ satisfies the following:*

(1) $(f'')^*(c') \cup \mathrm{Ex}(g) \cup B''$ *is an snc divisor for each $c' \in C'$.*
(2) $(f'')^*(c')$ *is reduced for each $c' \in C'$.*

This is slightly more precise than the one in [KKMSD73]. Let us indicate how the argument goes.

By (0.2) and (0.3), we can perform a blow up $g : X_1 \to X$ to obtain $f_1 : X_1 \to C$ such that X_1 is smooth and $\mathrm{Ex}(g) \cup g^{-1}(B) \cup f_1^{-1}(c)$ is an snc divisor for all $c \in C$. Unfortunately, $f_1^{-1}(c)$ is usually non-reduced. Nevertheless, (7.17) for $f_1 : X_1 \to C$ and the closed subset $B_1 := \mathrm{Ex}(g) \cup g^{-1}(B)$ obviously implies (7.17) for $f : X \to C$ and B. So we can replace X with X_1 and B with B_1, and assume that $f^*(c) \cup B$ snc for all $c \in C$.

The next step is to take a ramified cover $C' \to C$. It is easy to ensure that all the fibers of $X'_n \to C'$ be reduced. X'_n becomes singular, but all the singularities are toric (cf. (7.23)). The trick is to find a very careful resolution of these singularities. [KKMSD73, Theorem 11*] takes care of everything including B. However it uses the combinatorial theory of torus embeddings and requires a systematic treatment.

Our aim is to explain that, at least in dimension 3, a $\mathbb{Q}$-factorial terminalization of X'_n produces the desired semi-stable resolution, while avoiding a systematic use of the torus embeddings theory.

Remark 7.18. Our theorem (7.19) holds in the algebraic and analytic setting as well. We simply use the relevant version of terminalization (6.23) and $\mathbb{Q}$-factorialization (6.25) in the construction.

Theorem 7.19 (Semi-stable reduction for threefolds). *Notation as in (7.17). Assume that X is a non-singular threefold, $B \subset X$ is a divisor and that $f^{-1}(c) + B$ is snc for all $c \in C$ as above. Let $p : C' \to C$ be a finite morphism from a non-singular curve with the following property.*

> *For every $c \in C$, the multiplicities of the irreducible components of $f^*(c)$ divide the ramification index of p at every $c' \in p^{-1}(c)$.*

Then the normalization X'_n is canonical with only quotient singularities, and there exists a smooth projective terminalization X'_{tf} of X'_n (cf. (6.23), (6.25)) such that the conclusion of (7.17) is satisfied by $X'' := X'_{tf}$.

Remark 7.20. The result (7.19) and also the proof here are a natural generalization of the 2-dimensional case [DM69] where a 'smooth projective terminalization' is the minimal resolution.

7.21. Steps of the proof and notation. Let $f'_{tf} : X'_{tf} \to C'$ be the induced morphism and B'_n, B'_{tf} the pull back of B to X'_n, X'_{tf} respectively. X'_n is $\mathbb{Q}$-factorial by (5.15) and (7.21.1). So $\mathrm{Ex}(X'_{tf} \to X'_n)$ to be constructed below is of pure codimension 1 (2.63). We assume that B contains all the singular fibers since adding a fiber to B has no effect. Hence f is smooth on $X \setminus B$ and $X'_n \setminus B'_n$ is smooth. Thus $X'_{tf} \setminus B'_{tf} \simeq X'_n \setminus B'_n$. Hence by $\mathrm{Ex}(X'_{tf} \to X'_n) \subset B'_{tf}$, it is enough to check that $(f'_{tf})^*(c') \cup B'_{tf}$ is snc for each $c' \in C'$ for (7.17.1).

We prove (7.19) in the following steps:

(1) Check that X'_n is canonical with only quotient singularities (7.23).
(2) Let $X'_t \to X'_n$ be the standard terminalization (6.24).
(3) Write $q_t : X'_t \to X$. Express $\mathrm{red}(q_t^{-1}(B)) = \sum_\ell B_t'^{(\ell)}$, where $B_t'^{(\ell)}$ are reduced divisors determined by:

> Two prime divisors G_1, G_2 are in the same $B_t'^{(\ell)}$ iff $\mathrm{mult}_{G_1}(q_t^* E) = \mathrm{mult}_{G_2}(q_t^* E)$ for each divisor $E \subset B$.

By (6.32), we make $B_t'^{(\ell)}$ $\mathbb{Q}$-Cartier one by one in an arbitrary order. Thus we arrive at a terminalization $X'_{tf} \to X'_t$ on which the birational transforms $B_{tf}'^{(\ell)}$ are all $\mathbb{Q}$-Cartier.
(4) Check that all the fibers of $f'_{tf} : X'_{tf} \to C'$ are reduced (7.24).

(5) Check that X'_{tf} and $B'_{tf}{}^{(\ell)}$ are all smooth.

(6) Check that B'_{tf} is locally analytically an nc divisor.

The following explains why we introduced $B'_t{}^{(\ell)}$ in Step (3).

Lemma 7.22. *To do the steps of (7.21), we can work on the algebraic germs in (7.23) with $m = 3$, $f(x_1, x_2, x_3) = \prod x_i^{a_i}$ and $p(t) = t^d$.*

Proof. Over $X \setminus f^{-1}(c)$ there is nothing to prove. Let $W \subset X$ be a small analytic neighbourhood of a point $\xi \in f^{-1}(c)$. Then $X'|_W = W'$ and $W'_n|_W = W'_n$. We have $X'_t|_W = W'_t$ by (6.24). Since B is snc, any irreducible component of B restricts to one of $B|_W$ on W. Thus the decomposition $\{B'^{(\ell)}_t\}_\ell$ restricts to $\{(B|_W)'^{(\ell')}_t\}_{\ell'}$ except that $B'^{(\ell)}_t \cap W$ may be empty for some ℓ. Therefore $X'_{tf}|_W = W'_{tf}$ (6.33) if we choose the order of blow ups in Step (3) properly. Since we only need to check analytic local properties of X'_{tf}, we may as well check those of W'_{tf}. W'_{tf} remains analytically isomorphic if we replace the germs $\xi \in X$, $\xi \in B$, f at ξ and p at $p^{-1}(c)$ with an analytically isomorphic set of germs by (6.24) and (6.33). $\qquad\square$

Step (1) is taken care of by the following.

Proposition 7.23. *Let $X = (0 \in \mathbb{C}^m)$ be a germ and B a divisor with $\{\prod x_i^{a_i} = 0\} \subset B \subset \{\prod x_i = 0\}$, where $a_i \geq 0, d > 0$ are integers such that $\sum a_i > 0$ and d is a multiple of every non-zero a_i. Let $q_n : X'_n \to X$ be the projection, where X'_n is the normalization of*

$$X' := (\prod x_i^{a_i} = t^d) \subset (0 \in X \times \mathbb{C}^1),$$

$f'_n : X'_n \to X' \to \mathbb{C}^1$ the coordinate projection and $B'_n = q_n^* B$. Then:

(1) $K_{X'_n}$ *is Cartier, and $D'_n := ((f'_n)^* t = 0)$ is reduced.*

(2) $X'_n \setminus D'_n$ *is smooth and X'_n has only quotient singularities.*

(3) $K_{X'_n} + D'_n = q_n^*(K_X + \mathrm{red}(q_n(D'_n)))$.

(4) $(X'_n, 0)$ *is canonical and (X'_n, D'_n) is lc.*

Proof. Set $b_i = 1$ if $a_i = 0$ and $b_i = d/a_i$ otherwise. For any d-th root ϵ of 1, let $\pi : (0 \in \mathbb{C}^m) \to (0 \in X')$ be given by

$$\pi(u_1, \cdots, u_m) = (u_1^{b_1}, \cdots, u_m^{b_m}, \epsilon \prod_{a_i > 0} u_i).$$

Then π induces a finite morphism π_n onto the germ of an irreducible component of X'_n, and $\mathrm{im}\,\pi_n$ covers X'_n if we vary ϵ. Since $\pi_n^*(f'_n)^* t =$

$\pi^* t = \epsilon \prod_{a_i > 0} u_i$, D'_n is reduced and π_n is étale in codimension 1. X'_n has only quotient singularities since $\operatorname{im} \pi_n \cong \mathbb{C}^m / G$ for some finite group G.

Since $q_n : X'_n \to X$ is unramified on $X'_n \setminus D'_n$, we see that $X'_n \setminus D'_n$ is smooth, which is (2). We get (3) from the obvious $K_{\mathbb{C}^m} + (\pi^* t = 0) = (pr_1 \circ \pi)^* (K_X + \operatorname{red}(q_n(D'_n)))$, since π_n is étale in codimension 1. Hence $K_{X'_n}$ is Cartier and (X'_n, D'_n) is lc (5.20) because $(X, \operatorname{red}(q_n(D'_n)))$ is lc (2.31). To see that $(X'_n, 0)$ is canonical, let E be any exceptional divisor over X'_n. If $\operatorname{center}_{X'_n}(E) \not\subset D'_n$ then $a(E, X'_n) > 0$ by (2). If $\operatorname{center}_{X'_n}(E) \subset D'_n$ then $a(E, X'_n) \geq a(E, X'_n, D'_n) + 1 \geq 0$ since D'_n is Cartier and (X'_n, D'_n) is lc. $\qquad \square$

Steps (2) and (3) of (7.21) are automatic. The following takes care of Step (4).

Lemma 7.24. *Let Y be canonical, $D \subset Y$ a Cartier divisor and assume that (Y, D) is lc. Let $g : Y' \to Y$ be any crepant birational morphism. Then $g^* D \subset Y'$ is reduced.*

Proof. Let $g^* D = \sum b_i D_i$. Since $K_{Y'} \equiv g^* K_Y$, we obtain that

$$K_{Y'} \equiv g^* (K_Y + D) + \sum (-b_i) D_i.$$

Since (Y, D) is lc, $b_i \leq 1$ for every i and $g^* D$ is reduced. $\qquad \square$

Next we prove that X'_{tf} is smooth. This is not a formal consequence of the general theory, and one needs to use the $(\mathbb{C}^*)^3$-action. However, we work here only with its maximal torsion subgroup μ_∞ (the group of all roots of unity) to avoid any theoretical preparation. (In dealing with $(\mathbb{C}^*)^3$, one needs to use either that it is a topological group or that it is an algebraic group.) We note that μ_∞ is a divisible group and that the μ_r (the group of all r-th roots of unity) are all the finite subgroups of μ_∞.

Lemma 7.25. *The action of $(\mu_\infty)^m$ on X' in (7.23), defined by*

$$(\lambda_1, \cdots, \lambda_m) : (x_1, \cdots, x_m, t) \mapsto (\lambda_1^d x_1, \cdots, \lambda_m^d x_m, t \prod \lambda_i^{a_i}),$$

lifts to an action of $(\mu_\infty)^m$ on X'_n with finite kernel (that is, the kernel of the homomorphism $(\mu_\infty)^m \to \operatorname{Aut}(X'_n)$ is a finite group) such that B'_n is $(\mu_\infty)^m$-invariant. $\qquad \square$

Corollary 7.26. *The $(\mu_\infty)^3$-action on X'_n defined in (7.25) lifts to a $(\mu_\infty)^3$-action on X'_{tf} with finite kernel, and the $B'^{(\ell)}_{tf}$ are $(\mu_\infty)^3$-invariant.*

Proof. $(\mu_\infty)^3$ acts on the standard terminalization X'_t (6.24), and B'_t is $(\mu_\infty)^3$-invariant. Since $(\mu_\infty)^3$ is divisible, its action on any finite set is trivial. Thus every irreducible component of B'_t is invariant and the blow ups in $X'_{tf} \to X'_t$ are equivariant.   □

The singularities of X'_{tf} are terminal with a $(\mu_\infty)^3$-action. There are index 1 terminal singularities even with a $(\mathbb{C}^*)^3$-action; for instance $(x_1 x_2 = x_3 x_4) \subset \mathbb{C}^4$. These will be excluded by the conditions on $B'_{tf}{}^{(\ell)}$ (7.30). The following is the key lemma.

Lemma 7.27. *For every irreducible component $E \subset B'_{tf}$, let T_E be the subgroup of $(\mu_\infty)^3$ given by*

$$T_E = \{(t^{v_E(x_1)}, t^{v_E(x_2)}, t^{v_E(x_3)}) \mid t \in \mu_\infty\} \subset (\mu_\infty)^3,$$

where $v_E(\ )$ denotes the order of vanishing of a function along E. Then $T_E \cong \mu_\infty$ and E is fixed by T_E pointwise.

Proof. Consider the $(\mu_\infty)^3$-equivariant rational map

$$h : E \dashrightarrow \mathbb{P}_2 \quad \text{induced by} \quad (x_1^{v_2 v_3} : x_2^{v_3 v_1} : x_3^{v_1 v_2}),$$

where $v_i = v_E(x_i)$ and the $(\mu_\infty)^3$-action on $\mathbb{P}^2$ is defined by

$$(\lambda_1, \lambda_2, \lambda_3) : (y_1 : y_2 : y_3) \mapsto (\lambda_1^{d \cdot v_2 v_3} y_1 : \lambda_2^{d \cdot v_3 v_1} y_2 : \lambda_3^{d \cdot v_1 v_2} y_3).$$

The only $(\mu_\infty)^3$-invariant closed irreducible algebraic subsets of $\mathbb{P}^2$ are the three points $(1 : 0 : 0), (0 : 1 : 0), (0 : 0 : 1)$, the three lines $(y_i = 0)$ and $\mathbb{P}^2$.

The exponents are chosen such that all three coordinate functions have the same order of vanishing along E. Thus $h(E)$ is not contained in the coordinate lines of $\mathbb{P}^2$. The closure of $h(E)$ is a $(\mu_\infty)^3$-invariant subset of $\mathbb{P}^2$, thus h is dominant.

We note that T_E acts trivially on $\mathbb{P}^2$. Thus T_E acts on the general fibers of h which are finite. Since T_E is a divisible group, the T_E-action on E is trivial.   □

Let us check the smoothness of $B'_{tf}{}^{(\ell)}, X'_{tf}$ in a more general setting.

Lemma 7.28. *Let $(0 \in Y)$ be the germ of a 3-dimensional hypersurface with a μ_∞-action of finite kernel. Assume that $Y \setminus \{0\}$ is smooth and there is a Weil divisor $0 \in E \subset Y$ fixed by μ_∞ pointwise. Then E is smooth. Furthermore, E is $\mathbb{Q}$-Cartier iff Y is smooth.*

Proof. We have $\mu_a = \ker[\mu_\infty \to \operatorname{Aut} Y]$ for some $a > 0$. Thus the action will be assumed faithful, i.e. $\mu_\infty \subset \operatorname{Aut} Y$ via $\mu_\infty \cong \mu_\infty/\mu_a$.

Let $\mu_r \subset \mu_\infty$ with $r \gg 1$. Embed $Y \subset \mathbb{C}^4$ by (7.29) so that

$$\lambda : (y_1, \ldots, y_4, \phi) \mapsto (\lambda^{c_1} y_1, \ldots, \lambda^{c_4} y_4, \lambda^e \phi) \quad (\lambda \in \mu_r, c_i, e \in \mathbb{Z})$$

is the μ_r-action, where ϕ is the equation of Y. For some k, we may assume $c_i \equiv 0$ (r) iff $i \leq k$. The fixed point set of the μ_r-action is $(y_{k+1} = \cdots = y_4 = 0)$. This has dimension at least 2, so $k \in \{4, 3, 2\}$. $k = 4$ is impossible since the μ_∞-action is faithful. If $k = 3$ then $\phi \in (y_4)$ or $\phi \in \mathbb{C}\{y_1, y_2, y_3, y_4^r\}$. We have the latter since the action is faithful. If we set $\phi_i = \partial_{y_i}\phi$, the scheme-theoretic singular locus $\operatorname{Sing} Y \subset Y$ defined by the ideal $(\phi, \phi_1, \cdots, \phi_4)$ is independent of the choice of the coordinates $y_1, \cdots, y_4$, the equation ϕ and r. If Y is singular, $\operatorname{Sing} Y$ is an Artin scheme $\ni 0$ and it is easy to see that $(\phi, \phi_1, \cdots, \phi_4) \subset (y_1, \cdots, y_3, y_4^{r-1})$. Hence $\operatorname{Sing} Y$ is of length $\geq r - 1$ for every r, a contradiction. Thus Y is smooth, and the surface $(\phi = y_4 = 0)$ is smooth by $\phi \in \mathbb{C}\{y_1, y_2, y_3, y_4^r\}$. Hence E is smooth, and we are done if $k = 3$.

If $k = 2$, then the fixed point set is $L = (y_3 = y_4 = 0)$. Thus $E = L$ ($\subset Y$) is smooth and the equation of Y can be written as $y_3 f_3 + y_4 f_4 = 0$. Let $\pi : B_L Y \to Y$ denote the blow up of Y along L. $B_L Y \subset B_L \mathbb{C}^4$, thus the fibers of π have dimension at most 1. L is Cartier outside the origin, so π is an isomorphism outside the origin. By explicit computation, $\pi^{-1}(0) \cong \mathbb{P}^1$ iff $\operatorname{mult}_0 f_3 \geq 1$ and $\operatorname{mult}_0 f_4 \geq 1$. Thus X is singular at 0 iff E is not $\mathbb{Q}$-Cartier by (6.2). $\qquad\square$

We used the following in the proof above.

Lemma 7.29. *Let (R, M) be a local ring with an action of a finite commutative group G. Then every G-invariant ideal I has a minimal set of generators consisting of G-eigenvectors.*

Proof. Let $v \in I$ be such that $v + IM \in I/IM$ is a G-eigenvector with character χ. Set $w = \sum_{g \in G} \chi(g)^{-1} g(v)/|G| \in I$. Then $h(w) = \chi(h) \cdot w$ and $w \equiv v \mod IM$. We apply this to each generator of I. $\qquad\square$

The following settles Step (5).

Corollary 7.30. $B'_{tf}{}^{(\ell)}$ *and* X'_{tf} *are smooth.*

Proof. If E is an irreducible component of $B'_{tf}{}^{(\ell)}$, then T_E fixes the $\mathbb{Q}$-Cartier divisor $B'_{tf}{}^{(\ell)}$ pointwise (7.21.(3)). The singularities of X'_{tf} are terminal of index 1 (7.23.1) and hence isolated cDV points (5.38).

Thus we can apply (7.28) to get the smoothness of $B'_{tf}{}^{(\ell)}$ and X'_{tf} at any point of $B'_{tf}{}^{(\ell)}$. We are done since X'_{tf} is smooth outside $\cup_\ell B'_{tf}{}^{(\ell)}$ $\quad\square$

We now finish Step (6) and hence the proof of (7.19).

Lemma 7.31. B'_{tf} *is locally analytically an nc divisor.*

Proof. Pick any point $P \in B'_{tf}$. Let E_i be the irreducible components ($\ni P$) of B'_{tf}. Let $g_i \in T_{E_i}$ be an element acting non-trivially on X'_{tf}. Let G be the subgroup generated by the g_i's. By (7.29), the E_i are coordinate hyperplanes of the germ $(P \in X'_{tf})$. $\quad\square$

7.3 Special Semi-stable Flips

The aim of this section is to prove the existence of special semi-stable flips in dimension 3. The result and its proof are closely related to [K$^+$92, 20.8], which in turn is taken from [Sho92].

There are several variants of this result, differing slightly in the assumptions. The current choice is dictated by the needs of the proof in the next section.

Theorem 7.32. *Let* $(X, S + B_X)$ *be a* $\mathbb{Q}$*-factorial 3-dimensional pair and* $f : X \to Y$ *an extremal flipping contraction with respect to* $K + B_X$ *such that* S *is Cartier and* f*-trivial. Set* $C = \mathrm{Ex}(f)$*. Assume that*

(1) $(X, S + B_X)$ *is dlt,*
(2) $(X \setminus (S \cup \lfloor B_X \rfloor), 0)$ *is terminal, and*
(3) S *has an irreducible component* S_1 *such that* $(S_1 \cdot C) < 0$.

Then f *has a flip.*

Remark 7.33. The above theorem holds equally well in the projective, the open algebraic and the analytic settings. The arguments of this section work for the algebraic and analytic settings. See (7.37) on how to work only with projective varieties.

The key step of the proof is (7.34) which we state next. Then we prove that it implies (7.32), and finally we establish (7.34).

Proposition 7.34. *Let* $(X, S_1 + S_2 + B)$ *be a* $\mathbb{Q}$*-factorial 3-dimensional pair with* S_1, S_2 *irreducible and* $f : X \to Y$ *an extremal flipping contraction with respect to* $K + S_1 + S_2 + B$*. Assume that*

(1) $(X, S_1 + S_2 + B)$ *is dlt,*
(2) $\lfloor B \rfloor = 0$,

(3) $(S_1 \cdot C) < 0$ and $(S_2 \cdot C) > 0$, where $C = \mathrm{Ex}(f)$.

(4) S_1 is Cartier in codimension 2 on X.

Then, if we shrink X to a small neighbourhood of C, there is a Weil divisor $D^+ \subset X$ such that $(X, S_1 + S_2 + D^+ + \lfloor 2B \rfloor)$ is lc and $K_X + S_1 + S_2 + D^+ + \lfloor 2B \rfloor \sim 0$.

Remark 7.35. D^+ is closely related to the 1-complement defined in [Sho92]. The proof suggests that the natural object is not D^+ but a certain other divisor (denoted by $D^\sharp$) living on a suitable resolution.

Proof of (7.34) $\Rightarrow$ (7.32). $(S_1 \cdot C) < 0$, hence $C \subset S_1$. Since $(S \cdot C) = 0$, there is an irreducible component $S_2 \subset S$ such that $(S_2 \cdot C) > 0$, hence S_2 is f-ample. Set $S_3 := S - S_1 - S_2$ and $B := (1 - \delta)(S_3 + B_X)$ for some $0 < \delta \ll 1$. Since S is a reduced Cartier divisor and $(X, S + B_X)$ is dlt, S_1 is Cartier in codimension 2 on X by (5.55). The conditions of (7.34) are satisfied, hence we obtain $D^+ \subset X$ after we shrink X around C. Note that we can do shrinking here (and later as well) by (6.7) if we do not assume that X is $\mathbb{Q}$-factorial or f is extremal any longer. We note that $\mathrm{Supp}(S + \lfloor B_X \rfloor) \subset \mathrm{Supp}(S_1 + S_2 + \lfloor 2B \rfloor)$.

Since $(S_1 \cdot C) < 0$ and $(S_2 \cdot C) > 0$, we can choose an f-trivial Cartier divisor T which is a positive linear combination of components of $\mathrm{Supp}(S_1 + S_2 + \lfloor 2B \rfloor)$. By shrinking Y, we may assume that $T \sim 0$. Let m be the least common multiple of the coefficients of T. Let $\pi : X^* \to X$ be an irreducible component of the cyclic cover obtained by taking the mth root of T (2.50). Then $T^* := \pi^*(T)/m$ is a reduced Cartier divisor and $T^* \sim 0$ (7.23.1), $X^* \setminus T^*$ is smooth (7.32.2) and

$$K_{X^*} + T^* + D' = \pi^*(K_X + S_1 + S_2 + D^+ + \lfloor 2B \rfloor)$$

with $D' := \pi^*(D^+)$ follows from (7.23.3) because it is an assertion in codimension 1. Thus $(X^*, T^* + D')$ is lc by (5.20).

Claim 7.36. Let $B^* \in |2D'|$ be a general member. Then B^* is reduced, $(X^*, (1/2)B^*)$ is canonical and $2K_{X^*} + B^* \sim 0$.

Proof. Since $X^* \setminus T^*$ is smooth, after shrinking Y the complete linear system $|2D'|$ has no basepoints outside T^*, and so $(X^* \setminus T^*, B^*)$ is canonical by (5.17). $(X^*, T^* + (1/2)B^*)$ is lc by (2.33).

To see that $(X^*, (1/2)B^*)$ is canonical, let E be an exceptional divisor in some birational morphism $h : Z \to X^*$. If $h(E) \not\subset \mathrm{Supp}\, T^*$ then $a(E, X^*, (1/2)B^*) \geq 0$ since $(X^* \setminus T^*, B^*)$ is canonical. If $h(E) \subset \mathrm{Supp}\, T^*$ then

$$a(E, X^*, (1/2)B^*) \geq 1 + a(E, X^*, T^* + (1/2)B^*) \geq 1 - 1 = 0. \quad \square$$

$(X^*, (1/2)B^*)$ satisfies the assumptions of (6.11), thus the existence of the flip of $f : X \to Y$ is reduced to the existence of a 3-dimensional canonical flop (6.9). By (6.45), 3-dimensional canonical flops exist. This proves (7.32). ∎

Remark 7.37. If Y is a projective threefold in (7.32), we can manage to work only on projective varieties (without passing to open subsets) in this section with small changes.

In (7.34), add the condition that Y is projective with an ample divisor L and change the conclusion to

> For $\nu \gg 0$, there is a Weil divisor $D^+ \subset X$ such that $(X, S_1 + S_2 + D^+)$ is lc and $K_X + S_1 + S_2 + D^+ \sim \nu f^* L$

Thus $K_X + S_1 + S_2 + D^+ \sim_f 0$ instead. If $\nu \equiv 0 \mod (m)$ in the above proof, we can choose a general $V \sim \nu f^* L - T$ intersecting T transversally and disjoint from C, and then take the mth root of $T + V \sim \nu f^* L$. Thus (2.49) allows us to take the ramified covering globally and the argument for (7.34) $\Rightarrow$ (7.32) works. For the modified (7.34), the same argument works if (7.38) is changed similarly.

Proof of (7.34). Let $g : X' \to X$ be a log resolution of $(X, S_1 + S_2 + B)$ and $S := g_*^{-1} S_1$. We note that $(S_1, (S_2 + B)|_{S_1})$ is plt by (5.61). Thus $D := (g|_S)_*^{-1}(S_2|_{S_1})$ is a reduced divisor on S, the $\mathbb{Q}$-divisor Δ on S defined by $K_X + D + \Delta \equiv (g|_S)^*(K_{S_1} + (S_2 + B)|_{S_1})$ satisfies $\lfloor \Delta \rfloor \leq 0$, $(g|_S)_* \Delta = B|_{S_1} \geq 0$ and $(S, D + \Delta)$ is plt (2.30).

By (2.39), if we increase the coefficients of B a little, the assumptions and the conclusions of (7.34), as well as the assertions about $\lfloor \Delta \rfloor$ above, do not change. Thus by doing so, we can arrange that none of the coefficients of B, Δ are in $(1/2)\mathbb{Z}$ (the set of discontinuity points of the round down function $x \mapsto \lfloor 2x \rfloor$).

As the first step, we apply (7.38) with T the normalization of $f(S_1)$ and get a divisor $D_S^+ \in |-K_S - D - \lfloor 2\Delta \rfloor|$ with $(S, D + \lfloor 2\Delta \rfloor + D_S^+)$ lc.

The next step is to lift D_S^+ to X'. Let $\epsilon \in \mathbb{Q}$ be such that $0 < \epsilon \ll 1$. Then $-(K_X + S_1 + (1 - \epsilon)(S_2 + B))$ is f-ample. Let B'_ϵ be a $\mathbb{Q}$-divisor on X' defined by

$$K_{X'} + S + B'_\epsilon \equiv g^*(K_X + S_1 + (1 - \epsilon)(S_2 + B)).$$

Note that B'_ϵ is a monotone increasing sequence of divisors as $\epsilon \searrow 0$ and that $\lim g_* B'_\epsilon = S_2 + B$ and $\lim B'_\epsilon|_S = D + \Delta$. Since $(X, S_1 + (1 - \epsilon)(S_2 + B))$ is plt (5.51), we have $\lfloor B'_\epsilon \rfloor \leq 0$. Thus $\lfloor B'_\epsilon|_S \rfloor \leq 0$ and

$\lfloor 2B'_\epsilon |_S \rfloor = D + \lfloor 2\Delta \rfloor$. By (7.39), $D_S^+ \in |-K_S - D - \lfloor 2\Delta \rfloor|$ lifts to $D^\sharp \in |-K_{X'} - S - \lfloor 2B'_\epsilon \rfloor|$.

The last step is to show that $D^+ := g_* D^\sharp$ satisfies the conditions of (7.34). By $\lfloor B'_\epsilon \rfloor \leq 0$ and $\lim g_* B'_\epsilon = S_2 + B$, we have $\lfloor 2g_* B'_\epsilon \rfloor = S_2 + \lfloor 2B \rfloor$. Thus $K_X + S_1 + S_2 + D^+ + \lfloor 2B \rfloor \sim 0$ follows from $K_{X'} + S + \lfloor 2B'_\epsilon \rfloor + D^\sharp \sim 0$. Furthermore $(X, S_1 + S_2 + D^+ + \lfloor 2B \rfloor)$ is lc iff $(X', S + \lfloor 2B'_\epsilon \rfloor + D^\sharp)$ is lc (2.30) iff

$$(S, (\lfloor 2B'_\epsilon \rfloor + D^\sharp)|_S) = (S, D + \lfloor 2\Delta \rfloor + D_S^+)$$

is lc (5.50.2). The latter was checked to be lc in the first step. □

Proposition 7.38. *Let S be a smooth surface and $p : S \to T$ a proper birational morphism with exceptional set C; T normal. Let $D \subset S$ be a Weil divisor and Δ a (not necessarily effective) $\mathbb{Q}$-divisor such that*

    (1) *Supp D and Supp Δ have no common irreducible components,*
    (2) *$(S, D + \Delta)$ is plt and $D + \Delta$ is snc,*
    (3) *$p_* D \neq 0$ near each point of $p(C)$,*
    (4) *$\lfloor \Delta \rfloor \leq 0$, and $p_* \Delta$ is effective,*
    (5) *$-(K_S + D + \Delta)$ is p-nef.*

Then, if we shrink S around C, there is a Weil divisor $D_S^+ \subset S$ such that $(S, D + \lfloor 2\Delta \rfloor + D_S^+)$ is lc and $K_S + D + \lfloor 2\Delta \rfloor + D_S^+ \sim 0$.

Proof. D is smooth by (5.51). Take any point $t \in p(C)$ and a small open neighbourhood U of t. By (3.38), $(T, p_*(D + \Delta))$ is plt, hence $p(D)|_U$ is smooth (5.51) and non-empty (7.38.3), that is $D \cap p^{-1}(U)$ has exactly one irreducible component which is not contained in C. On the other hand, $D \cap p^{-1}(U)$ is connected by (5.49). Hence $D \cap p^{-1}(U)$ is irreducible and has exactly one point (say, x_t) in $p^{-1}(t)$. Since $t \in p(C)$ was arbitrary, $p : D \to T$ is finite.

By (7.39) we have a surjection

$$p_* \mathcal{O}_S(-K_S - D - \lfloor 2\Delta \rfloor) \twoheadrightarrow p_* \mathcal{O}_D(-K_D - \lfloor 2\Delta|_D \rfloor).$$

D is finite over T, hence any line bundle on D is generated by global sections if we shrink T around $p(C)$. Thus we can choose a general $D_S^+ \in |-K_S - D - \lfloor 2\Delta \rfloor|$ such that $D_S^+ \cap C \cap D = \emptyset$. That is, D_S^+ is disjoint from D near C. Hence over any $U \ni t$ as above, we have $D \cdot (D_S^+ + \lfloor 2\Delta \rfloor) = D \cdot \lfloor 2\Delta \rfloor \leq 1$ because D and Δ meet only at x_t, $D + \Delta$ is snc and the coefficients a_i of Δ satisfy $a_i < 1$, i.e. $\lfloor 2a_i \rfloor \leq 1$.

By construction, $0 \sim K_S + D + D_S^+ + \lfloor 2\Delta \rfloor$ and $(S, D + \lfloor 2\Delta \rfloor + D_S^+)$ is lc near C by (5.58). □

Proposition 7.39. *Let* $q : Y \to Z$ *be a proper birational morphism,* Y *smooth. Let* $S \subset Y$ *be an irreducible divisor and* B *a (not necessarily effective)* $\mathbb{Q}$*-divisor such that* $S+B$ *has snc. Assume that* $-(K_Y+S+B)$ *is* q*-nef. Then there is a surjection*

$$q_*(\mathcal{O}_Y(-K_Y - S - \lfloor 2B \rfloor)) \twoheadrightarrow q_*(\mathcal{O}_S(-K_S - \lfloor 2B|_S \rfloor)).$$

Proof. Notice that

$$-K_Y - 2S - \lfloor 2B \rfloor \equiv K_Y - 2(K_Y + S + B) + \{2B\}.$$

Thus $R^1 q_*(\mathcal{O}_Y(-K_Y - 2S - \lfloor 2B \rfloor)) = 0$ by (2.68). Pushing forward the exact sequence

$$
\begin{aligned}
0 \;\to\;\; & \mathcal{O}_Y(-K_Y - 2S - \lfloor 2B \rfloor) \;\;\to\;\; \mathcal{O}_Y(-K_Y - S - \lfloor 2B \rfloor) \\
\to\;\; & \mathcal{O}_S(-K_S - \lfloor 2B|_S \rfloor) \;\;\to\;\; 0
\end{aligned}
$$

to Z gives the result. $\qquad\square$

7.4 Semi-stable Flips

The following is the main result of this section.

Theorem 7.40. *Let* (X, B) *be a 3-dimensional klt pair with* B *effective. Let* $f : X \to Y$ *be a flipping contraction with respect to* $K_X + B$. *Assume that there exists a flat morphism* $s : Y \to C$ *to a smooth curve such that* $s \circ f$ *is lc. Then* f *has a flip.*

Remark 7.41. The arguments of this section work in the projective, the open algebraic and the analytic settings as well, except that one needs one small change (7.55) in the projective setting.

Corollary 7.42. *Let* (X, B) *be a 3-dimensional* $\mathbb{Q}$*-factorial dlt pair with* B *effective. Let* $f : X \to Y$ *be an extremal flipping contraction with respect to* $K_X + B$. *Assume that there exists a flat morphism* $s : Y \to C$ *such that* $s \circ f$ *is lc. Then* f *has a flip.*

Proof. If we replace B with $(1 - \epsilon)B$ for $0 < \epsilon \ll 1$, the flip does not change (6.5). This reduces the corollary to (7.40). $\qquad\square$

7.43. The proof of (7.40) consists of two main steps (7.44) and (7.51) followed by an easy reduction step (7.54). In the main steps, we create situations in which one can run and complete the MMP with only special semi-stable flips and divisorial contractions. This idea was first used in [Sho92].

The following is the first main step, called 'subtracting H'.

Theorem 7.44. *Let $(X, B + H)$ be a 3-dimensional $\mathbb{Q}$-factorial pair with effective $\mathbb{Q}$-divisors B, H and $f : X \rightarrow Y$ a projective birational morphism. Let $T \subset Y$ be a Cartier divisor, $S = f^*T$ and S_i the irreducible components of S. Assume the following:*

(1) $(X, S + B + H)$ *is dlt.*
(2) $(X \setminus S, 0)$ *is terminal.*
(3) *There exist $a > 0$ and $b_j \in \mathbb{Q}$ such that*

$$H \equiv_f a(K_X + B) + \sum b_j S_j.$$

(4) $K_X + B + H$ *is f-nef.*

Then (X, B) has a minimal model over Y.

Proof. We give a proof in the form of several lemmas by running a MMP over Y guided by H. The notation and the assumptions of (7.44) are assumed in these lemmas.

Lemma 7.45. *There exists a rational number $\lambda \in [0, 1]$ such that*

(1) $K_X + B + \lambda H$ *is f-nef, and*
(2) *if $\lambda > 0$ then there exists a $(K_X + B)$-negative extremal ray R over Y such that $R \cdot (K_X + B + \lambda H) = 0$.*

The lemma follows from (3.6). If $\lambda = 0$ then the theorem is proved. Assume $\lambda > 0$ and let $\phi : X \rightarrow V$ be the contraction of R.

Lemma 7.46. $K_X + B$ *is f-nef over $Y \setminus T$ and ϕ induces an isomorphism $X \setminus S \simeq V \setminus \phi(S)$.*

Proof. The first assertion follows from the conditions (3) and (4) and the second is a corollary to the first. $\square$

Lemma 7.47. *If ϕ contracts a divisor E, then the above conditions (1–4) still hold if we replace $f : X \rightarrow Y$ with $V \rightarrow Y$ and B, S, H with $\phi_* B, \phi_* S, \lambda \phi_* H$.*

Proof. The lemma follows from the general properties of divisorial contractions. $\square$

Lemma 7.48. *If ϕ is a flipping contraction, then ϕ is special. If $p : X \dashrightarrow X^+$ is the flip of ϕ (given by (7.32)), then the above conditions (1–4) still hold if we replace $f : X \rightarrow Y$ with $X^+ \rightarrow Y$ and B, S, H with $p_* B, p_* S, \lambda p_* H$.*

Proof. One has to prove that ϕ is special. Note first that $(X \setminus S, 0)$ is terminal by the condition (2). By the hypothesis $R \cdot (K_X + B + \lambda H) = 0$ and $R \cdot (K_X + B) < 0$, one sees that $R \cdot H > 0$. Hence by Condition (3) there exists j' such that $R \cdot S_{j'} \neq 0$ and by $R \cdot (\sum S_j) = 0$ there exists j'' such that $R \cdot S_{j''}$ has the opposite sign. This means that ϕ is special. $\square$

Lemma 7.49. *We can apply the above procedure to the new set up in the cases (7.47) and (7.48) if $H \neq 0$. After repeating this finitely many times, H becomes 0, and one obtains a minimal model of (X, B) over Y. In particular Theorem (7.44) holds.*

Proof. It is obvious that (7.47) does not occur infinitely many times. The flip in the assertion (7.48) is a $(K_X + B)$-flip with the flipping curve on S. Hence it cannot repeat infinitely many times (7.11). $\square$

7.50. The second main step says that if a good resolution exists then we can avoid non-special semi-stable flips during the modified MMP by adding an extra boundary. It is to be followed by the subtraction of the extra boundary (7.44), to reach the goal of the original MMP. The good resolution can be found (7.54) in the semi-stable case (7.40).

Theorem 7.51. *Let (X, B) be a klt threefold pair with B effective, $f : X \to Y$ a projective birational morphism with Y normal and $s : Y \to C$ a flat morphism such that $s \circ f$ is lc. Assume the following.*

(1) *$(s \circ f)(\mathrm{Ex}(f))$ is 0-dimensional.*
(2) *There are a projective resolution $g : Z \to X$ and a free linear system $|L|$ on Z such that*

   (a) *every fiber F of $Z \to C$ is reduced and the divisorial part of $F \cup g_*^{-1} B \cup \mathrm{Ex}(g)$ is snc,*
   (b) *$L \equiv_{f \circ g} ag^*(K_X + B) + \sum b_i G_i$ for some $a > 0$, $b_i \in \mathbb{Q}$ and irreducible divisors G_i in some fibers of $Z \to C$.*

Then (X, B) has a canonical model over Y.

Remark 7.52. The reader may be curious to see so many reduced boundary components thrown in during the proof. It should be emphasized that this made it possible to avoid non-special flips and at the same time it imposed the condition on $X \setminus (S \cup \mathrm{Supp}\lfloor B \rfloor)$ rather than $X \setminus S$ in (7.32.2).

Proof. Let $\Sigma \supset (s \circ f)(\mathrm{Ex}(f) \cup g(\cup_i G_i))$ be any reduced divisor of C such that $Z \to C$ is smooth over $C \setminus \Sigma$. Set $T = s^*(\Sigma)$, $S = f^*T$ and

$G = g^*S$. Then S, T and G are reduced by (7.51.2.a), and the G_i in (7.51.2.b) are irreducible components of G.

Let us write $g^*(K_X + B) = K_Z + P - N$ using effective P, N with no common components. Since $(s \circ f) : (X, B) \to C$ is lc, $(s \circ f \circ g) :$ $(Z, P - N) \to C$ is also lc. Hence P, G have no common components (7.2.5). Then $N \sim_g K_Z + P$ and N is g-exceptional. Let $M \subset Z$ be a reduced divisor such that $\operatorname{Supp} M = \operatorname{Supp} N \setminus \operatorname{Supp} G$. Then $(Z, G + P + M + (s \circ f \circ g)^*(\xi))$ is dlt for every $\xi \in C \setminus \Sigma$ (7.51.2.a). Let $L_1, \cdots, L_4$ be general members of $|L|$ so that $(Z, G + H + P + M)$ is dlt, where $H = \sum L_i$.

We run a $(K_Z + H + P + M)$-MMP over Y. For simplicity, we denote by the same symbols Z, B, H, etc. the variety to work on and the birational transforms during the MMP. Set $h : Z \to Y$.

Claim 7.53. During the MMP, we have

(1) $a(K_Z + P + M) \equiv_h H + a(M + N) + \sum b_i G_i$ $(a > 0, b_i \in \mathbb{Q})$,
(2) H is free, and $G \sim_h 0$,
(3) $(Z, G + H + P + M)$, $(Z, F + M)$ are dlt for each fiber F of $s \circ h$,
(4) every extremal contraction $\phi : Z \to W$ is flipping and special with $\operatorname{Ex}(\phi) \subset G$, or divisorial with $\operatorname{Ex}(\phi) \subset M \cup G$,
(5) $(Z \setminus (G + M), 0)$ is terminal.

Furthermore, the flips exist and terminate during the MMP. Hence the MMP ends, and then we have $M = 0$.

Proof of the claim. (7.53.1) follows from (7.51.2.b). Since (Z, H) remains dlt, $L_1 \cap \cdots \cap L_4 = \emptyset$ by (5.63). Hence $|L|$ remains free and so does H. $G = h^*(T)$. Hence (7.53.2). The MMP is both a $(K_Z + G + H + P + M)$-MMP by $G \sim_h 0$ and a $(K_Z + F + P + M)$-MMP by H nef. Hence (7.53.3).

For (7.53.4), let $\ell \subset \operatorname{Ex}(\phi)$ be an arbitrary irreducible curve. We have $a\ell \cdot (M + N) + \sum b_i (\ell \cdot G_i) < 0$ by (7.53.1) and (7.53.2).

Assume first $\ell \cdot G_i = 0$ for all G_i. Then $\ell \cdot M_k < 0$ whence $\ell \subset M_k$ for some normal irreducible surface $M_k \subset M$ (5.52). Let F be the fiber of $s \circ h$ through ℓ and $F_j \supset \ell$ an irreducible component of F. Then $(\ell \cdot F_j) = 0$ by $F_j = F$ if $F \not\subset G$ or by $F_j = G_i$ for some G_i if $F \subset G$. F_j is Cartier in codimension 2 (5.56) by $(Z, F + M)$ dlt, and $F_j|_{M_k} = \ell$ as a set in a neighbourhood of ℓ (5.62). Thus ℓ is a $\mathbb{Q}$-Cartier divisor on M_k with $(\ell)^2 = 0$. Hence $M_k \to \phi(M_k)$ is not biratonal (3.39) and $\operatorname{Ex}(\phi) = M_k$.

Assume $\ell \cdot G_i \neq 0$ for some G_i. Since $G = \sum G_i \sim_h 0$, we have another

G_j with $(\ell \cdot G_i) \cdot (\ell \cdot G_j) < 0$. Thus $\mathrm{Ex}(\phi) \subset G_i \cup G_j \subset G$. ϕ is special if flipping. Thus (7.53.4).

By (7.53.4), $Z \setminus (G+M)$ remain isomorphic during MMP. Since Z was originally smooth, (7.53.5) follows. The special flip exists by (7.32), and the termination holds (7.7). When the MMP ends, $K_Z + H + P + M$ is h-nef. $H|_{Z \setminus G} \equiv_h 0$ by (7.51.2.b). Hence $(M+N)|_{Z \setminus G} \equiv_h (K_Z + P + M)|_{Z \setminus G}$ is h-nef over $Y \setminus T \simeq X \setminus S$. Since it is an effective exceptional divisor over $Y \setminus T$, we see $M = 0$ (3.39).   □

Let $p : W \to Y$ be the minimal model. By G_W, H_W, etc. we denote the birational transforms of G, H, etc. By (7.53), one has $G_W = p^*T$, $(W \setminus G, 0)$ is terminal, $(W, G_W + P_W + H_W)$ is dlt, $K_W + P_W + H_W$ is p-nef and H_W is free. These satisfy the conditions of (7.44).

Thus by (7.44), (W, P_W) has a minimal model over Y. Since (X, B) is klt, $\lfloor P \rfloor = 0$ and $\lfloor P_W \rfloor = 0$. Hence applying the Relative Basepoint-free Theorem (3.24) to the minimal model, we get the canonical model of (W, P_W). Note that the $(K_Z + H + P + M)$-MMP was a part of a $(K_Z + P + M)$-MMP by (7.53.2). Thus (W, P_W) and $(Z, P + M)$ have the same canonical model over Y. Since $K_Z + P + M = g^*(K_X + B) + M + N$ and $N \subset \mathrm{Ex}(g)$, it is also the canonical model of (X, B) over Y (3.53).   □

7.54 (Proof of (7.40)). Let $a > 0$ be an integer such that $Q = a(K_X + B)$ is Cartier. We shrink Y to a neighbourhood of $f(\mathrm{Ex}(f))$ so that $f_* \mathcal{O}_X(Q)$ is generated by global sections. Let $\mathrm{Bs}|Q|$ be the scheme-theoretic intersection of all the members of $|Q|$. Note that $\mathrm{Supp}\, \mathrm{Bs}|Q| \subset \mathrm{Ex}(f)$ and $\mathrm{Supp}\, \mathrm{Bs}|Q|$ is of codimension ≥ 2.

Let $g_1 : X_1 \to X$ be the normalization of the blow up of X along $\mathrm{Bs}|Q|$. Note that there is a Cartier divisor $E_1 \subset \mathrm{Supp}\, (g_1)^{-1}\mathrm{Ex}(f)$, and $|L_1| = g_1^*|Q| - E_1$ is free.

Let $r : C' \to C$ be a finite covering with sufficient ramification to apply the semi-stable reduction (7.17), that is (7.19), to the morphism $X_1 \to C$ and the subset $\mathrm{Ex}(g_1) \cup (g_1)^{-1}(B)$. Let Y', X', X_1' be (the normalization of) the base change of Y, X, X_1 by r, respectively. Let Z' be the resolution of X_1' given in (7.17). We have the induced morphisms

$$
\begin{array}{ccccccccc}
Z' & \xrightarrow{g_2'} & X_1' & \xrightarrow{g_1'} & X' & \xrightarrow{f'} & Y' & \xrightarrow{s'} & C' \\
 & & \downarrow p_1 & & \downarrow p & & \downarrow q & & \downarrow r \\
 & & X_1 & \xrightarrow{g_1} & X & \xrightarrow{f} & Y & \xrightarrow{s} & C
\end{array}
$$

Set $g' : Z' \to X', B' = p^*B$ and $|L'| = (p_1 \circ g_2')^*|L_1|$. We check that (X', B'), $f' : X' \to Y'$ and $|L'|$ satisfy the conditions of (7.51).

Since (X, B) is klt and $(s \circ f) : (X, B) \to C$ is lc, we have (X', B') is klt (7.2) and $(s' \circ f') : (X', B') \to C'$ is lc (7.6). We have $\text{Ex}(f') \subset p^{-1}(\text{Ex}(f))$ and (7.51.1) follows. (7.51.2.a) follows from (7.17). $|L'|$ is free since $|L_1|$ is also free. By the construction, we have

$$L' \sim a(p \circ g')^*(K_X + B) - (p_1 \circ g_2')^* E_1.$$

We note that $(p_1 \circ g_2')^* E_1$ and $(g')^*(K_{X'} + B') - (p \circ g')^*(K_X + B)$ are equivalent to linear combinations of irreducible components of fibers of $Z' \to C'$ (7.23.3). Thus (7.51.2.b) follows and we can apply (7.51).

Now (X', B') has a canonical model over Y', which is the flip of f' (6.4) and f has a flip (6.9). $\qquad\qquad\qquad\qquad\qquad\qquad\qquad\qquad\qquad\square$

Remark 7.55. If X, Y and C are all projective and one wants to work only on projective varieties, only the shrinking argument in (7.54) has to be modified in this section.

Here is what you can do. Before defining Q in (7.54), take an ample divisor A on Y such that $f_* \mathcal{O}_X(a(K_X + B)) \otimes A$ is generated by global sections and set $Q = a(K_X + B) + f^* A$ without shrinking Y. The rest works without any changes.

7.5 Applications to Families of Surfaces

The aim of this section is to discuss three applications of the semi-stable MMP. First we consider the compactification of the moduli space of surfaces of general type. It is this and related problems that led to the investigation of semi-stable MMP historically.

Next we turn to a study of the miniversal deformation spaces $\text{Def}(H)$ for surface singularities. For quotient singularities one can use the semi-stable MMP to get a rather complete picture of $\text{Def}(H)$.

Finally we study simultaneous resolution for families of surfaces. The main result gives a necessary and sufficient condition for their existence.

7.56 (Compactification of moduli spaces).

It has been known for a long time that the set of all smooth projective algebraic varieties with fixed numerical invariants is in a natural one–to–one correspondence with the set of points of a variety, the so-called moduli space. The simplest example is elliptic curves, where the j-invariant establishes a correspondence between the set of elliptic curves over $\mathbb{C}$ and the points of $\mathbb{C}$. It has also been known that the set of all curves of a fixed genus g naturally corresponds to the points on a certain variety, usually denoted by M_g.

The spaces M_g are not proper, and there has been considerable interest in finding a natural compactification. This raises the question: What kind of objects should correspond to the boundary of M_g?

An answer was established in [DM69]:

Definition 7.57. A *stable curve* of genus g over a field k is a 1-dimensional proper scheme D over k which has only ordinary nodes as singularities such that ω_D is ample and $\chi(D, \mathcal{O}_D) = 1 - g$.

A *family of stable curves* is a proper and flat morphism $f : X \to S$ such that every fiber is a stable curve.

One can prove that the set of all stable curves of genus g naturally corresponds to the points on a variety $\bar{M}_g$, which contains M_g as an open and dense subvariety [DM69].

It should be emphasized that there are other solutions to the compactification problem. For instance, one can consider curves D which have only ordinary nodes and cusps as singularities such that ω_D is ample and $\chi(D, \mathcal{O}_D) = 1 - g$. We have to assume in addition that D does not have any irreducible components of genus 1 which intersect the rest of D in a single point. This definition also leads to a good theory.

We would like to have an analogous theory for higher dimensional varieties. Experience suggests that the following special case of (7.56) is the most important:

Let C be a smooth curve and $C^0 \subset C$ an open subset. Assume that we have a family $f^0 : X^0 \to C^0$, whose fibers are smooth curves of genus g. How can one extend this to a family $f : X \to C$?

In some sense, a satisfactory solution in case C is the spectrum of a DVR implies a solution of (7.56) in general. (This should be compared with the valuative criteria of separatedness and properness.)

The following lemma shows how this problem is related to the MMP. (The easy proof is left to the reader; cf. (7.59).)

Lemma 7.58. *Let $g : S \to C$ be a morphism from a surface to a smooth curve whose generic fiber is smooth. The following are equivalent:*

(1) *Every fiber has at worst ordinary nodes as singularities.*
(2) *If C' is any smooth curve and $h : C' \to C$ a non-constant morphism then $S \times_C C'$ is canonical.*
(3) *$g : S \to C$ is log canonical.*

Conditions (7.58.2–3) make sense in all dimensions, and they are also equivalent:

Lemma 7.59. *Let $g : X \to C$ be a morphism from a variety to a smooth curve whose generic fiber is canonical. The following are equivalent:*

(1) *If C' is any smooth curve and $h : C' \to C$ a non-constant morphism then $X \times_C C'$ is canonical.*

(2) *$g : X \to C$ is log canonical.*

Proof. (2) $\Rightarrow$ (1) follows from (7.2). To see the converse, pick a point $c \in C$ and choose $h : C' \to C$ such that h has a unique ramification point $c' = h^{-1}(c)$ and the ramification index is r. Then

$$K_{X'} = h^*(K_X + (1 - \tfrac{1}{r})g^{-1}(c)).$$

By assumption X' is canonical hence klt. Thus by (5.20) $(X, (1 - \tfrac{1}{r})g^{-1}(c))$ is klt. This holds for every $r > 0$, hence $(X, g^{-1}(c))$ is lc by (2.35). $\square$

7.60. It remains to find the higher dimensional analogue of (7.58.1). Inversion of adjunction (5.50) suggests that, at least for normal fibers, the right condition should be: Every fiber is lc.

In the non-normal case one needs to develop a suitable generalization of lc. This notion is called semi-log canonical, abbreviated as slc. Their theory is very similar to the log canonical case, once some foundational problems are settled. See [KSB88, K$^+$92] for details.

It turns out, however, that (7.58.1) does not have a higher dimensional analogue. Lc morphisms cannot be characterized by their fibers. This is shown by the following example.

Example 7.61. The degeneration of a variety to a cone over its hyperplane section is constructed as follows.

Let $Z \subset \mathbb{P}^n$ be a projectively normal variety with a hyperplane section $D = (x_n = 0)$. Let $C_Z \subset \mathbb{P}^{n+1}$ be the cone over Z with vertex at $(0 : \cdots : 0 : 1)$. The projection $\pi : (x_0 : \cdots : x_{n+1}) \mapsto (x_n : x_{n+1})$ gives a morphism $f_Z : X_Z \to \mathbb{P}^1$ where $X_Z := B_D C_Z$ is the blow up of C_Z along D. $f_Z^{-1}(s : t) \cong Z$ if $s \neq 0$ and $f_Z^{-1}(s : t) \cong C_D$ if $s = 0$.

Apply this construction to the Veronese surface $V \subset \mathbb{P}^5$ and to the degree 4 rational normal scroll $S \subset \mathbb{P}^5$. Both have the degree 4 rational normal curve $D \subset \mathbb{P}^4$ as their general hyperplane section. We can construct two families of surfaces $f_V : X_V \to \mathbb{P}^1$ (resp. $f_S : X_S \to \mathbb{P}^1$) such that $f_V^{-1}(0 : 1) \cong C_D$ and $f_V^{-1}(1 : t) \cong V$ (resp. $f_S^{-1}(0 : 1) \cong C_D$ and $f_S^{-1}(1 : t) \cong S$).

C_D is the cone over the degree 4 rational normal curve and its minimal resolution is the ruled surface F_4. Near the vertex C_D is isomorphic to the quotient singularity $\mathbb{C}^2/\mu_4(1,1)$, so it is log terminal.

Observe that $(K_V^2) = 9$, $(K_S^2) = 8$ and $(K_{C_D}^2) = 9$. In the first family the self-intersection of the canonical class of the fiber is constant but in the second family it jumps. Thus we see that the fibers alone do not adequately describe which families are 'good' and which families are 'bad'.

We get some insight into the nature of the problem if we notice that X_V has canonical singularities (its only singular point is the cone over the Veronese which is isomorphic to the terminal quotient singularity $\mathbb{C}^3/\mu_2(1,1,1)$), but K_{C_S} is not even $\mathbb{Q}$-Cartier.

This example shows that for families $f : X \to C$ over a smooth curve it is not enough to assume that every fiber is lc. We need to impose additional restrictions to guarantee that X is canonical. Essentially by (5.50), we only need to assume that $K_{X/C}$ is $\mathbb{Q}$-Cartier.

For morphisms to a general base $f : X \to Y$ this condition becomes more subtle. As long as Y is smooth, it is sufficient to assume only that $K_{X/Y}$ is $\mathbb{Q}$-Cartier. In general, especially when Y is not reduced, the corresponding condition is rather technical (cf. [Kol94b]).

The semi-stable MMP shows that, at least for surfaces, every 1-parameter family can be compactified:

Theorem 7.62. *Let C be a smooth curve over $\mathbb{C}$ and $C^0 \subset C$ an open subscheme. Let $f^0 : (X^0, B^0) \to C^0$ be a proper lc morphism such that $K_{X^0} + B^0$ is f^0-ample.*

Then there is a smooth curve C' and a finite and surjective morphism $h : C' \to C$ such that $(X^0, B^0) \times_C C' \to h^{-1}(C^0)$ extends to a proper lc morphism $f' : (X', B') \to C'$ with $K_{X'} + B'$ being f'-ample:

$$
\begin{array}{ccccc}
(X^0, B^0) & \leftarrow & (X^0, B^0) \times_C C' & \subset & (X', B') \\
f^0 \downarrow & & \downarrow & & f' \downarrow \\
C^0 & \leftarrow & C^0 \times_C C' & \subset & C'.
\end{array}
$$

Proof. Extend $f^0 : (X^0, B^0) \to C^0$ to a projective morphism $f : (X, B) \to C$. Apply (7.19) to get $h : C' \to C$ such that $(X^0, B^0) \times_C C'$ has a semi-stable log resolution. Now apply (7.10) to obtain $f' : (X', B') \to C'$.   □

7.63 (Deformation spaces of quotient singularities).   Let $(0 \in H)$ be a singularity. As we mentioned in (4.60), $(0 \in H)$ has a miniversal deformation space Def(H), unique up to local analytic isomorphism. (4.61) gives an explicit construction for hypersurface singularities.

In general, however, it is very difficult to describe the miniversal deformation space. Next we explain how one can use the semi-stable MMP to determine the number of irreducible components of Def(H) when H is a surface quotient singularity. Even this case turns out to be quite subtle. We start with an auxiliary result.

Lemma 7.64. *Let* $0 \in X$ *be a 3-dimensional terminal singularity and* $0 \in S \subset X$ *a Cartier divisor. The following are equivalent*

(1) (X, S) *is plt,*
(2) $0 \in S$ *is a quotient singularity,*
(3) $0 \in S$ *is either a Du Val singularity or analytically isomorphic to* $(xy + z^{dn} = 0)/\mu_n(1, -1, r)$ *for some* $(r, n) = 1$.

Proof. The equivalence of (1) and (2) follows from (5.50) and (4.18). (3) implies (2). The converse is not hard to establish using (5.43). See [KSB88, 3.10] for details. □

Theorem 7.65. *[KSB88, 3.9] Let* $(0 \in H)$ *be a surface quotient singularity. There is a one–to–one correspondence between the following two sets:*

(1) *irreducible components of* Def(H);
(2) *proper birational morphisms* $f : H' \to H$ *(up to isomorphism) such that* $K_{H'}$ *is* f-*ample and every singularity of* H' *is among those listed in (7.64.3).*

[KSB88] provides a rather tedious algorithm to enumerate the set (7.65.2). Very efficient algorithms to determine this set and further information about Def(H) can be found in [BC94, Ste93]. These ideas can be further generalized to describe the irreducible components of Def(H) for many more surface singularities, see [Kol91b, Sec. 6].

Sketch of proof. We just explain how to construct a birational morphism $f : H' \to H$ from an irreducible component $Z \subset$ Def(H). We need the result that every irreducible component of Def(H) contains smoothings (cf. [Art74]). That is, there is a flat deformation

$$
\begin{array}{ccc}
H & \subset & Y \\
\downarrow & & \downarrow g \\
0 & \in & C
\end{array}
$$

such that $g : Y \setminus H \to C$ is smooth and the image of the corresponding morphism $C \to \mathrm{Def}(H)$ is contained in Z.

We may also assume (possibly after a base change) that $g : Y \to C$ admits a semi-stable resolution. By (7.10) Y has a canonical model $f : X \to Y$. By (7.66), $H' := f^*H$ satisfies the requirements of (7.65.2).

Quite a few things remain to be established. We need to check that H' does not depend on the choice of $g : Y \to C$ and that every H' is obtained this way. These turn out to be consequences of properties of deformations of rational surface singularities and their partial resolutions. The reader should consult [KSB88, Sec. 3] for details. $\qquad\Box$

Lemma 7.66. *Let Y be a normal variety and $H \subset Y$ a Cartier divisor. Let $f : X \to Y$ be a proper birational morphism. Assume that*

(1) $(H, 0)$ *is normal and klt,*
(2) X *is $\mathbb{Q}$-factorial and K_X is f-nef,*

Then $f^*H = f_*^{-1}H$ and (X, f^*H) is plt in a neighbourhood of f^*H.

Proof. We can write $f^*H = f_*^{-1}H + E$ where E is an f-exceptional divisor. Let $y \in H$ be a codimension 1 point. Then H and hence also Y are smooth at y. Since K_X is f-nef, this implies that f is an isomorphism over y (3.39). This shows that $f_*^{-1}H \cap E$ is $f|_{f_*^{-1}H}$-exceptional.

$f_*^{-1}H$ is Cartier in codimension 2 by (5.56), hence

$$K_{f_*^{-1}H} \equiv (K_X + f_*^{-1}H)|_{f_*^{-1}H} \equiv_f (K_X - E)|_{f_*^{-1}H}.$$

On the other hand, $K_{f_*^{-1}H} \equiv f^*K_H + F$ where F is f-exceptional and $\lceil F \rceil$ is effective.

Putting these together we obtain that $F + E|_{f_*^{-1}H} \equiv_f K_X|_{f_*^{-1}H}$, in particular, $F + E|_{f_*^{-1}H}$ is f-nef. By (3.39) this implies that $-F - E|_{f_*^{-1}H}$ is effective. In $-F$ every divisor appears with coefficient < 1 and if $E \neq 0$ then at least one irreducible divisor in $E|_{f_*^{-1}H}$ appears with coefficient ≥ 1. Thus $E = 0$, $F \leq 0$ and so $f^*H = f_*^{-1}H$.

$K_{f_*^{-1}H} - F \equiv f^*K_H$, hence $(f_*^{-1}H, -F)$ is klt. Since $-F$ is effective, this implies that $(f_*^{-1}H, 0)$ is klt (2.35). Thus (X, f^*H) is plt in a neighbourhood of f^*H by (5.50). $\qquad\Box$

7.67 (Simultaneous resolution of families of surfaces). We have seen in section 4.3 that any flat family of Du Val singularities has a simultaneous resolution, at least after a finite and surjective base change. For an arbitrary flat family of surfaces $g : Y \to S$ it is of interest to find conditions which guarantee the existence of simultaneous resolutions.

The problem is essentially local on Y, but it turns out that a natural answer can be best obtained for proper families. The criterion can be made local, but it needs some care (cf. [KSB88, 2.25]).

Theorem 7.68. *[KSB88] Let S be a scheme over $\mathbb{C}$ and $g : Y \to S$ a flat and proper morphism whose fibers are surfaces with isolated singularities only. For every $s \in S$ let $\bar{Y}_s \to Y_s$ denote the minimal resolution of the fiber. Then*

(1) $s \mapsto (K^2_{\bar{Y}_s})$ *is lower semi-continuous,*

(2) g *has a simultaneous resolution after a finite and surjective base change $S' \to S$ iff $s \mapsto (K^2_{\bar{Y}_s})$ is locally constant.*

Sketch of proof. We prove only the case when $S = C$ is a smooth curve. This easily implies (7.68.1). (7.68.2) also turns out to be a consequence once some general results about the simultaneous resolution functor have been established (cf. [KSB88, Sec. 2]).

There is a semi-stable resolution after a base change $C' \to C$. To simplify notation we assume that $g : Y \to C$ itself has a semi-stable resolution. We can thus apply (7.9) to obtain

$$X \xrightarrow{f} Y \xrightarrow{g} C$$

where $X \to C$ is dlt and K_X is f-nef. Let X_c denote the fiber of $g \circ f$ over $c \in C$. X_c is the minimal resolution of Y_c for general $c \in C$, but for certain values, say $0 \in C$, the fiber may be reducible or singular. Write $X_0 = X'_0 + E_0$ where X'_0 is the birational transform of Y_0 and E is f-exceptional. Then

$$
\begin{aligned}
(K^2_{\bar{Y}_c}) &= (K^2_{X_c}) = (K^2_{X_0}) = ((K_{X_0}|_{X'_0})^2) + ((K_{X_0}|_E)^2) \\
&\geq ((K_{X_0}|_{X'_0})^2) = ((K_{X'_0} + E|_{X'_0})^2) \geq (K^2_{\bar{Y}_0}),
\end{aligned}
$$

the last inequality by (7.69). This proves (1). Equality holds iff $E = 0$ and X_0 has at worst Du Val singularities. In this case there is a simultaneous resolution after a further base change by (4.28). $\square$

Lemma 7.69. *Let S be a proper surface, $\bar{S} \to S$ its minimal resolution. Let $f_1 : S_1 \to S$ be a proper birational morphism such that $(S_1, 0)$ is klt. Let Δ_1 be an effective f_1-exceptional divisor such that $K_{S_1} + \Delta_1$ is f_1-nef. Then $((K_{S_1} + \Delta_1)^2) \geq (K^2_{\bar{S}})$ and equality holds iff $\Delta_1 = 0$ and S_1 has only Du Val singularities.*

Proof. Let $g_1 : S_1 \to S_2$ be the minimal model of $(S_1, 0)$ over S. Then $K_{S_1} \equiv g_1^* K_{S_2} + \Delta_2$ and Δ_2 is effective. Thus $K_{S_1} + \Delta_1 \equiv g_1^* K_{S_2} + (\Delta_1 + \Delta_2)$, hence $((K_{S_1} + \Delta_1)^2) \geq (K_{S_2}^2)$ by (7.70) applied to $S_1 \to S$.

Let $g_3 : S_3 \to S_2$ be the minimal resolution. Then $K_{S_3} \equiv g_3^* K_{S_2} - \Delta_3$ and Δ_3 is effective (4.3). Hence $(K_{S_2}^2) \geq (K_{S_3}^2)$ by (7.70).

Finally let $g_4 : S_3 \to \bar{S}$ be the minimal model of S_3 over S. As before, we obtain that $(K_{S_3}^2) \geq (K_{\bar{S}}^2)$.

Equality holds iff $\Delta_1 = 0$, $S_2 = S_1$ has Du Val singularities and $S_3 = \bar{S}$. $\qquad\square$

Lemma 7.70. *Let $g : U \to V$ be a birational morphism of normal and proper surfaces. Let N be a g-nef divisor and E an effective g-exceptional divisor such that $N + E$ is also g-nef. Then $((N + E)^2) \geq (N^2)$ and equality holds iff $E = 0$.*

Proof. $((N + E)^2) = (N^2) + (N \cdot E) + (N + E \cdot E) \geq (N^2)$. If equality holds then $(N \cdot E) = (N + E \cdot E) = 0$, hence also $(E^2) = 0$. By (3.40) this implies that $E = 0$. $\qquad\square$

7.6 A Survey of Further Results

The purpose of this section is to survey related results and further developments along the lines discussed in the book. We aim to give a reasonably complete list of references, rather than a detailed explanation of the precise statements.

7.71 (Flips). One of the main results, mentioned already several times, is the existence of flips in dimension 3. The classical case of threefolds with terminal singularities is treated in [Mor88]. Later this method was developed to obtain a classification of flips with terminal singularities [KM92]. This result has several applications, for instance the deformation invariance of the plurigenera for smooth projective threefolds and the existence of a coarse moduli space for threefolds of general type up to birational equivalence [KM92, Chap. 12].

The more general case of flips for threefold lc pairs (X, Δ) is treated in [Sho92]. Many of the ideas of this paper are developed further in the seminar notes [K+92, Chaps. 16–22]. These notes also contain another proof of log canonical threefold flips in [K+92, Chaps. 4–8], by reducing their existence to the classical case treated in [Mor88]. The reduction is very similar to the one we used in section 5.4.

The existence of flips in higher dimensions is still unknown. This is probably the most important open problem of the theory. The most significant special cases settled are toric varieties [Rei83a] and varieties with certain other algebraic group actions [BK94].

7.72 (Flips and GIT). Since the discovery of flips it has been realized that flips appear in many contexts. One of the most important of these is the realization that flips appear naturally in the geometric invariant theory of [Mum65]. The geometric invariant theory quotient of a scheme by the action of a group depends on the choice of an ample line bundle, called a polarization. Different choices of the polarization lead to different quotients. These quotients are birational to each other and the birational maps are frequently compositions of flips and inverses of flips. The general theory is developed in [Tha96, DH98, MW97].

7.73 (Kodaira dimension). Let X be a smooth projective variety. For m sufficiently large and divisible, the image of $|mK_X| : X \dashrightarrow \mathbb{P}$ does not depend on m up to birational equivalence. The dimension of the image is called the *Kodaira dimension* of X. It is denoted by $\kappa(X)$. The investigation of the Kodaira dimension, and especially a study of its behavior in fiber spaces was initiated by Iitaka. Most of the results in this direction predate the minimal model theory. [Uen75] is a good introduction while [Mor87] is a survey of the later results.

7.74 (Abundance and the Canonical Rings). We have already mentioned the finite generation of canonical rings and abundance in section 6.2. The older version is the finite generation question:

Is the canonical ring $\oplus_{m=0}^{\infty} H^0(X, \mathcal{O}_X(mK_X))$ of a smooth projective variety finitely generated?

The first significant result for threefolds is due to [Fuj86], who proved finite generation under the assumption that $\kappa(X) \leq 2$. His method, reducing the 3-dimensional question to a 2-dimensional lc pair (S, Δ), is one of the early successes of the theory of log canonical pairs.

As explained in section 6.2, assuming the appropriate MMP, abundance implies finite generation (but probably not the other way around). Abundance for threefolds turned out to be quite hard, the case $\kappa = 1$ and $\kappa = 2$ requiring different methods. The proof was established in a series of papers by Kawamata and Miyaoka. A simplified version can be found in [K+92, Chaps. 9–15], which also contains many references. Abundance for threefold lc pairs (X, Δ) is proved in [KMM94a].

Another interesting question is the following. Find a constant $c(n)$

such that if X is a smooth projective variety of dimension n and $\kappa(X) \geq 0$ then $|mK_X| \neq \emptyset$ for some $0 < m \leq c(n)$.

For threefolds the $\kappa = 0$ case is treated in [Kaw86], the $\kappa = 1$ case was settled by Mori (unpublished) and the $\kappa = 2$ case is in [Kol94a]. The $\kappa = 3$ case is still open, despite some partial results by [Kol86a, Luo94].

7.75 (Fujita's Conjecture). Let X be a smooth projective variety and L an ample divisor on X. Fujita conjectured that $|K_X + mL|$ is free for $m \geq \dim X + 1$. More generally, one can ask which linear combinations of K_X and L are nef, free, ample or very ample. There have been many results in this direction, with significant contributions by Reider, Demailly, Ein–Lazarsfeld, Kawamata, Angehrn–Siu, Helmke and Smith. The lectures of [Laz96] provide a very good introduction.

The strongest current results say that Fujita's conjecture is true in dimensions ≤ 4 [Kaw97]. In any dimension, $|K_X + mL|$ is basepoint free for $m > \frac{1}{2} \dim X (\dim X + 1)$ [AS95]. [Kol97, secs. 5–6] contains a survey of recent results.

The methods developed in these works can also be applied to the study of Abelian varieties. This was first noticed in [Kol95, Chap.17] while the strongest results in this direction are contained in [EL97]. We mention one, in order to give a flavour of these applications:

Let (A, Θ) be a principally polarized Abelian variety, which is not a product. Then (A, Θ) is canonical. Equivalently, Θ is normal with only rational singularities.

7.76 (Birational geometry of Fano fiber spaces). We saw in section 3.8 that canonical models are unique and minimal models are unique in codimension 1. By contrast, if an MMP starting with X produces a Fano fiber space, then different choices during the MMP frequently result in very different Fano fiber spaces. It seems very hard to tell when two Fano fiber spaces are birational to each other.

The MMP for pairs helps with this problem in two ways. First, by a suitable choice of the boundary Δ, one can frequently run the MMP for (X, Δ) which results in an especially simple Fano fiber space. This was explored in the papers [Ale91, Cor96]. Another application of this approach is in [CF93].

Second, by combining the techniques of the MMP with earlier ideas of Sarkisov, [Cor94] developed a method that provides a factorization of any birational map between Fano fiber spaces into elementary steps. Unfortunately these steps are not very easy to understand, and the complete analysis has been carried out in only a few cases.

7.77 (Fano threefolds). The MMP suggests that one should study in detail the varieties obtained as the end result of the program. One such class is Fano threefolds, where we allow terminal singularities. Smooth Fano threefolds have been classified earlier, see [Isk80, MM81]. There are 104 distinct deformation types. Much less is known about singular Fano threefolds. There are only finitely many deformation types of $\mathbb{Q}$-factorial terminal Fano threefolds with $\rho = 1$ by [Kaw92a]. The same holds for Fano threefolds with canonical singularities, without assuming $\mathbb{Q}$-factoriality or $\rho = 1$ [KMM94b]. So far there is no practical way of obtaining the actual list. The examples developed in [Fle89] suggest that there may be very many cases.

7.78 (Calabi–Yau threefolds). Calabi–Yau varieties are normal varieties whose canonical class is numerically trivial. (Different authors pose different restrictions on the singularities.) These form an important class of varieties in their own right since the ones with terminal singularities are minimal models with Kodaira dimension 0. (Assuming the abundance conjecture, the converse also holds.) Most of the recent interest in them, however, stems from their connection with string theory (see, for instance, [Mor97] and the extensive list of references there). The number of deformation types is enormous, possibly infinite. The conjectures of [Rei87a] attempt to relate all Calabi–Yau threefolds to each other. A significant step in this direction is the smoothablility of many Calabi–Yau threefolds established in [NS95]. Calabi–Yau threefolds related to Abelian threefolds are characterized in [SBW94].

7.79 (Elliptic threefolds). A variety X is called *elliptic* if it admits a morphism $f : X \to Z$ whose general fiber is an elliptic curve. Then $\kappa(X) < \dim X$ and every variety with $\kappa(X) = \dim X - 1$ is birational to an elliptic one. As in Kodaira's canonical bundle formula for elliptic surfaces, the key point seems to be to understand the behaviour of the singular fibers. The papers [Nak91, DG94] give an almost complete description in dimension three.

7.80 (MMP for complex manifolds). There are several difficulties in trying to formulate the MMP for proper varieties or for compact complex manifolds. The basic problem is that the cone of curves is probably not the right object to start with, since an effective curve can be numerically trivial.

It is, however, possible, that all these unusual cases can be described in geometrically clear ways, and there may be a corresponding contraction morphism. Some special cases are discussed in [Kol91a].

For Kähler manifolds these problems do not appear, and a rather close analogue of the MMP is likely to work. The first steps in this direction are developed in [CP97].

7.81 (MMP in positive and mixed characteristic). It is believed that the MMP works in positive and in mixed characteristic cases.

The first steps for threefolds in positive characteristic were taken in [Kol91a]. [Kee98] develops a new approach to basepoint freeness in positive characteristic which proves the contraction theorem in many cases. Very little is known about flips.

For semi-stable families of surfaces in positive and in mixed characteristic the whole MMP has been carried out in [Kaw94].

7.82 (Deformation invariance of plurigenera). Recently [Siu98] proved that if $f : X \to S$ is a smooth family of varieties of general type then the plurigenera of the fibers $h^0(X_s, \mathcal{O}_{X_s}(mK_{X_s}))$ are locally constant functions on S for $m \geq 1$. This was generalized by [Kaw98] to the case when the fibers have canonical singularities and to reducible fibers by [Nak98]. [Kaw98] also shows that a small deformation of a canonical singularity is again canonical and [Nak98] proves that a small deformation of a terminal singularity is again terminal.

Bibliography

[Abh56] S. S. Abhyankar. On the valuations centered in a local domain. *Amer. J. Math.*, 78:321–348, 1956.

[AdJ97] D. Abramovich and A. J. de Jong. Smoothness, semistability and toroidal geometry. *Jour. Alg. Geom.*, 6:789–802, 1997.

[AGZV85] V. I. Arnold, S. M. Gusein-Zade, and A. N. Varchenko. *Singularities of Differentiable Maps I–II.* Birkhäuser, 1985.

[AK97] D. Abramovich and K. Karu. Weak semistable reduction in characteristic 0. Preprint, 1997.

[Ale91] V. A. Alexeev. General elephants on Q-Fano 3-folds. *Comp. Math.*, 91:91–116, 1991.

[Art62] M. Artin. Some numerical criteria for contractibility of curves on algebraic surfaces. *Amer. J. Math.*, 84:485–496, 1962.

[Art66] M. Artin. On isolated rational singularities of surfaces. *Amer. J. Math.*, 88:129–136, 1966.

[Art69a] M. Artin. Algebraic approximation of structures over complete local rings. *Publ. Math. IHES*, 36:23–58, 1969.

[Art69b] M. Artin. Algebraisation of formal moduli I. In D. C. Spencer and S. Iyanaga, editors, *Global Analysis*, pages 21–72. Tokyo Press - Princeton Univ. Press, 1969.

[Art70] M. Artin. Algebraisation of formal moduli II. *Ann. of Math.*, 91:88–135, 1970.

[Art74] M. Artin. Algebraic construction of Brieskorn resolutions. *J. Alg.*, 29:330–348, 1974.

[Art76] M. Artin. *Deformations of Singularities.* Tata Lecture Notes, 1976.

[Art86] M. Artin. Néron models. In G. Cornell and J. Silverman, editors, *Arithmetic Geometry*, pages 213–230. Springer Verlag, 1986.

[AS95] U. Angehrn and Y.-T. Siu. Effective freeness and point separation for adjoint bundles. *Invent. Math.*, 122:291–308, 1995.

[AW97] M. Andreatta and J. Wiśniewski. A survey on contractions of higher dimensional varieties. In *Algebraic Geometry, Santa Cruz 1995*, volume 62. Amer. Math. Soc., 1997.

[BC94] K. Behnke and J. Christophersen. M–resolutions and deformations of quotient singularities. *Amer. J. Math.*, 116:881–903, 1994.

[Ben83] X. Benveniste. Sur l'anneau canonique de certaines variétés de dimension 3. *Inv. Math.*, 73:157–164, 1983.

242   *Bibliography*

[BG71] S. Bloch and D. Gieseker. The positivity of the Chern classes of an ample vector bundle. *Inv. Math.*, 12:112–117, 1971.

[BK94] M. Brion and F. Knop. Contraction and flip for varieties with group action of small complexity. *J. Math. Sci. Univ. Tokyo*, 1:641–655, 1994.

[Bla56] M. A. Blanchard. Sur les variétés analytiques complexes. *Ann. Sci. Ec. Norm. Sup.*, 73:157–202, 1956.

[BP96] F. A. Bogomolov and T. G. Pantev. Weak Hironaka theorem. *Math. Res. Lett.*, 3:299–308, 1996.

[BPdV84] W. Barth, C. Peters, and A. Van de Ven. *Compact Complex Surfaces*. Springer, 1984.

[Bre97] G. Bredon. *Sheaf Theory*. Springer, 1997.

[Bri66] E. Brieskorn. Über die Auflösung gewisser Singularitäten von holomorphe Abbildungen. *Math. Ann.*, 166:76–102, 1966.

[Bri68] E. Brieskorn. Rationale Singularitäten komplexer Flächen. *Inv. Math.*, 4:336–358, 1968.

[Bri71] E. Brieskorn. Singular elements of semisimple algebraic groups. In *Proc. Int. Congr. Math. Nice*, pages 279–284. Gauthier-Villars, 1971.

[BS76] C. Bănică and O. Stănăşilă. *Algebraic Methods in the Global Theory of Complex Spaces*. J. Wiley and Sons, 1976.

[CF93] F. Campana and H. Flenner. Projective threefolds containing a smooth rational surface with ample normal bundle. *J. f.r.u.a. Math.*, 440:77–98, 1993.

[CKM88] H. Clemens, J. Kollár, and S. Mori. *Higher Dimensional Complex Geometry*. Astérisque, 1988.

[Cor94] A. Corti. Factoring birational maps of threefolds after Sarkisov. *J. Alg. Geom.*, 4:223–254, 1994.

[Cor96] A. Corti. Del Pezzo surfaces over Dedekind schemes. *Ann. Math.*, 144:641–683, 1996.

[CP97] F. Campana and Th. Peternell. Towards a Mori theory on compact Kähler threefolds. *Math. Nachr.*, 187:29–59, 1997.

[Cut88a] S. D. Cutkosky. Elementary contractions of Gorenstein threefolds. *Math. Ann.*, 280:521–525, 1988.

[Cut88b] S. D. Cutkosky. Weil divisors and symbolic algebras. *Duke Math. J.*, 57:175–183, 1988.

[DG94] I. Dolgachev and M. Gross. Elliptic threefolds I: Ogg–Shafarevich theory. *J. Alg. Geom.*, 3:39–80, 1994.

[DH98] I. Dolgachev and Y. Hu. Variation of geometric invariant theory quotient. *Publ. Math. IHES*, to appear, 1998.

[DM69] P. Deligne and D. Mumford. The irreducibility of space of curves of given genus. *Publ. Math. IHES*, 36:75–110, 1969.

[Dur79] A. Durfee. Fifteen characterizations of rational double points and simple critical points. *L'Ens. Math.*, 25:131–163, 1979.

[DV34] P. Du Val. On isolated singularities of surfaces which do not affect the conditions of adjunction I–III. *Proc. Camb. Phil. Soc.*, 30:453–465, 483–491, 1934.

[EL97] L. Ein and R. Lazarsfeld. Singularities of theta divisors and the birational geometry of irregular varieties. *Jour. AMS*, 10:243–258, 1997.

[Elk78] R. Elkik. Singularités rationnelles et déformations. *Inv. Math.*, 47:139–147, 1978.

[Elk81] R. Elkik. Rationalité des singularités canoniques. *Inv. Math.*, 64:1–6, 1981.

[EV92] H. Esnault and E. Viehweg. *Lectures on Vanishing Theorems.* Birkhäuser, 1992.

[Fle81] H. Flenner. Rational singularities. *Arch. Math.*, 36:35–44, 1981.

[Fle89] A. Fletcher. Working with weighted complete intersections. preprint, 1989.

[Fuj83] T. Fujita. Semipositive line bundles. *J. Fac. Sci. Tokyo Univ.*, 30:353–378, 1983.

[Fuj85] T. Fujita. A relative version of Kawamata-Viehweg's vanishing theorem. preprint, 1985.

[Fuj86] T. Fujita. Zariski decomposition and canonical rings of elliptic threefolds. *J. Math. Soc. Japan*, 38:19–37, 1986.

[Ful84] W. Fulton. *Intersection Theory.* Springer, 1984.

[Fur86] M. Furushima. Singular Del Pezzo surfaces. *Nagoya Math. J.*, 104:1–28, 1986.

[GH78] P. Griffiths and J. Harris. *Principles of Algebraic Geometry.* John Wiley and Sons, Inc., 1978.

[God58] R. Godement. *Topologie algébrique et théorie des faisceaux.* Hermann, Paris, 1958.

[GR70] H. Grauert and O. Riemenschneider. Verschwindungssätze für analytische Kohomologiegruppen auf komplexen Räumen. *Invent. Math.*, 11:263–292, 1970.

[GR84] H. Grauert and R. Remmert. *Coherent analytic sheaves.* Springer, 1984.

[Gra62] H. Grauert. Über Modifikationen und exzeptionelle analytische Mengen. *Inv. Math.*, 146:331–368, 1962.

[Gra72] H. Grauert. Über die Deformationen isolierter Singularitäten analytischer Mengen. *Inv. Math.*, 15:171–198, 1972.

[Gro68] A. Grothendieck. *Cohomologie Locale des Faisceaux Cohérents et Théorèmes de Lefschetz Locaux et Globaux — SGA 2.* North Holland, 1968.

[Har66] R. Hartshorne. *Residues and Duality.* Springer Lecture Notes vol. 20, 1966.

[Har70] R. Hartshorne. *Ample Subvarieties of Algebraic Varieties.* Springer Lecture Notes vol. 156, 1970.

[Har77] R. Hartshorne. *Algebraic Geometry.* Springer, 1977.

[Hen11] A. Henderson. *The Twenty–seven Lines upon the Cubic Surface.* Cambridge Univ. Press, 1911.

[Hir60] H. Hironaka. *On the theory of birational blowing up.* PhD thesis, Harvard Univ., 1960.

[Hir64] H. Hironaka. Resolution of singularities of an algebraic variety over a field of characteristic zero. *Ann. of Math.*, 79:109–326, 1964.

[Hir71] F. Hirzebruch. *The Hilbert Modular Groups, Resolution of the Singularities at the Cusps and Related Problems*, pages 275–288. Springer Lecture Notes vol. 244, 1971.

[HS71] P. Hilton and U. Stammbach. *A course in Homological Algebra.* Springer, 1971.

[Iit77] S. Iitaka. On logarithmic Kodaira dimension of algebraic varieties. In *Complex Analysis and Algebraic Geometry*, pages 175–190. Cambridge Univ. Press, 1977.

[Isk80] V. A. Iskovskikh. Anticanonical models of three-dimensional algebraic varieties. *J. Soviet Math.*, 13:745–814, 1980.

[K⁺92] J. Kollár et al. *Flips and Abundance for Algebraic Threefolds.* Soc. Math. France, Astérisque vol. 211, 1992.

[Kaw82] Y. Kawamata. A generalisation of Kodaira-Ramanujam's vanishing theorem. *Math. Ann.*, 261:43–46, 1982.

[Kaw84a] Y. Kawamata. The cone of curves of algebraic varieties. *Annals of Math.*, 119:603–633, 1984.

[Kaw84b] Y. Kawamata. On the finiteness of generators of the pluri-canonical ring for a threefold of general type. *Amer. J. Math.*, 106:1503–1512, 1984.

[Kaw86] Y. Kawamata. On the plurigenera of minimal algebraic threefolds with $K \approx 0$. *Math. Ann.*, 275:539–546, 1986.

[Kaw88] Y. Kawamata. The crepant blowing-up of 3-dimensional canonical singularities and its application to the degeneration of surfaces. *Annals of Math.*, 127:93–163, 1988.

[Kaw91] Y. Kawamata. On the length of an extremal rational curve. *Inv. Math.*, 105:609–611, 1991.

[Kaw92a] Y. Kawamata. Boundedness of Q-Fano threefolds. In *Proc. Int. Conf. Algebra*, pages 439–445. Contemp. Math. vol. 131, 1992.

[Kaw92b] Y. Kawamata. Termination of log flips for algebraic 3-folds. *Internat. J. Math*, 3:653–660, 1992.

[Kaw94] Y. Kawamata. Semistable minimal models of threefolds in positive or mixed characteristic. *J. Alg. Geom.*, 3:463–491, 1994.

[Kaw97] Y. Kawamata. On Fujita's freeness conjecture for 3-folds and 4-folds. *Math. Ann*, 308:491–505, 1997.

[Kaw98] Y. Kawamata. Deformations of canonical singularities. *(to appear)*, 1998.

[Kee98] S. Keel. Basepoint freeness for nef and big line bundles in positive characteristic. *Annals of Math.*, to appear, 1998.

[KKMSD73] G. Kempf, F. Knudsen, D. Mumford, and B. Saint-Donat. *Toroidal Embeddings I.* Springer Lecture Notes vol. 339, 1973.

[Kle66] S. Kleiman. Toward a numerical theory of ampleness. *Annals of Math.*, 84:293–344, 1966.

[KM87] Y. Kawamata and K. Matsuki. The number of the minimal models for a 3-fold of general type is finite. *Math. Ann.*, 276:595–598, 1987.

[KM92] J. Kollár and S. Mori. Classification of three dimensional flips. *Jour. AMS*, 5:533–703, 1992.

[KMM87] Y. Kawamata, K. Matsuda, and K. Matsuki. *Introduction to the Minimal Model Problem*, volume 10 of *Adv. Stud. Pure Math*, pages 283–360. Kinokuniya — North-Holland, 1987.

[KMM94a] S. Keel, K. Matsuki, and J. McKernan. Log abundance theorem for threefolds. *Duke Math. J.*, 75:99–119, 1994.

[KMM94b] J. Kollár, Y. Miyaoka, and S. Mori. Boundedness of Q-Fano threefolds. (unpublished), 1994.

[Kol83] J. Kollár. *Toward moduli of singular varieties.* PhD thesis, Brandeis Univ., 1983.

[Kol84] J. Kollár. The cone theorem. *Annals of Math.*, 120:1–5, 1984.

[Kol86a] J. Kollár. Higher direct images of dualizing sheaves I. *Annals of Math.*, 123:11–42, 1986.

[Kol86b] J. Kollár. Higher direct images of dualizing sheaves II. *Annals of Math.*, 124:171–202, 1986.

[Kol89] J. Kollár. Flops. *Nagoya Math. J.*, 113:15–36, 1989.

[Kol91a] J. Kollár. Extremal rays on smooth threefolds. *Ann. Sci. ENS*, 24:339–361, 1991.

[Kol91b] J. Kollár. Flips, flops, minimal models, etc. *Surv. in Diff. Geom.*, 1:113–199, 1991.

[Kol92] J. Kollár. Cone theorems and bug-eyed covers. *J. Alg. Geom.*, 1:293–323, 1992.

[Kol94a] J. Kollár. *Log Surfaces of General Type; Some Conjectures*, volume 162 of *Contemp. Math.*, pages 261–275. AMS, 1994.

[Kol94b] J. Kollár. Push forward and base change for open immersions. Salt Lake City summer school on Moduli of Surfaces, 1994.

[Kol95] J. Kollár. *Shafarevich Maps and Automorphic Forms*. Princeton Univ. Press, 1995.

[Kol96] J. Kollár. *Rational Curves on Algebraic Varieties*, volume 32 of *Ergebnisse der Math.* Springer Verlag, 1996.

[Kol97] J. Kollár. Singularities of pairs. In *Algebraic Geometry, Santa Cruz 1995*, volume 62 of *Proc. Symp. Pure Math.* Amer. Math. Soc., 1997.

[KSB88] J. Kollár and N. Shepherd-Barron. Threefolds and deformations of surface singularities. *Inv. Math.*, 91:299–338, 1988.

[Lam86] K. Lamotke. *Regular Solids and Isolated Singularities*. Vieweg, 1986.

[Lau73] H. Laufer. Taut two-dimensional singularities. *Math. Ann.*, 205:131–164, 1973.

[Lau77] H. Laufer. On minimally elliptic singularities. *Amer. J. Math.*, 99:1257–1295, 1977.

[Laz96] R. Lazarsfeld. Lectures on linear series. In *Complex Algebraic Geometry – Park City*, volume 3 of *Park City/IAS Math. Ser.*, pages 161–220. Amer. Math. Soc., 1996.

[Lip69] J. Lipman. Rational singularities with applications to algebraic surfaces and unique factorization. *Publ. Math. IHES*, 36:195–279, 1969.

[Loo84] E. Looijenga. *Isolated singular points on complete intersections.* Cambridge Univ. Press., 1984.

[Luo94] T. Luo. Plurigenera of regular threefolds. *Math. Zeitschr.*, 217:37–46, 1994.

[Mat69] H. Matsumura. *Commutative Algebra.* Benjamin/Cummings, 1969.

[Mat86] H. Matsumura. *Commutative Ring Theory.* Cambridge Univ. Press, 1986.

[MM81] S. Mori and S. Mukai. Classification of Fano threefolds with $b_2 \geq 2$. *Manuscr. Math*, 36:147–162, 1981.

[MM86] Y. Miyaoka and S. Mori. A numerical criterion for uniruledness. *Annals of Math.*, 124:65–69, 1986.

[Mor79] S. Mori. Projective manifolds with ample tangent bundles. *Annals of Math.*, 110:593–606, 1979.

[Mor82] S. Mori. Threefolds whose canonical bundles are not numerically effective. *Annals of Math.*, 116:133–176, 1982.

[Mor85] S. Mori. On 3-dimensional terminal singularities. *Nagoya Math. J.*, 98:43–66, 1985.

[Mor87] S. Mori. Classification of higher-dimensional varieties. In *Algebraic Geometry, Bowdoin 1985*, volume 46 of *Proc. Symp. Pure Math.*, pages 269–332. Amer. Math. Soc., 1987.

[Mor88] S. Mori. Flip theorem and the existence of minimal models for 3-folds. *Jour. AMS*, 1:117–253, 1988.

[Mor97] D. Morrison. Mathematical aspects of mirror symmetry. In J. Kollár, editor, *Complex Algebraic Geometry*, pages 265–340. AMS IAS/Park City Series, 1997.

[MP97] Y. Miyaoka and T. Peternell. *Geometry of Higher Dimensional Algebraic Varieties*. Birkhäuser, 1997.

[Mum65] D. Mumford. *Geometric Invariant Theory*. Springer, 1965.

[MW97] K. Matsuki and R. Wentworth. Mumford–Thaddeus principle on the moduli space of vector bundles on a surface. *Intern. J. Math*, 8:97–148, 1997.

[Nag60] M. Nagata. On rational surfaces I–II. *Mem. Coll. Sci. Univ. Kyoto*, 32, 33:351–370; 271–329, 1960.

[Nak87] N. Nakayama. The lower semi-continuity of the plurigenera of complex varieties. In *Algebraic Geometry, Sendai,1985*, volume 10 of *Adv. Studies in Pure Math*, pages 551–590. Kinokuniya - North-Holland, 1987.

[Nak91] N. Nakayama. Elliptic fibrations over surfaces. In *Algebraic Geometry and Analytic Geometry*, pages 126–137. Springer Verlag, 1991.

[Nak98] N. Nakayama. Invariance of the plurigenera of algebraic varieties. *RIMS Preprint*, 1998.

[NS95] Y. Namikawa and J. Steenbrink. Global smoothing of Calabi–Yau threefolds. *Inv. Math.*, 122:403–419, 1995.

[Par98] K. H. Paranjape. Bogomolov–Pantev resolution, an expository account. In *New trends in Algebraic Geometry, Warwick, July 1996*. Cambridge Univ. Press, 1998.

[Pin80] H. Pinkham. Résolutions simultanée de points doubles rationels. In *Séminaire sur les Singularités des Surfaces*, pages 179–204. Springer Lecture Notes vol. 777, 1980.

[PS97] É. Picard and G. Simart. *Théorie des fonctions algébriques*. Gauthiers–Villars, 1897.

[Ree58] D. Rees. On a problem of Zariski. *Ill. J. Math.*, 2:145–149, 1958.

[Rei76] M. Reid. Elliptic Gorenstein singularities of surfaces. Preprint, 1976.

[Rei80] M. Reid. Canonical threefolds. In A. Beauville, editor, *Géométrie Algébrique Angers*, pages 273–310. Sijthoff & Noordhoff, 1980.

[Rei83a] M. Reid. Decomposition of toric morphisms. In *Arithmetic and Geometry II.*, volume 36 of *Progress in Math*. Birkhäuser, 1983.

[Rei83b] M. Reid. Minimal models of canonical threefolds. In S. Iitaka, editor, *Algebraic Varieties and Analytic Varieties*, volume 1 of *Adv. Stud. Pure Math.*, pages 131–180. Kinokuniya and North-Holland, 1983.

[Rei83c] M. Reid. Projective morphisms according to Kawamata. Preprint, 1983.

[Rei87a] M. Reid. The moduli space of threefolds with $K = 0$ may nevertheless be irreducible. *Math. Ann.*, 278:329–334, 1987.

[Rei87b] M. Reid. Young person's guide to canonical singularities. In *Algebraic Geometry Bowdoin 1985*, volume 46 of *Proc. Symp. Pure Math*. Amer. Math. Soc., 1987.

[SBW94] N. I. Shepherd-Barron and P. Wilson. Singular threefolds with numerically trivial first and second Chern classes. *J. Alg. Geom.*, 3:265–281, 1994.

[Ser56] J.-P. Serre. Géometrie algébrique et géometrie analytique. *Ann. Inst. Fourier*, 6:1–42, 1956.

[Sha94] R. I. Shafarevich. *Basic Algebraic Geometry*. Springer Verlag, 1994.

[Sho85] V. Shokurov. The nonvanishing theorem. *Izv. A. N. SSSR Ser. Mat.*, 49:635–651, 1985.

[Sho92] V. Shokurov. 3-fold log flips. *Izv. Russ. A. N. Ser. Mat.*, 56:105–203, 1992.

[Sie69] C. L. Siegel. *Topics in Complex Function Theory*, vols. I–III. Wiley, 1969.

[Siu98] Y.-T. Siu. Deformation invariance of plurigenera. *Invent. Math. (to appear)*, 1998.

[Ste88] J. Stevens. On canonical singularities as total spaces of deformations. *Abh. Math. Sem. Univ. Hamburg*, 58:275–283, 1988.

[Ste93] J. Stevens. Partial resolutions of quotient singularities. *Manuscr. Math.*, 79:7–11, 1993.

[Sza95] E. Szabó. Divisorial log terminal singularities. *J. Math. Sci. Univ. Tokyo*, 1:631–639, 1995.

[Tak94] K. Takegoshi. Higher direct images of canonical sheaves. *Math. Ann.*, 303:389–416, 1994.

[Tha96] M. Thaddeus. Geometric invariant theory and flips. *Jour. AMS*, 9:691–729, 1996.

[Tyu70] G. Tyurina. Resolution of the singularities of flat deformations of rational double points. *Funct. Anal. and Applic.*, 4:77–83, 1970.

[Uen75] K. Ueno. *Classification Theory of Algebraic Varieties and Compact Complex Spaces*. Springer Lecture Notes vol. 439, 1975.

[vdW91] B. L. van der Waerden. *Algebra (English translation)*, volume I. 7th ed, vol II. 5th ed. Springer Verlag, 1991.

[Vie82] E. Viehweg. Vanishing theorems. *J. f. r. u. a. Math.*, 335:1–8, 1982.

[Zar62] O. Zariski. The theorem of Riemann-Roch for high multiples of an effective divisor on an algebraic surface. *Annals of Math.*, 76:560–615, 1962.

Index

:=, is defined by, 6

$A \geq B$, $A - B$ is effective, 4

$L^{[i]}$, the double dual of $L^{\otimes i}$, 64

$(L_1 \cdots L_d \cdot Z)$, intersection number, 29

$(L^d \cdot Z)$, d-fold intersection number, 29

$\cong_{et}$, étale equivalence, 149

$\equiv$, numerical equivalence, 4

$\equiv_f$, numerical f-equivalence, 5

$D_{\geq 0} = \{Z \mid (D \cdot Z) \geq 0\}$, 19

$D^{\perp} = \{Z \mid (D \cdot Z) = 0\}$, 19

$\phi_{|D|} : X \dashrightarrow \mathbb{P}(H^0(X, \mathcal{O}(D)))$, 68

$\phi_{|D|/Y} : X \dashrightarrow \mathbb{P}_Y(f_*\mathcal{O}_X(D))$, 94

$\lfloor D \rfloor$, the round down of divisor D, 5

$\lfloor d \rfloor$, the round down of $d \in \mathbb{R}$, 5

$\lceil D \rceil$, the round up of divisor D, 5

$\lceil d \rceil$, the round up of $d \in \mathbb{R}$, 5

$\sim$, linear equivalence, 4

$\sim_f$, linear f-equivalence, 5

$\{D\}$, the fractional part of divisor D, 5

$\{d\}$, the fractional part of $d \in \mathbb{R}$, 5

$f_*(Z)$, the birational transform of Z by f, 5

$g_*^{-1}(Z) = (g^{-1})_*(Z)$, 5

(-1)-Curve, 8

1-Complement, 221

1-Cycle, 11, 18
  effective , 18
  relative, 45

$1/r(a_1, \cdots, a_m)$, a μ_r-action, 171

Abelian surface, 22

Abundance conjecture, 81

Adjunction, 173
  formula, 21, 182
  inversion of, 173, 174

Algebra, symbolic power, 189

Algebraic equivalence, 10, 11

Algebraic function, 148

Algebraic space, 148

Algebraic valuation, 50

Almost all, 14

Ample, f-, 34, 35

A_n, a type of a Du Val point, 122

Analytic case, 47

Analytic singularity type, 193

Analytic surface, 25

Aut, the automorphism group, 165, 217

Automorphism, 22, 165

Basepoint-free theorem, 75, 78
  relative, 94

Bend and break, 9, 11

Big, 67

Big, f-, 94

Bimeromorphic map, 6

Birational map, 6

Birational transform, 5

Blow up
  symbolic, 199
  weighted, 142, 143, 167, 170

Branch, 114

Bundle
  canonical, 47
  conic, 28
  lattice, 10

C, a type of an extremal ray, 28

Calabi–Yau threefold, 239

Canonical, 42, 56, 165, 238
  bundle, 47
  divisor, 47
  flop, 191
  ring, 80, 237
  singularity, 42

Canonical bundle formula, 52
  Kodaira's, 49

Canonical model, 107
  weak, 107

Cartier,$\mathbb{Q}$-, 4

249